PUBERTAL MATURATION
in
FEMALE DEVELOPMENT

PATHS THROUGH LIFE

A series of volumes edited by David Magnusson

Volume 1
Individual Development from an Interactional
Perspective: A Longitudinal Study

Volume 2
Pubertal Maturation
in Female Development

PATHS THROUGH LIFE
Volume 2

PUBERTAL MATURATION in FEMALE DEVELOPMENT

Håkan Stattin
David Magnusson
University of Stockholm

LAWRENCE ERLBAUM ASSOCIATES, PUBLISHERS
1990   Hillsdale, New Jersey   Hove and London

Copyright © 1990 by Lawrence Erlbaum Associates, Inc.
All rights reserved. No part of the book may be reproduced in
any form, by photostat, microform, retrieval system, or any other
means, without the prior written permission of the publisher.

Lawrence Erlbaum Associates, Inc., Publishers
365 Broadway
Hillsdale, New Jersey 07642

Library of Congress Cataloging-in-Publication Data

Stattin, Håkan.
   Pubertal maturation in female development / Håkan Stattin and
David Magnusson.
     p. cm.—(Paths through life ; v. 2)
   Bibliography: p.
   Includes indexes.
   ISBN 0-8058-0595-8
   1. Puberty. 2. Menarche. 3. Developmental psychobiology.
4. Teenage girls—Psychology. I. Magnusson, David. II. Title.
III. Series.
   [DNLM: 1. Human Development. 2. Puberty—psychology. 3. Women—
psychology. W1 PR963FL v. 2 / WS   450 S797p]
QP84.4.S73   1990
612.6'61—dc20
DNLM/DLC
   for Library of Congress   89-11743
Printed in the United States of America   CIP
10 9 8 7 6 5 4 3 2 1

Contents

	Preface	vii
Chapter 1	The Issue of Biological–Psychosocial Interaction	1
Chapter 2	The General Approach and the Basic Model	43
Chapter 3	Data	78
Chapter 4	Psychological Adaptation and Self-Concept	99
Chapter 5	Interpersonal Relations	130
Chapter 6	Social and Emotional Adjustment	179
Chapter 7	The Short-Term Consequences	224
Chapter 8	Developing Girls in a Developing Environment	248
Chapter 9	Mediators of the Influence of Pubertal Timing	262
Chapter 10	The Long-Term Consequences	301
Chapter 11	Some Final Reflections	347
	References	370
	Author Index	391
	Subject Index	399

Preface

This volume is concerned with female development, particularly during adolescence. A biosocial model for female development during this age period is presented as a framework for empirical studies, focusing on the role of individual differences in biological maturation. The prospective, longitudinal design of the research program offers possibilities to investigate both the short- and the long-term consequences for individual variations in pubertal development. The issue of the long-term consequences is a major issue that is looked into here in considerably more detail than has been done in previous research in this area.

Two colleagues particularly have contributed to the production of this volume. Professor Joan McCord read an earlier version of the manuscript and generously offered her sharp criticism, comments, and suggestions. Dr. Sigrid Gustafson not only checked the English, but also contributed considerably to improving the final version of the manuscript by her knowledgeable and clear insights in the problems that are central in the volume. We thank both of them and our colleagues in the research group for continuous stimulating discussions.

This volume is the second in a series that has the title Paths Through Life. The volumes published in the series present research performed within the theoretical frame of and built on data from a longitudinal research program—Individual Development and Adjustment (IDA)—planned and carried through under the auspices of the second author at the Department of Psychology, University of Stockholm since the mid-1960s.

The first volume in the series (Magnusson, 1988) presented the theoretical background of the longitudinal program, discussed the methodological and research strategy implications of the theoretical framework and reported empirical studies illustrating the potentiality of a longitudinal approach to the study of individual development.

For stimulating and constructive discussions and for a creative and enjoyable climate we thank all our collaborators in the research group.

The research presented in this volume was made possible by financial support from the Bank of Sweden Tercentenary Fund, The Swedish Council for Planning and Coordination of Research and the Swedish Council for Social Research.

1

The Issue of Biological–Psychosocial Interaction

FACING THE ISSUE

Adolescence is the phase in the life cycle that is associated with reaching sexual and social maturation, consequently designated as a biosocial stage of development (Ausubel, 1954). It has few definite demarcations (Brooks-Gunn & Petersen, 1984). At its lower end some measure of puberty can mark the entry into this period, but no more circumscribed end point can be established (English, 1957). During this period, thorough shifts occur in endocrine systems; in other bodily systems; and in emotional, cognitive, moral, social, and interpersonal development. The period is one of heightened introspection, of increased awareness of personal and physical qualifications. As a social marker of development, the period is customarily associated with a number of developmental tasks, such as acquiring an appropriate sex-role and identifying oneself as a mature member of society. During this period the person should behaviorally and emotionally establish independence from parents and achieve a sense of autonomy and individuation. New reference groups are formed, mainly through interaction and identification with peers. During this period most teenagers establish their first contacts with the opposite sex. The period is one of preparing for higher education and for a vocational career. Also, adolescence is intimately connected with achieving a stable individual identity and a view of life, ideologically and socially.

These processes operating in adolescence are complex and multifaceted, involving both childhood reminiscences and future adult behaviors. They emerge as the consequence of the interplay of biological factors, cognitive–experiential processes, feelings and motivational factors, social relations, and environmental factors, both proximal and distal. Often it is the specific constellation or pattern of factors operating in the biological, psychological, and the social domains that will determine the effectiveness of coping with transition tasks in adolescence, not isolated experiences, however strong they might seem.

Beyond the complexity of human behavior in this period, the developmental aspect adds one more factor: the factor of time. The time scale used almost exclusively in research on human growth is chronological age. As psychologists we are accustomed to the age-graded perspective, (i.e., to viewing individual behavior against the background of the behavior repertoire of same-age subjects). We count, for example, prevalence of a particular behavior on an age basis. As methodologists we are accustomed to basing our examinations on age-homogeneous samples, and the instruments we use are standardized by chronological age. The result is an assumption that subjects mature in monotonically similar ways. Thus, the interindividual differences that occur at a certain chronological age refer to the same systems' functioning in the same way across individuals. This assumption is particularly critical for individual predictions over time. In order for a variable to constitute an effective prognostic factor, or a criterion variable, at a given point in time, it must be demonstrated to be interindividually equivalent (i.e., to involve, across persons, systems that exhibit similar logical and meaningful relationships to the chronological age point at which the measure is taken; Goldstein, 1979).

Research on physical maturity has demonstrated conclusively that the assumption of an age-homoegenous development does not always hold true. Hence, chronological age cannot be used as the only meaningful reference scale for development (Goldstein, 1979; Magnusson & Allen, 1983b; Peskin, 1967; Young, 1963). The age-related perspective, which appears to posit a unitary progression through which teenagers change in similar ways across time, might well disguise the existence of quite different developmental progressions that would become evident if social transition was viewed from a maturational timing perspective. In fact, it is not uncommon nowadays to take the position that "biological age" is a central determinant for transition behavior in adolescence, in some respects, perhaps, a more sensitive index of psychosocial changes in the individual than other time-linked measures.

This volume is concerned with the maturational timing issue and particularly with the significance of pubertal development among females. We direct attention to the role of variations in the timing of biological maturation in females for progression in various areas in the adolescent life phase. The purpose is to systematically analyze, from an individual difference perspective, the psychosocial consequences that take place as an effect of entering puberty earlier or later in females. What is the significance of pubertal change in adolescence? To what extent do emotional and social adjustment during this period have a maturation basis? How are variations in the timing of pubertal maturation interwoven in the complex network of factors underlying the completion of normal developmental tasks over time? The issue before us is to explore how interindividual differences in pubertal development are linked to new social behavior and emotional reactions in adolescence and how they relate to social support systems and to self-concept. In the pursuit of an understanding of the role of physical maturation for psychosocial transition in adolescence, we propose and test, within the framework of an interactional model, various developmental progressions through adolescence that are affected by variations in pubertal timing. In summary, the dual aim of this research is to map the domains related to differential pubertal development in individuals and to attempt to map the processes operating in these domains.

This endeavour involves attempting to crystallize the primary meaningful connections between the physical growth component and behavior: to determine the *configuration of change* on different levels of organization and how changes in one part of the total organization are linked with changes in other parts. We are attentive to those factors in the girls' environment that are affected by their bodily development, to how significant others respond to the change in the girls, and to how the girls themselves are affected by other persons' views and behaviors directed toward them. Particular attention is paid to the mediators of the impact of biological maturation on social and emotional behavior. In the end we hope to say something more definite about which areas are and are not likely to be affected by interindividual differences in physical growth in girls, and, in particular, about how these connections are formed.

Characteristic of physical growth in adolescence is the rapidity with which it occurs and the wide range of effects it has on the body. The physical growth process is completed within a relatively short space of time. This fact does not necessarily imply that interindividual differences in physical growth lose their significance for psychosocial functioning later on. An important task before us is to examine the evidence for the possible long-term implications of timing of pubertal maturation. Given this objective, the only meaningful way in which to address the long-term consequences of variations in pubertal development for the adult life course is to follow one and the same group of subjects through adolescence to adulthood. The present investigation constitutes a prospective longitudinal analysis of Swedish females over two decades, from 1965 when they were 10 years old to the present time when they are about 30.

Some of the questions we attempt to answer are the following:

- Are variations in the timing of pubertal development among girls related to their psychological and social life situation in the adolescent years?
- If so, *when* is the relation most prominent?

- In *what areas* is the relation most prominent?
- *How* does this relation come about?
- Does the relation *sustain over time* in adolescence?
- Do interindividual differences in physical maturation have any *long-term consequences* for adult life? If so, in what areas, for which girls, and through which developmental processes does pubertal development operate?

The data for this analysis consist of a broad range of biological, mental, psychological, behavioral, and social factors collected for the sample of females over a considerable age span. Data are reported on the relation between biological maturity and factors connected with:

- *The person herself:* self-image, self-esteem, emotionality, personality constellation, etc.,
- *her family:* mother and father relations, family conflicts, family orientation, etc.,
- *her peers:* emotional relations to peers, popularity in the peer group, peer–parent preferences, characteristics of her peer network, peers as norm transmitters, etc.,
- *her opposite-sex relations:* popularity among boys, dating, sexual intercourse, attitudes toward sex, etc.,
- *her school adjustment:* school interest and aspirations, school achievement, teacher relations, school problem behaviors, etc.,
- *her general social adjustment:* mental problems, social deviancy, delinquency, alcohol patterns, abortions, etc.
- *her adult life situation:* marriage, children, level of education, work life, etc.

The scientific goal is to explain and understand if, to what extent, and why interindividual differences in pubertal development have short- and long-term impact on the life situation in various areas. Not only may such an exploration give insight into the role of timing of pubertal maturation for the development from a teenager to an adult, it could also contribute to a deeper understanding of the nature and the direction of adolescent transitions in general.

An Interactionist Approach

The abrupt changes in bodily proportions and functions accompanying puberty are a most salient change characteristic in adolescents. One enters this life phase with childish bodily proportions but leaves it with adult-like physical shape and with the capacity to reproduce. However, even though the emergence of this new pubertal force creates a sort of discontinuity in individual development, there are few reasons to believe that pubertal timing, except in some few obvious ways, has a unidirectional influence on behavior. There are even fewer reasons to assume that its influence generally should be interpreted in a casual way. As Brooks-Gunn and Petersen (1984) have pointed out, the mere fact that there exists a relationship between pubertal growth and social behavior does not imply a cause–effect relationship. What we witness might rather be the co-occurrence of different developmental factors. If so, the biological determinism model, which infers that observed changes in feelings, motivations, social relations, and so on are casually related to changes in the physical-bodily domain, has limited applicability.

Although the point in time when puberty takes place is largely beyond the control of the individual's purposive behavior and although physical growth itself follows a more or less predetermined course once its onset occurs in individuals, pubertal development is not a process that emerges suddenly, without connection with prior aspects of the person or her social environment. Interindividual differences in pubertal timing are not determined randomly. They are connected with general cultural environmental conditions, with genetic determinants and with specific individual experiences, like health history, protein intake, and amount of daily exercise, to take a few examples. Birth order and the number of siblings have been found to be associated with age of menarche (James, 1973; Roberts & Dann, 1967; Roberts, Rozner, & Swan, 1971), such that the larger the family, the later in time does puberty occur. This relationship has been attributed to less nutritional intake per person and higher likelihood of diseases in larger families (Tanner, 1965). Malnutrition (as evidenced in anorexia nervosa), diabetes, illness, and intensive athletic activity prior to puberty have been found to delay the puberty entry. Stress, in a wide sense, can also affect the age at which the puberty sequence starts. In a Swedish longitudinal study, Miserandino (1986) reported effects of eating habits and sleeping problems at age 8 on the point in time for menarche.

Stockier-built pre-teens reach their point for menarche at an average earlier time (Adams, 1981; Bruch, 1941; Garn, 1980; Greulich, 1944; Kantero & Widholm, 1971; McNeill & Livson, 1963). It has been suggested that a "critical weight," or height/weight ratio, reflecting the percentage of body fat, determines when girls reach their menarche (Frisch & Revelle, 1970). Alternative hypotheses, associating the menarche more with skeletal maturation, have also been advanced (Ellison, 1982; Tanner, 1978). Among nonexercising females, those who reach menarche early are, on the average, heavier than those who reach menarche late; however, this heaviness is not necessarily connected with an absolute body fat standard (Garn, 1980). Among exercising females on the élite level, the amount of exercise seems to be a vital factor for when puberty occurs. Futhermore, menstrual disturbances among teenage girls are often considerably more common among hard-exercising females than among females in general. In a recent Swedish study of training patterns and menstrual disturbances among hard-exercising teenage girls (Lindahl & Bergh, 1988), girls whose menstrual cycle were totally turned off trained on an average of 477 hours per year, in comparison with 395 hours for females with irregular menstruations and 366 hours for females with normal menstruation. Only 4 out of 10 hard-training girls had regular menstrual cycles. Menstrual problems were most common among gymnasts and cross-country skiers.

Studies conducted in various cultures show rather wide differences in growth rates. More than 6 years' difference in median age at menarche was, for example, reported by Tanner (1966) for subjects from Cuba and the Bundi tribe in New Guinea. These population differences have been attributed to factors affecting the standards of living and health, such as protein intake and availability of medical facilities, and to genetic differences. Similar types of explanations have been offered for the differences in growth rate observed between subjects from urban and rural areas, with subjects from urban areas having an accelerated growth rate.

Undoubtedly, menarcheal age is under genetic influences (Garn, 1980; Hiernaux, 1972; Jensen, 1969; Kantero & Widholm, 1971; Roberts & Dann, 1967; Tanner, 1962). Studies have been performed investigating the correspondence of menarcheal age between girls and their mothers, as well as between girls and their sisters. The early developed girl tends to have an early developed mother and an early developed sister. The correlation for age at menarche between mothers and daughters has been estimated at about .20–.25 and that between siblings at .35–.40 (Damon, Damon, Reed, & Valadian, 1969; Garn, 1980; Zacharias, Rand, & Wurtman, 1976). Menarcheal age also has been studied for monozygotic twins, diozygotic twins, and nonrelated subjects. Tanner (1962) cited data showing correlations at the level of .70 between identical twins with regard to skeletal maturity, at around .30 between siblings and no correlation between unrelated subjects. Genetic influences also are found with respect to height and weight and height and weight gain in adolescence (Fischbein, 1977). The casual relationships involved among the various hereditary and environmental factors are, however, to a great extent still unclear. One hypothesis that has been put advanced is that growth *rate*, as manifested by earlier menarche, is a function primarily of environmental factors, such as nutrition, whereas the ultimate growth *outcome*, like adult height, is affected more by genetic determinants (see Jensen, 1969; Tanner, 1965).

Even if pubertal change were totally under the control of fundamental biological drives, its significance in females' development could not be determined without due consideration of environmental conditions. Margaret Mead's (1952) ethnological studies of adolescencs in Samoa made clear that the impact of variations in biological maturation on behavior is highly dependent on overriding cultural expectations and demands. Its consequences for the life course will depend on the societal conditions for development in their members. Many differences can be found in institutionalized and informal upbringing conditions from society to society, presumably connected with the issue on how behavior is affected by pubertal growth. For example, the concept of dating is a social phenomenon that American, but not Swedish, teenagers are quite familiar with. This fact cannot be bypassed when attempts are made to compare the effect of biological maturation in the two countries in the realm of heterosexual relations.

In addition, there are history-graded factors. The cultural impact on the relationship between biological maturation and behavior will apply to the historical era in which studies of physical growth are conducted. Findings obtained in longitudinal growth studies that started in the late 1920s and the early 1930s do not necessarily hold for today's generation of teenagers.

Societal norms, values, and customs are transformed into the real-life situations that teenagers confront, into the demands and expectations of the immediate environment. Therefore, the study of physical growth becomes the study of how physical growth is connected with the particular features of the environment in which individuals are raised. It concerns how the individual copes with her own physical and sexual maturation and how she adjusts to the demands and the expectations made on her biological development from the social and cultural environment. From the perspective of the individual girl it is a dual, reciprocal process: an adjustment to one's own physiological change, including the recognition of how this change is related to the general development toward adult status, and an adjustment to others' reaction to this change.

Because the effect of physical maturation is connected with the individual's own reaction to his or her own physical development, other significant persons' reactions to this change, the conformity to the role as early or late developer, and so on, the impact of physical maturation on the functioning of individuals in different areas is most fruitfully approached from a perspective of an integrative theoretical model involving specific outlines of the interconnection between physical, cognitive, emotional, social, and behavioral factors. In fact, there are probably very few domains of research where it is so obviously necessary to study behavioral change from the perspective of reciprocal, interactive influences as in research on the effects of physical growth. This is a challenging task, and its very challenge is perhaps one of the reasons we have recently witnessed an upsurge of interest in the study of processes accompanying pubertal change (Lerner & Foch, 1987).

Our research approach in this book can be characterized as basically a holistic and an interactionist approach (Magnusson, 1988). The fundamental outlook is that individuals function as total integrated organisms. Development does not take place in one single component in isolation from the rest. It contributes to the whole.

One way to illustrate what goes on in the individual over time is to use a systems metaphor. From that perspective the person can be considered as a totality or as a total system whose functioning depends on the functioning of subsystems of feelings, thoughts, attitudes, interests, behaviors. From the viewpoint of a system with interrelated parts, isolated changes are unlikely to occur. The constituent parts of the configuration are bound to each other, and the total system tends to maintain its configuration (Weiss, 1969). From another perspective, structures and processes occurring in one part of the system can be used as input for investigation of what occurs at other subsystems or systems on a more superordinate level. In Koestler's (1978) vocabulary, we deal with "holons," having one face directed toward higher order levels and one face toward lower order levels. A model for development that investigates maturation from one particular viewpoint is not "better" or more "true" than another model specifying the developmental parameters from another viewpoint. On the contrary, they approach one and the same phenomenon from different angles. Ultimately, the thorough understanding of an element in development involves taking into account the relevant factors operating on superordinate and subordinate levels as well as those functioning on the same organizing niche.

The mere fact that development seldom occurs in one single component or at one level of the organization, has direct research strategy consequences.

It has direct consequences for data collection. To obtain an effective description of development over time requires covering a broad range of biological, psychological, and social factors in development. Likely or appropriate behavior at one age might be absent at another. Therefore, the selection of data to be collected must involve a careful analysis of the structures and processes operating at different age levels. Administrating "the same" questions to the same individuals from one time to another, although perhaps ideal from a psychometric viewpoint, may be less meaningful from a psychological perspective. From earlier developmental research (Bergman & Magnusson, 1984a; 1984b, 1986, 1987; Magnusson, 1985, 1987, 1988; Magnusson, Dunér, & Zetterblom, 1975; Magnusson, Stattin, & Dunér, 1983; Olofsson, 1971; Stattin, 1979; Stattin & Magnusson, 1984, in press; Stattin, Magnusson, & Reichel, in press) we have become convinced that adolescence is not a simple way station between childhood and adulthood. Rather, the adolescent period is a period that, in some sense, has its own social rituals and rules. The desire for experimentation is a primary characteristic among teenagers, and they are engaged in activities that are not particularly lasting, nor, for that matter, functionally related to adult behavior. Adolescence must be described as a substantive period in its own right and understood from the developmentally unique social conditions prevailing during this life phase. Data must reflect these unique features.

Second, that development seldom occurs only in one part of the organization has consequences for data treatment. Because development operates simultaneously in different structures and on different levels, it is only by conducting simultaneous analyses at different levels of this organization that we arrive at comprehensive information about the developmental process. The "multiple level of analysis" is the approach presented in this volume.

The postulate that individuals develop as integrated wholes does not exclude the factors operating outside of them. Rather, interaction occurs as much between different features "within" the individual as between the individual and her environment. In systems terms, the person–situation unit is the combining element of the person and situation components. This person–situation unit has become the focal point in much recent psychological research (cf. Magnusson & Endler, 1977). The emphasis on this unit leads to the second postulate: The individual functions and develops in an ongoing, dynamic, and reciprocal process of interaction with her environment.

The use of the world "reciprocal" is not merely lip service but emphasizes that the person interacts with an environment that constantly changes. One of the ideas that has been stressed in the physical growth literature is that the social development accompanying girls' pubertal growth and parents' reactions to their daughter's physical status are complementary processes. The female herself contributes to these environmental changes. Attitudes and behavior of significant others are likely to change as a consequence of physical changes in the female. Perceiving that their daughter is approaching adult status physically and personally, her parents give her more privileges and treat her as more mature than before. By virtue of her biological maturation, "development" in this case has occurred in the girl's immediate environment. Thus, it makes sense intuitively that girls' development cannot be seen to occur within a stable environment. The perspective of development as a reciprocal process whereby the individual is both the producer and the product of her own development and through which individual attributes, behaviors, and emotions are shaped in an bi-directional fashion relative to significant others, is becoming a central focus of the research on physical development (Lerner & Foch, 1987).

PUBERTY IMPACT: THE STATUS OF THE FIELD

Before going into the issue of pubertal development from the interindividual point of view, we briefly address this issue from an intergenerational perspective.

The Secular Trend and its Impact on Behavior

Children grow faster and enter puberty at an earlier age today than they did in years past. This earlier maturation from generation to generation, over more than the last century, has been said to reflect a *secular* trend. This downward tendency in the onset of puberty over generations has been observed in all cultures. As far as can be judged by the literature, the trend toward earlier puberty has been observed since the middle of the 18th century (Diers, 1974; Tanner, 1962). Our generation of teenagers approach adult size perhaps as much as 7 to 8 years earlier than did those raised in the beginning of this century. In Sweden 100 years ago, the average age at menarche was 15.75 years (Tanner, 1966). In the beginning of the 1970s girls attained menarche at about 13 years (Ljung, Bergsten-Brucefors, & Lindgren, 1974). This means that the menarcheal age was lowered more than 3 months per decade. Whether the secular trend with respect to age of menarche will continue at the same rate is an open question (Brundtlang & Walloe, 1973; Garn, 1980; Poppleton & Brown, 1966; Tanner, 1962, 1966).

The secular trend has been attributed to factors such as improved nutrition, first and foremost, and to increased standards of living resulting in better health (Tanner, 1962). Furthermore, we cannot underestimate the effect of the increased knowledge of factors affecting the growth of the infant and the better social and nutritional care of infants today. Early protein intake might be an important determinant for the earlier maturation (Tanner, 1978).

Another feature of the secular trend is the final outcome of maturity. Not only do children today reach maturity earlier than before, but, on the average, they will also ultimately surpass their parent generation in adult height. One interpretation that has been offered assigns the increased size of offsprings to the consequence of outbreeding (Tanner, 1965).

The secular trend has opened discussions on the connection of physical maturation to mental, social, and emotional development over generations. The case of teenage pregnancies is an obvious example:

> Even where contraception is widely practiced, the increase in the number of teenage pregnancies may be attributed in part to the secular decrease in the age at sexual maturation. Despite the transient period of adolescent infertility, there are now many 12-year-olds and even 11-year-olds who are capable of maintaining a conceptus. As a result of the secular decrease in the age at menarche, there is an increasing number of mothers who are ill prepared for the task of mothering, thus creating both biological and psychological hazards to themselves and their progeny. (Garn, 1980, p. 125)

An opposite view has been declared by Bullough (1983), who claimed that the secular trend in menarcheal age has been overestimated, and who found that the rate of pregnancy per person among adolescents actually decreased from 1960 to the middle of the 1970s.

The increase in incidence and prevalence of alcohol and drug use, smoking, sexual intercourse, autonomy problems, and so on, among increasingly younger teenagers, has, from the secular perspective, been attributed to an earlier physical maturation that has no parallel in earlier emotional and moral development. Problem behavior has also been attributed to a case of intergenerational misperception. Muuss (1970) claimed that

> the secular trend in physical growth has its correlates in many areas of development in that certain interest patterns and attitudinal changes as well as social sexual interests and social sophistication seem to occur at an earlier age today than a generation ago. . . . Some of the contemporary adolescent problems and conflicts, especially the present concern with the "generation gap," may be better understood if one were to consider the earlier biological maturation of youth as contrasted with their chronological age which usually serves as the frame of reference for parents and teachers. (pp. 276–277)

This intergenerational misperception implies that adults systematically misinterpret the prevalence of some social behaviors among teenagers when comparing these with the prevalence in their own generation.

The adolescent role is longer today than it once was. Children become "physically adult" at younger ages but are not treated as adults until much later. The prolonged education holds them as minors, without economic independence and adult responsibility. Adolescent problems have been interpreted in the light of this lack of societal demands on social responsibility and lack of adult status recognition in their young members (cf. Ausubel, 1954). In this marginal situation between childhood and adulthood and deprived of adult-like tasks, problem behavior and earlier onset of conduct problems have been thought to result from more freedom to select among different lifestyles. Indeed, dating occurs earlier (Rice, 1975), and teachers complain that discipline problems are not uncommon even in the first grades. Some writers have expressed great concern over what society will eventually become if the trend toward earlier biological maturation and later adult social status continues. Hebbelinck (1977), for one, has proposed that "the adolescent in year 2000 will live in a period of great social and psychological turmoil" (p. 154) due to this gap.

The consequences of the secular trend for social behavior, however, are still speculative. They are based on considerations of the interrelation of physical maturity, social development, and the attainment of adult status among teenagers of former times compared with teenagers today.

Individual Differences in Pubertal Development

In the within-generation basis for comparison, we limit the scope to a particular teenage cohort and start asking questions about the co-occurrence of physical, social, and emotional growth. What patterns of physical, social, and affective phenomena can be discerned, and is the timing of maturation an important factor for psychosocial adjustment and development in adolescence?

Because the systematic empirical research on these issues goes back to the 1920s, it is now more than half a century since the first investigations were initiated with regard to the effect of the timing of puberty on teenagers. The chief issue dealt with at that time concerned whether or not development ran parallel in different areas, such that early physical development was also associated with early social, emotional, and mental development in the same individual. Some writers argued that correspondences among physical, psychological, and social indicators were inevitable, that individuals who physically developed earlier than their peers also had advantages in other areas. Late maturing, on the other hand, would be associated with feelings of inferiority, lingering parental dependence, and low self-esteem. Other researchers maintained that growth was specialized, with measures of association from different areas being so small as to be without practical value. For extreme early and late developers, on the other hand, correspondences in social behavior, social adjustment, and health were more likely (see Abernethy, 1925; Gates, 1924).

The overall conclusion from studies conducted in the 1920s, 1930s, 1940s, and the 1950s was that it was advantageous, particularly for a boy, to enter puberty early. Many favorable consequences seemed to be connected with the more mature physique of the early developed boys. A more masculine mesomorphic appearance among boys would be beneficial in the interpersonal context: being positively appraised by people in general and beneficial for heterosexual contacts. Early maturation also meant acquiring greater strength relative to later developed boys, an advantage that would particularly benefit the early maturing boy in athletic activities. The early developer was supposedly treated as older than his actual age by parents and peers, was expected to behave more maturely, and was given more privileges, which made gaining independence from his parents easier, in contrast to the experience of his late maturing peers.

Up to the present time, much empirical support for the positive impact of early maturation among boys has been gathered, particularly as it relates to the social realm. Jones and Bayley (1950) reported that the early maturing boy was rated by adults as more attractive, relaxed, interested in girls, and appeared more reserved and assured, in contrast to the later developed boys, who had a more childish or "little-boy" type of motor activity (eager, peppy, animated, active, busy, talkative, uninhibited, restless, and attention-seeking). Jones (1958) found that the early-maturing boys scored higher in public recognition while Duke et al. (1982) reported that the late-maturing boys were rated lower in educational aspirations and achievement, as measured by the boys themselves, their parents, and their teachers. In a Finnish outpatient clinic sample, investigated by Frisk (1968), the late-developed boys were characterized by feelings of physical inferiority, enuresis, and nail-biting, and they were badly treated by peers. Also common were compensatory self-assertions, pilfering, defiance, and smoking. Their troubles were most intensive around the age of 15. Crocket and Petersen (1987) found better body image among the early developers, who also had more comfortable heterosexual relationships. Tobin-Richards, Boxer, and Petersen (1983) reported that boys who rated themselves as early in pubertal development perceived themselves as more attractive and had more positive body image than did boys who perceived themselves as late developed relative to same-age peers. Among the different growth indices involved, facial hair was most closely connected with attractiveness and body image.

Comparing early- and late-developing boys in late adolescence, Jones and Mussen (1958) reported more negative self-concepts among the late developed boys. These boys also tended to have poorer parental relations and higher dependency needs. The picture of less favorable personal and social adjustment among the late developers in late adolescence was supported in a study by Weatherley (1964). College males who rated themselves as late developed showed higher anxiety and less dominance compared to average matured and early-developed boys. Also, it should be noted that boys generally experience their pubertal development positively (Gaddis & Brooks-Gunn, 1985).

Support has been presented that indicate that the favorable development for the early-maturing boys persists into adult life. Jones (1957) reported the early developers to be more responsible, higher in dominance, socialization, and self-control, capable of making a good impression, poised, responsible, and goal-directed as adults. Ames (1956, cited in Eichorn, 1963) reported higher social participation at adult age among males who were early developers. Of particular interest was the finding of more frequent high occupational positions among the early-developed males. Kinsey, Pomeroy, and Martin (1948) found the highest rate of sexual activity among the early developed males, who also married earlier.

Jones (1965) reported a study comparing the males in a longitudinal growth study in a number of respects with regard to personality at the average age of 33. The early developer, according to skeletal age measures, scored higher in adulthood with regard to capacity for status, responsibility, socialization, self-control, good impression, and achievement via conformity, whereas the late-developed males were portrayed as relatively more flexible and achieving via independence. From retrospective reports at the age of 38, the early developers described their social experiences as well as their developmental status more positively than did the late developers. Other measures portrayed the late developers as more perceptive and insightful and the early developer as more conventional. The results were interpreted as showing a continuation over time of the beneficial social situation among early developers, which had been found in adolescence, with, however, some evidence of cognitive rigidity. A capacity for coping and tolerance of ambiguity were the main developmental themes for the late-developed males.

Another perspective on these research findings from adult age, placing emphasis on subjects' preparations and reactions to the internal properties of puberty, rather than on society's and significant others' reactions to the individual's physical maturation, was offered by Peskin and Livson (Peskin, 1967, 1973; Livson & Peskin, 1980). The socially more favorable situation among the early-developed males relative to late developed and the earlier adult responsibility among the early developers were reinterpreted by Peskin (1973) to represent social conforming in order to escape confrontation with threatening pubertal impulses, with which the early developer was neither ready nor prepared to cope. Comparing early and late developers equated for the time of pubertal onset (Peskin, 1967), the early-matured males were reported after pubertal onset, among other things, to show lower activity level, less exploratory behavior, social submission, and lower intellectual curiosity, whereas late-matured males showed more activity, exploration, and curiosity. Peskin concluded that the early-developed male controls his pubertal drive by conforming socially, and "engages in a socially sanctioned 'flight' into adulthood from his less tolerable puberty" (p. 13).

With the exception of Peskin's findings, much data from different areas converge in a more favorable developmental transition for the early relative to the late-developed male. Empirical data for girls have shown more contradictory results across studies, with a fewer number of domains systematically documented as related to interindividual differences in physical growth in adolescence, fewer long-term effects, and an absence of simple, straightforward explanations of how differences in pubertal timing are related to female development generally. It is commonly recognized nowadays that the field of research as it pertains to maturational effects for females is still in a highly speculative and fragmentary state (Brooks-Gunn, Petersen, & Eichorn, 1985).

Most researchers acknowledge that puberty timing does have some impact for transition behaviors among girls in adolescence, but at the same time they claim that the issue is complex and that there are no easy answers (see Greif & Ulman, 1982, for a review of the literature on the impact of menarche). Futhermore, although it is a current commonplace to acknowledge the importance of the time for pubertal entry among females, few inferences are formulated regarding the possible processes operating from physiology to behavior.

The general favorable social development associated with early male puberty has few counterparts in empirical findings for females. There are both positive and negative consequences of entering puberty early versus late. A widely held opinion is that early pubertal development for females, because they will deviate in physical characteristics from their same-age peers, is detrimental for these girls' social status in early adolescence; however, as time proceeds they will have an advantage over other, later-developed girls through being more the focus of boys' attention, thereby heightening their self-esteem (cf. Rice, 1978). On the other hand, boyfriend relations may give rise to concern on the part of the early-developed girls' mothers, who fear for the consequences of these girls' early heterosexual contacts. Actually, whereas a more mature physical status is met by positive adult attitudes for boys, it might evoke worries from adults in the immediate environment for girls. Besides, whereas early physical maturation in boys generally co-occurs with what is culturally desired, early physical maturation in females does not necessarily harmonize with the cultural ideal (Simmons, Blyth, & McKinney, 1983), both with regard to social development and physical development per se. The cultural bodily ideal emphasizes slenderness. However, with physical maturation follows an increase in body fat, giving the early-developed girl a more stocky bodily figure than the later developer.

The typical position heralded in the literature tends to connect differences between the early- and the late-developed girls to the immediate bodily changes of puberty. Golub (1983), for example, concluded that

> the early-maturing girl presents us with a specific set of potential problems. She is less likely to be prepared for puberty and menarche. Hence, she is more likely to be frightened. Her early sexual maturation—the development of breasts and curves—will influence the way in which others react to her and she may be subjected to overtures for which she is not psychologically or socially ready. She is also likely to experience some ambivalent or negative feelings about herself. On the one hand she enjoys having a mature body and being admired; on the other, she would like to be like everyone else. Generally the early maturing girls has lower self-esteem and more behavioral problems than her peers. Conversely, the late-maturing girls is troubled by her lack of breast development and sex appeal. She worries about not having started to menstruate and thinks that perhaps she is not normal. (pp. 317–318)

Reviewing the literature in the field, one is struck by the different conclusions researchers have drawn regarding the existing empirical material. Let us take some examples. Comparing the results for personal and social adjustment between the sexes, Weatherley (1964) found essentially satisfactory implications of developing early for males, but mixed results, with some suggestions for negative implications, of maturing early for girls. Weatherley concluded that

> what is perhaps noteworthy in the findings for girls is the very fact that they were so much less dramatic than for boys. Since previous research also produced less definite results for girls than for boys, one is drawn to the conclusion that for girls as opposed to boys rate of physical maturation is a much less influential variable mediating personality development. This is not surprising. In our

society the cultural sex-role prescription for males is relatively unambigious and is one which places a high value upon attributes associated with physical strength and athletic prowess, especially in the adolescent and young adulthood years . . . the feminine sex-role prescription is much less definite and stereotyped; consequently it is less likely to be closely tied to any specific pattern of physical attributes. (p. 1209)

The same conclusion, with a different interpretation, was arrived at by Burns (1979), who reviewed correspondences between self-concept and maturity timing:

In contrast to the dramatic effects bodily change has on the male self concept in adolescence, physical change, whether early or late, is much less a potent influence on the self-concepts of adolescent girls. This difference in effects may be due to the male cultural norm of tall, brawny masculinity, whereas early maturing for girls contain no prestigeous advantage . . . it is possible to make a tentative speculation that physical maturation in adolescence has a less dramatic effect on girls than on boys because the former have greater flexibility for altering or changing their looks through sensible use of cosmetics, padding, etc., but the latter can do little to alter their performance. (p. 183)

In opposition to these conclusions, More (1953) found more far-reaching effects of pubertal timing in girls than in boys, and also that early physical maturity was more beneficial for a girl than late maturation:

It appears from the data that the American female goes through a period of abrupt changes in emotional life and the pattern of her social behaviors at or near puberty. The relative suddenness with which her body functions shift presents her with a series of new emotions, or at least old emotions in new strengths or configurations, which must be taken care of rapidly if she is to continue to get along in her society. By and large, girls make this turnover when it is required of them by their bodies; and, in this sense, they exhibit concordance of physical development with emotional maturation and social behavior forms. . . . Among the boys studied, puberty appears to have been a more gradual affair. . . . It may be that the personality of the boys is less subject to external pressures which force it into a uniform mold than are girls. They, as a group, present us with more diversity, less similarity, of personality development than do girls. (pp. 117–119)

Parenthetically, More's conclusions were based on empirical data from a limited subject sample and exhibiting limited variations in pubertal development.

There are also examples where variations in pubertal timing influence social behavior, but viewed against the background of what is age normative, the factor is of secondary concern. In the words of Neugarten and Datan (1973):

> Many of the major punctuation marks of the life cycle are not only orderly and sequential, but many are social rather than biological in nature, and their timing is socially regulated. These concepts point to one way of structuring the passage of time in the life span of the individual; and in delineating a social clock that can be superimposed upon the biological clock. (p. 62)

For example, comparing the relative influence of sexual maturation in girls with the expected behavior from knowledge of the chronological ages of a sample of 12- to 17-year-old girls, Dornbusch, et al. (1981) reported that chronological age was the main determinant for dating. After this factor entered the prediction of dating, the factor of physical maturation did not contribute additionally, although seen in isolation it had a quite high relationship with the dating variable:

> The social system thus may take biological development into account as it develops and promulgates institutionalized images of appropriate behavior. But what is striking in this study is that individual differences in sexual development seem not to influence a behavior, dating, which, though not a direct measure of sexual behavior, is commonly thought of as reflecting the development of sexual drives. Social pressures appear to overwhelm the individual whose rate of biological development deviates from the norm. Here is a fascinating example of social reactions to biological processes in which standards for the social group reduce dramatically the impact of individual biological processes. (p. 184)

These examples have been given in order to highlight the various interpretations of empirical data on the impact of puberty in females. Other examples can readily be found picturing contrary positions in overall theoretical conceptualizations of the importance of physical growth for adolescent development generally: which girls are most affected, whether early or late maturation is advantageous or disadvantageous, whether early or late maturation is positively or negatively related to self-esteem, and so on—even among advocators of the same personality orientation. For example, whereas some psychodynamic theorists regard the menarche as being essentially a traumatic experience for girls (Deutsch, 1944), others view the event as a positive turning point for subsequent experiences among teenage girls (Kestenberg, 1961).

Domain Specificity of Pubertal Impact

On the surface, the field of research on the pubertal-timing impact in females seems to be in a state of confusion, and in contrast with boys, a somewhat circumscribed and domain-specific role of pubertal timing is postulated in present day research on females:

> It is clear that pubertal status affects fewer aspects of early adolescent development than might be expected. This general conclusion is in agreement with the conclusions of other researchers who report that pubertal effects are specific rather than general . . . It is quite clear at this point that attributions of pervasive influence to puberty are inappropriate. (Crocket & Petersen, 1987, p. 186)

Petersen (1987) delineated the important areas connected with pubertal development to body image, heterosexual relationships, and dating. Brooks-Gunn (1987) proposed that effects related to maturational timing in early adolescence would be found, temporarily, in just a few domains such as dating, menarcheal attitudes, and sex-role expectations. More pervasive consequences would eventually be found with respect to body image and menstrual attitudes.

A number of considerations strengthen the position that the impact of individual differences in pubertal development is specific rather than general. First, whether early or late physical maturation is favorable or not has been found to be quite dependent on the area investigated. Garwood and Allen (1979), hypothesized and found some support for the idea that the attainment of menarche simultaneously led to an increase in self-esteem, due to a clearer view of oneself as a woman, but also to increased problems, due to not having enough experience to deal with impending transition tasks such as autonomy, dating, and career plans. Simmons, Blyth, and McKinney (1983) reported that early maturation was advantageous for heterosexual relations and independence but disadvantageous for body image and school adjustment.

Second, the importance of the factor of pubertal timing is dependent on the time in adolescence when observations are made. To be taller and weigh more are not prestigious qualities for the early-matured girl in the early teens. These attributes will make her stand out in the peer group (Stolz & Stolz, 1944), and even when accompanied by more womanly bodily proportions, they have no heterosexual stimulus value in the context of same-age undeveloped boys. Later, when the feminine-defining physical attributes are responded toward positively in the dating situation, the early-matured girl's social status as well as her self-esteem may increase. Faust (1960) showed that early pubertal maturation was detrimental to a girl's social status in the early teenage years but more advantageous to her prestige later on. Thus, girls' developmental status may be beneficial at one point in time during adolescence but disturbing at another. That relationships change across time implies, in terms of research strategy, that it might be futile to look for general outcomes of maturity timing.

Third, whether it can be beneficial or disturbing for a girl to enter puberty early or late will also depend on attributes of the individual, the conditions of her upbringing prior to puberty, and the adequacy of her preparation for the event. It has been suggested, and there are some supporting data, that physically attractive girls have more to lose through the rapid changes that occur in puberty than do less attractive girls. The self-esteem of the former girls, presumably more closely connected with their physical appearance, makes the puberty period a more distressing experience (Zakin, Blyth, & Simmons, 1984). Dreyer, Hulac, and Rigler (1971) found some support for the hypothesis that the change in body image from pre to postmenarcheal status in girls would be dependent on the field dependence–independence of the girls, such that it would be more marked changes among the field independent girls.

Fourth, the influence on behavior of individual differences in biological maturity may be more indirect than direct. Simmons, Carlton-Ford, and Blyth (1987) found no direct relationship between pubertal development and self-esteem. However, early pubertal development was associated with greater school problem behaviors, and these, in turn, were negatively connected with self-esteem. Simmons, Blyth, Van Cleave, and Bush (1979) examined whether or not early pubertal development was associated with low self-esteem at Grades 6 and 7. No such association was found. However, in conjunction with a major environmental transition, changing school, pubertal development was related to low self-esteem. In fact, the girls with "multiple transitions"—early developers (physiological) who changed school (environmental) and had started to date (social)—displayed the lowest self-esteem in Grade 7. Consequently, although early or late physical maturation in itself was not a significant factor for self-esteem, it became important when it simultaneously co-occurred with other factors affecting the same individual.

Fifth, not only might one and the same measure of maturation have domain-specific effects on behavior, different growth features might have different implications (Apter, Galatzer, Beth-Halachmi, & Laron, 1981; Benedek, Poznanski, & Mason, 1979; Brooks-Gunn, 1984, 1987; Brooks-Gunn & Warren, 1985a; Jenkins, 1983; Livson & Bronson, 1961; Tobin-Richards, Boxer, & Petersen, 1983). For example, menarche might be particularly connected with issues like adult status, social transition, sexual interests, and heterosexual contacts; breast development particularly related to opposite-sex contacts and to feelings to pride and shame in early adolescence; height gain in early adolescence associated with social ostracism; and weight being especially connected with overall body and self-image.

Some data indicate that menarche is a more powerful determinant for heterosexual relations and sexuality than are pubic hair or breast development (Westney, Jenkins, & Benjamin, 1983). A higher correspondence between pubertal development and physical similarity and disclosure of menarcheal status to friends has been reported for pre-postmenarcheal status and breast development than for pubic hair growth (Brooks-Gunn, Warren, Samelson, & Fox, 1986). Brooks-Gunn et al. proposed that "hair growth may not be as relevant to friendships as menarche or breast growth, possibly because it is not discussed (like menarche), is not commented upon by others (like breast development), had a different psychological meaning attached to it by the adolescent, and/or is perceived more negatively in the culture than breast growth or menarche" (p. 12). Although girls who mature "on time" tend to have more positive self-conceptions than girls who mature early or late, this curvilinear relationship is changed to a linear one when breast development is used as marker of pubertal development (Petersen, 1985). Brooks-Gunn (1984) reviewed some results from studies on the significance of different growth indices. She reported that fifth-grade girls with more advanced breast development considered marriage and children more important than the girls with less-advanced breast development. Furthermore, the more-advanced girls showed higher adjustment, better body image, and better peer relations. Finally, more of the girls whose breasts had begun to develop reported being teased—an experience that no one regarded as pleasant—than did girls whose breast development had not begun. No relationships between pubic hair growth and aspects of psychological and social functioning were found.

In a review of some literature on the concomitants and implications of breast development for adolescent females, Benedek, Poznanski, and Mason (1979) commented that the development of the breasts rather than the menarche is the overt signal to parents and to the adolescent herself that adolescence is about to set in. However, due to the continuous development of this aspect of physical maturation, females seldom remember accurately when their breasts started to develop. This is in contrast to the menarche, which is a discrete event and remembered in considerably more detail. Due to the intimate relation between breasts and sexuality—for the girl herself and for the surrounding social environment—sexual aspects of maturation can be supposed to enter at a much earlier time in the life of the teenage girls than the commonly investigated connection between menarche and genital sexuality.

In the end, it might even be the case that the long-range implications of pubertal timing might involve one type of maturation indicators for early developers and another type for late developers (cf. Faust, 1983). One might also speculate on the possibility of whether or not an individual's score on one type of physical growth measure is less determining for personality development and interpersonal relations than is her timing pattern for a set of growth measures, (i.e., that the ipsative profile rather than the nomothetic variation is the essential determining growth characteristic in development).

Sixth, whether early or late physical maturation is advantageous or not is related to the character of the environment and overriding demographic factors. Early pubertal maturation (or late) might be beneficial in one context but a drawback in another. Girls with early (late) pubertal timing might select certain environments over others, and the environment's response to these girls might be radically different from one context to another. To take one example, for ballet dancers, who must be able to combine constant dietary with heavy exercise, late physical maturation is an asset (Brooks-Gunn, 1987; Brooks-Gunn & Warren, 1985b), although not necessary so for girls in general. Although late-matured dancers had more positive body image than did dancers who developed on time, Brooks-Gunn and Warren (1985a) found that in the comparison group (girls in private schools) it was the on-time group of girls who had lower body image. Furthermore, whether or not early maturation is positively or negatively associated with self-confidence has been reported to be dependent on socioeconomic conditions (Clausen, 1975). When discussing reasons for the limited effect of pubertal status across studies, Petersen (1985) argued that fundamental characteristics of the U.S. society itself, such as the age- and grade-structured environments for adolescents, set limits for the influence of pubertal maturation on behavior.

All things considered, it makes sense to expect specific rather than global consequences of maturity timing in girls, and it seems inappropriate to say that early or late pubertal development is generally beneficial or disadvantageous for girls, making an overall evaluation of processes instigated by physical maturation difficult. Therefore, one conclusion that can be drawn from the reviewed studies regarding the impact of biological maturity on social transition behavior is that the understanding of social transition behaviors in adolescence requires more careful analyses of the role of pubertal development at different points in time than has heretofore been made. Such analyses will necessitate taking into consideration the cultural, psychological, and social factors that mediate the association (Brooks-Gunn & Petersen, 1984; Petersen, 1980; Petersen & Taylor, 1980). Simple cause–effect relations involving physical maturation are probably marginal in comparison to the cases involving multiple determinants, mediated relationships, and interactive effects.

CONCEPTUAL AND METHODOLOGICAL ISSUES

It appears that many of the inconsistencies across studies and the appearance of few main developmental progressions associated with early versus late physical maturation must be evaluated against the quality of the data presented. Contemporary knowledge of how physical development in girls is related to other areas of development is meager for the following three reasons:

1. there is a scarcity of studies on physical development in females,
2. existing studies are often hampered by weak designs, and
3. there is a gap between empirical results and their interpretation.

Partly for these reasons, we are far from a thorough understanding of how physical growth in females determines their psychosocial adjustment in both the short-term and in the long-term perspective.

Limited Empirical Data

Given the certainty with which many have expressed their opinions on the impact of pubertal timing for psychosocial functioning in females it is surprising to find how little empirical work has actually been presented in this field. There are considerably more speculations and general views on the subject than there are empirical data to support them, and there are almost as many attempts to survey research work in this area as there are empirical studies themselves. Up to the present time, fewer than 100 empirical studies have been reported relating measures in the domains of self-concept, behavior, achievement, emotions, and interpersonal relations to pubertal growth. A scarcity of studies predominantly characterized the period up to the middle of the 1970s. Thereafter, a renewed interest in issues associated with biopsychosocial development has resulted in a marked increase of investigations on the role of physical development in adolescence.

We hesitate to say that few empirical analyses have been performed. Probably considerably more empirical analyses have been done than what have shown up in scientific journals, but there is a common tendency to report positive findings rather than null-findings or contradictory results, and the latter types of investigations possibly are overrepresented among the "hidden" studies. For example, Clausen (1975), in his review of a considerable body of data on the social meaning of physique and the issue of early versus late physical maturation in two of the first longitudinal growth studies, the Oakland Growth Study and the Guidance Study, both of which have produced a wealth of knowledge on the role of physical maturation in mental, emotional, and social adjustment, commented, in the end, that he had "presented very little of the data on girls, partly because of the limits of space but primarily because of divergent findings." (p. 45)

Not only is there a scarcity of systematic research in this area, but because a great part of it is concerned with the immediate experience and concomitants of the menarche—how girls react to their first menstruation, feelings attached to the menarche and menstruation in general, knowledge, expectations, and preparations for menarche, sources of support and information and so on—there are subsequently even fewer studies that have dealt with the general issue of the connection between maturational timing and psychosocial development over longer periods of time. There are just a handful of studies that have related maturational timing in girls to their adult life situation. Replications are rare in this field of research, raising the question of whether the obtained findings were due to chance and questioning the generality of obtained relationships.

This state of affairs, of limited empirical data and the associated risk for sampling errors, means that we lack a solid empirical basis for drawing conclusions on the role played by physical maturation in both short- and long-term development. The normal development of science is to move from descriptive hypothesis generating to more detailed model testing and more exact hypotheses. A characteristic feature of the contemporary situation for research on physical growth in females is that we are far from having passed the descriptive stage. Exploratory research predominates, we lack integrative models, and we have only recently began to get a rough grasp of the particular domains that are connected with the timing factor.

Methodology

As reviewed by Greif and Ulman (1982), difficulties arising from limited control of chronological age when comparing early and late developers, limited size of subject samples, unreliability of measures, limited use of multiple criteria and scarce use of multimethod analyses of data, and the reliance on self-reports, obstruct a comprehensive knowledge of the role of pubertal maturation in development. Other serious limitations are restrictions of cross-sectional designs, inadequate subject samples, and lack of appropriate statistical methodology that matches underlying theoretical assumptions about the impact of pubertal timing.

Type of Data. As in many other fields in developmental psychology, research on physical maturity has been dominated by empirical investigations using cross-sectional data. This is unfortunate because the personal and social consequences of developing early or late probably are quite different at different ages. Besides, the nature of the developmental progressions, how the constellation of physical timing relative to psychosocial variables at one point in time changes to another constellation at another point in time, which are the primary determinants for change, and how the effects are mediated, cannot be tackled with cross-sectional data. These issues need longitudinal data to be examined. Probably one of the most serious fallacies found in the literature is the one of drawing inappropriate casual or time-linked conclusions based on cross-sectional information. Such examples can be found.

We are perhaps in a more fortunate position than are many other fields by having at least some longitudinal traditional. The longitudinal growth studies that were initiated at the Institute of Human Development at Berkeley, California, at the end of the 1920s and the beginning of the 1930s, were pioneer works based on solid theoretical ground and were conducted with methodological stringence. They are invaluable when making comparisons with present-day results on pubertal timing. The Berkeley Growth Study initiated by Nancy Bayley in 1928 encompassed about 60 children studied from birth up to 36 years of age. The Guidance Study that was started by Jean W. Macfarlane in 1928 followed 252 infants up to age 30. The Oakland Growth Study (the Adolescent Growth Study) was initiated by Harold E. Jones and Herbert Stoltz in 1931, and it involved 215 children ages 11–12 who were followed up to the age of 40. These longitudinal programs were broadly aimed, utilizing a wide spectrum of individual functioning—biological, psychological, cognitive, motivational, emotional, and social—studied at several age points.

It took a half a century before new longitudinal and follow-up investigations were initiated. Characteristically, these were more narrowly focused, as to both scope and time. Among them are the studies on self-image by Rierdan and Koff (Koff, Rierdan, & Silverstone, 1978), the Early Adolescent Study (Petersen, 1984), the Milwaukee Study (Simmons, Blyth, & McKinney, 1983), and research on the menarcheal experience by Brooks-Gunn and Ruble (1982a, 1982b, 1983).

The broad and the narrow longitudinal and follow-up studies, as well as good cross-sectional investigations on subjects at different ages (cf. Faust, 1960) contain the most useful data we have at the present time for drawing conclusions on the significance of maturital timing. Among them the broad longitudinal projects can most effectively set the factor of pubertal timing into its developmental context alongside psychological, interpersonal, and social changes over time.

Sample Characteristics. Characteristic for many empirical investigations of the impact of the timing of puberty are the small subject samples and their limited ecological generalizability. Our knowledge on the consequences of maturing early versus late can be said to be knowledge primarily derived from data on White, middle-class girls (Brooks-Gunn, Petersen, & Eichorn, 1985). The subjects are often school pupils, parent permission is often required, and the cooperativeness and interest of the school administration is a prerequisite.

Due to ethical considerations, the "we-take-what-we-can-get" rationale often cannot be avoided. However, absence of information on what constitute the target population is disturbing. If, for some reason, subjects in a population are prevented from participation, the minimal information attainable is the proportion of subjects who participated of the total number of subjects intended to be covered. Exclusion of this information provides the reader with no possibility of estimating the drop-out effect. If possible, researchers should attempt to ascertain whether the participating individuals constituted a reasonable random sample of the intended population or a selected "well-adjusted" group.

Loss of data on a reasonable level unfortunately tend to result in unproportionally greater loss of data for individuals from underpriviliged conditions and/or for subjects with serious personal and social problems. Investigators acquainted with longitudinal research are aware of this biased drop-out of exactly the individuals who display problems of inner and outer maladjustment. Researchers employing cross-sectional data have little possibility of estimating this effect with their data. Longitudinal designs offer more control options. An example of a pubertal timing study in which it is likely that the dropout has affected the results is Simmons et al.'s (1983) short-term longitudinal investigation of girls followed from Grade 6 to Grade 10—the Milwaukee longitudinal study.

Originally in Grade 6, Simmons, Blyth, and McKinney's sample constituted 237 girls: "Eighteen schools were randomly selected from specific strata, and all sixty grade students in these schools were invited to participate. Each of these children had signed parental consent. In total, we obtained signed parental consent from 82 percent of those from whom we originally sought this permission" (p. 236). Of these girls, 151, or 64% of the girls tested in Grade 6, could be traced in the community and could be reinterviewed in Grade 10.

Some of the main results in this study were lower school performance, more marked social problem behavior, and more school problems among the earliest-developed girls. The social and academic difficulties among these girls were concentrated to the earlier grades (Grades 6 and 7), and few differences between early- and late-matured girls appeared in the upper grades (Grades 9 and 10). If this study had been cross-sectional and had pertained to the 10th-grade girls, the nonsignificant relationship between the timing of puberty and academic interest, as well as between pubertal timing and social adjustment, would be reported with few additional comments. However, because Simmons, Blyth, and McKinney had relevant data at earlier ages for the dropout girls, they could compare these girls with the rest of the girls in Grades 6 and 7 in relevant respects and, through this rationale, characterize those who could not be retested in the upper grades. All previous knowledge from longitudinal studies tells us that we should expect the subjects with social adjustment problems to be overrepresented among the drop-outs. Indeed, whereas 45% of the early developers who were tested in the earlier grades were lost in Grade 10, the percentage figure for the "average" developers was 36%, and for the late developers 28%.

Not only were the loss of data greatest for the early developers, who were characterized as most socially maladapted in the early grades, the later dropouts among the early-matured groups of girls had had more school problems and lower grade point averages in the early grades than had been the case for the early-developed girls who did not drop out. No significant differences for these variables were observed between the remaining girls and the drop-outs among the average developers or among the late developers.

In summary, the drop-out effect was multiplicative. First, an unknown proportion of girls with social problems, probably greater than the 18% of girls for whom parental consent was not obtained originally in Grade 6, disappeared before the actual investigation started. Second, the drop-out was concentrated to the early-matured girls who were worse off with respect to school achievement and social adaptation in the early grades. Third, the drop-out was concentrated to those of the early-matured girls who had higher incidence of social and academic problems in the earlier grades. Thus, the nonsignificant differences among early-, average-, and late-matured girls with respect to social adaptation and academic behavior in the upper grades is likely to be an artifact of the biased drop-out.

The loss of data for underprivileged groups of subjects can be found in other investigations within the physical growth domain. The adult follow-ups of the subjects in the Guidance study and the Oakland Growth study involved about half of the original samples. In the Guidance Study (MacFarlane, Allen, & Honzik, 1954), observations of a group of children was originally done when they were 21 months old. At this time the control group contained 116 subjects. Forty-two percent of them belonged to lower socioeconomic categories. At the age of 14, and 41 (35.3% of the original sample) were traced and tested, and 29% of these subjects belonged to the low socioeconomic categories.

It is not unreasonable to assume that to the extent that early or late physical maturation is negative for females, these negative consequences might be especially focused for females who in other respects, socioeconomically or intellectually, are in an disadvantageous position. Therefore, the study of psychosocial consequences of pubertal maturation requires investigating the total range of girls, not only the "average" or the well-adjusted girls (cf. Simmons, et al., 1979). It is probably the case that the relationship of individual differences in pubertal development to other, particularly social, variables has gone unnoticed for the very reason that the subjects sampled often have been well-adjusted, middle-class girls willing and permitted to participate (i.e., constituted a sample group with an disproportional lack of data for the segment of the female population characterized by certain adjustment difficulties).

Methods for Data Treatment. To the extent that maturity timing can be treated as a continuous variable with a normal distribution, and to the extent that we could expect a monotonic positive or monotonic negative relationship with other variables of interest, the correlational strategy is the appropriate statistical test of the associations. This is also a methodology adopted frequently in research on the effects of maturing early or late. However, there is much that suggests that the necessary conditions for the use of linear correlations often cannot be met in the realm of physical maturation. Nonlinear and interactive associations may, in fact, be masked by employing the linear methodology (cf. Magnusson, 1985, 1988; Shader, Harmatz, & Tammerk, 1974).

Many studies of the relationship between pubertal timing and psychosocial development are based on the idea that it is primarily the girls who are "out of range" in their physical development, compared to the majority of same-age peers, who are susceptible to environmental pressures (Petersen & Taylor, 1980; Weatherley, 1964). Obviously, the correlation coefficient is not the appropriate association measure when the anticipated consequences are concentrated to subgroups of girls at both the extreme ends of the maturity continuum.

In other cases, the expected effect is thought to occur only among one of these extreme groups of girls. For example, the early-matured girls are commonly thought to have an exposed position in the early teenage years when their same-age female and male peers are still undeveloped. The application of a linear methodology when investigating the effects of pubertal timing in this case may mask the true underlying relationship, with its effect concentrated to the early developers.

In conclusion, theoretical expectations of curvilinear effects in both the cases mentioned set limits to the use of linear models as the appropriate methodology. Possibilities exist to specify interaction effects in regression models. However, as a general method, and given the hypothesis-generation stage of development of the field, this is not advisable. What is obviously needed is a measurement technique that can take account of nonlinear relationships and interactions.

An alternative, which also has been suggested by Shader et al. (1974), is the use of ANOVA and ANCOVA techniques. Although analysis of variance does not take full advantage of the rank ordering of individuals in physical timing, the loss of power by treating physical maturity as a categorial variable should be weighed against the possibility of conducting post hoc tests and thereby detecting nonlinear effects. Furthermore, in order to test a given theoretical model, planned comparisons could be conducted to specify the necessary comparisons among subgroups of girls. The analysis of variance method for data treatment, in addition to pattern analysis, are the methods that are used in the present study.

THEORETICAL APPROACHES

In some sense, the situation with respect to model building has not changed much from the time when the first systematic investigations of pubertal timing were initiated. We have passed the descriptive phase, involving estimation of the relevant measures of pubertal growth, their distributions, and their changes over time. However, we are still in an exploratory stage accumulating evidence in different areas on how the variations in pubertal timing affects the everyday life of girls throughout the adolescent years. The need to develop models linking the biological, the psychological, and the social facets in development have been expressed by many. Some hunches about which transition mechanisms are in operation have emerged. However, we still lack a comprehensive understanding of the developmental progressions or which factors underlie the relationships between maturity timing and psychosocial adjustment in different ages.

There is as yet no single conceptual framework that can account for the diverse findings or encompass the multitude of relationships presented in the empirical literature. Such a model, at the outset, must specify the general mechanisms responsible for how change in physical structure is translated into social and emotional growth, and it must specify the appropriate operational measures. What comes closest to such an integration is the psychodynamic claim that many of the common developmental themes in adolescence—individuation, parent–child relations, peer affiliation, heterosexual relations—are the primary and secondary consequences of the increase in sexual drive accompanying puberty. Following the latency period, characterized by gradual and continuous development of ego capacities, puberty with its intensification of libidinal impulses abruptly brings development into a state of disequilibrium where the building up of new and diverse object relations, a sexual identity, and adjustment to one's sexual feelings becomes immense.

This type of psychodynamic model is what Petersen and Taylor (1980) denote as a "biological deterministic model," which postulates a time-linked association between biology and psychosocial adjustment. The intensification of genital impulses is both the push and pull in development. No specific testing of the model has yet been made other than in limited areas, and operational definitions of utilized concepts are limited. At the present, the model is too broad and too vague to account for the shifts in effects of maturational timing observed throughout adolescence. In the historical perspective the psychodynamic viewpoint has moved from regarding puberty, as reflected in the menarcheal event, as a predominantly traumatic event to the present day emphasis on positive organization and accentuation of femininity.

Apart from more broad conceptual frameworks there are some approaches to research that have been advanced to clarify the role of pubertal timing in development. They are reviewed here. These approaches are not to be regarded as separate and mutually exclusive. Furthermore, they are not to be seen as theory-guided approaches in the ordinary sense of the word "theory." For convenience, we have dealt with them as three common frames of reference for dealing with the functional relationship between physical growth and psychosocial development.

The first approach, the case of *unitary development*, suggests an intimate relationship between physical maturation and psychosocial functioning in girls. With early pubertal development is assumed to follow an earlier entry into adult-type activities and more matured forms of behavior, perception of oneself as reproductively mature and identification as a mature female, as well as a more consolidated emotional status. Another set of research questions is raised by researchers who, rather than postulating a unitary development across various areas, assume the existence of uneven or *asynchronic development*. Just the fact that development does not proceed at the same pace in the physical domain as in other domains is thought to have specific and often negative consequences for some girls. The third approach has been coined *the "deviancy" hypothesis* in the literature. It makes no claims on the evenness or unevenness of development, but rather stresses the implications of being an early-developed or late-developed girl relative to other same-age girls.

Unitary Development

A major perspective in the field of pubertal impact among girls is the postulate that social and emotional development co-occur with physical maturation. The onset of puberty triggers or facilitates the onset of development in other areas which signify a more mature status. Being physically older than her chronological age, the early-maturing girl is expected to be sensitive to and develop the values, interests, and behavior appropriate for girls who are older than herself: heterosexual interest, female identification, motherhood, interest in childrearing, and so forth. She is generally expected to approach the adult way of functioning at an earlier time than her physically later-developed peers. This process is facilitated by the presumed support she received from those most akin to her, who encourage the girl to adopt a more adult-like behavior.

The first explicit attempt to test this approach empirically was made by Stone and Barker in the 1930s in the Oakland Growth Study. Empirically, they employed the Pressey and the Sullivan tests of developmental maturity to test the idea of earlier social maturity as a function of earlier biological maturity, and they confirmed that the interests of the early physically developed girls were more similar to those of chronologically older girls, than were the interests of the late developers, in areas such as heterosexual interests, adornment and bodily display, avoidance of strenuous activities, and engagement in imaginative and secretive activities. The early-maturing girls displayed interest patterns that were on the average 11 to 15 months more advanced than would be expected from their chronological age.

The expected accelerated social development among early-developing girls has been found in other empirical studies to occur in areas such as heterosexual relations, autonomy, and independence from the home, increased interests in children and motherhood, and aspirations to establish a family (Brooks-

Gunn, 1984; Davies, 1977; Dornbusch et al., 1981; Goldberg, Blumberg, & Kriger, 1982; Kiernan, 1977).

The concordance of development in the physical, emotional, and social domains was also an issue that More (1953) attempted to investigate. He hypothesized that development in social behaviors and attitudes, a firmer sex-role identity, and more emotionally matured behavior appeared concurrently with puberty. The early-matured girl was supposed to assume greater responsibility for herself, and for activities in the social networks she was engaged in, compared to the late-matured girl. She would often initiate activity, stimulate others, participate in social activity, and generally show better qualities of leadership. She would to a higher extent stand for her opinion against the pressures of peers and authorities, be able to maintain a warm interpersonal climate, display adult attitudes and responsibility in society. Compared to the late-matured girl she would be more satisfied with her sex role, and be able to rely on her feminine qualities. There would emerge more realistic self-perceptions and behaviors that were more governed by self-initiated standards. She also would be more emotionally stable than her later-matured peers. Results also confirmed that early-maturing girls were more mature emotionally and socially than were the late developers, and that the advantage that the early-maturing girls had was not limited only to the early teens, but appeared at the age of 16 as well. Of particular interest was the finding that emotional adjustment seemed to be the critical watershed if early physical maturation was to find its way in more matured social behavior.

Faust (1960) also presented evidence that early physical maturation was accompanied by the adopting of more matured behavior and social relationships. In all grades investigated—6, 7, 8, and 9—earlier-developed girls in her study were rated by their peers as more grown-up, and in the two upper grades they were also considered to engage more frequently with older peers than did later-developed girls. Finally, Jones and Mussen (1958) compared very early- and very late-matured 17-year-old girls on a variety of different TAT measures. The early-matured girls showed higher self-esteem and less need of recognition. There was also some evidence that the late-developed girls were more dominated by their parents at this age.

Before leaving the discussion of the unitary approach, certain additional comments should be made regarding the significance of the menarche. The first menstruation has in previous literature been pictured as a strong *maturity sensitizer*, having widespread effects on other areas of development. For psychodynamic theorists, the menarche constitutes one of several normative developmental crises with which females have to cope over the life course. If successful adaptation occurs, it will form the basis for a well-functioning female role and identity. If coping is unsuccessful, old psychosexual conflicts will be reactivated.

The first menstruation in girls constitutes a discrete and sharply delineated event around which substantial changes in self-conception are assumed to take place. Prior to menarche, girls have relatively unclear body images, vague cognitions and diffuse fantasies about sexuality and femininity (Hart & Sarnoff, 1971; Kestenberg, 1961, 1967a, 1967b). Hart and Sarnoff (1971) characterized the permenarcheal period as one of "diffusion" and Deutsch (1944) labeled it "the period of expectation." The menarche brings order and consolidation to the feminine identity, and more realistic body image, and better perceptual organization (Blos, 1962; Hart & Sarnoff, 1971; Kestenberg, 1961, 1967a, 1967b, 1968; Notman, 1983). If accepted and integrated with the personality structure, the self-describing qualities as feminine and woman should be beneficial for self-esteem.

Kestenberg (1961) has postulated a direct parallel between hormonal activity and perceptual functioning. In the premenarcheal girl, the diffuse inner hormonal activity and inner tensions have their counterparts in diffuse cognitions, in unclear body image, and in intense but short-lived affectional reactions. The menarche channelizes the unstructured inner sensations, and operates as a platform for organization, promoting reality orientation, structure, and regularity in the life of the teenage girl.

It is of course acknowledged that a number of factors are responsible for how a girl copes with and experiences the menarcheal event. Whether or not she will draw benefits from it is first and foremost dependent on her earlier life history, where prior emotional problems make for poorer adaptation (Kestenberg, 1961). The extent to which it promotes womanhood is also connected with her momentary relations to her parents, and their reactions to the event, as well as with the girl's preparation for the experience emotionally (Ritvo, 1977). However, in the normal course, progression should occur and the menarche should have a positive effect on the girl's self-image.

There is some empirical support that menarche sensitizes a girl to her female identity, as a potential mother, and to aspects associated with adult patterns of living, and that these changes may have an impact on her self-judgments (Brooks-Gunn & Petersen, 1984; Garwood & Allen, 1979; Simmons, Blyth, & McKinney, 1983). Rierdan, Koff, and co-workers have tested hypotheses concerning the role of menarche for the body image and female identification in girls. Employing different procedures to classify responses to the Draw-a-Person test, they found support for assumptions of stronger sexual identification among postmenarcheal girls than among premenarcheal (Koff, Rierdan, & Silverstone, 1978; Rierdan & Koff, 1980a, 1980b). In a study on fourth- through sixth-grade girls, most of whom were premenarcheal, Williams (1983) reported that the majority thought that their mothers would treat them more like an adult after their first menstruation, and about half thought that their fathers would do the same.

Asynchronic Development

In the context of physical maturation, a situation of asynchronic development in an individual can be said to occur when there are discrepancies between the physical maturity level of the individual and his or her development in other areas: cognitive, emotional, interpersonal, moral, and so on (see Eichorn, 1975). The prototype that has been discussed in the physical growth literature is the early-matured girl who has adopted an advanced social life pattern commensurate with her physical development even though her emotional developmental status is not mature enough to cope with the pressures and demands from her interpersonal milieu. It is this type of developmental asynchrony that researchers discussing the implications of the secular trend have also addressed to account for the increase in incidence and prevalence of problem behavior among young teenagers.

Whatever reasons are outlined for anticipating negative consequences of accelerated social development in some early-matured girls, basically it is claimed that these girls are too emotionally immature or have too little experience to cope with these new behaviors (Golub, 1984; Simmons, Carlton-Ford, & Blyth, 1987). Frisk (1975) has contended that the early-matured girls' premature step out of childhood often is accompanied by limited social support. Parental conflicts might develop due to these girls' striving for autonomy and engaging in more advanced behavior than is normative among girls their age. In a similar vein, disturbances in the contacts with parents immediately following pubertal onset, particularly with regard to the mother–daughter contact, have been posited by several researchers (Hill, Holmbeck, Marlow, Green, & Lynch, 1985; Steinberg, 1987). Additionally, Jones (1949) had advocated that because the early-developed girl has few friends who match her own physical status, she is forced toward association with older friends. This situation could eventually lead to negative consequences for her school adjustment (Davies, 1977). Adopting the values of older peers, the early-developed girl will identify with the less favorable attitudes toward school that characterize older groups of adolescents. In addition, it has been argued that to mature early is to be exposed to unreasonable expectations from peers and adults who pressure these girls into behavior for which they are not developmentally ready (Clausen, 1975). The new social life that is opened for the early developer—heterosexual experiences, independence from home, autonomy, career choice, and the like—emerges at a time when some of the girls have insufficient experience to handle it effectively (Frisk, Tenhunen, Widholm, & Hortling, 1966; Garwood & Allen, 1979; Simmons, Blyth, Van Cleave, Bush, 1979; Simmons, Carlton-Ford, & Blyth, 1987).

Heterosexual relations in particular seems to be a domain that is connected with particularly negative consequences for the early-matured girl. In her early contacts with boys she might be forced into sexual behavior, increasing her risk for becoming pregnant. Another obvious consequence of premature heterosexual contacts is family conflicts (Frisk, 1968; Frisk, Tenhunen, Widholm, & Hortling, 1966). Parents are concerned that their daughters do not engage in heterosexual contacts too early; rather than promoting a more mature social life and supporting the early-matured girl's entry into stable relations with boys, the parents of the early-developed girl often try to oppose their daughter's contacts with boys (Jones, 1949). According to Berzonsky (1981):

> Since girls, on the average, already mature faster than their male cohorts, the precocious girl is even more advanced than her male cohorts. She must consequently date boys considerably older than herself. A three- or four-year edge (or deficit in this case) in social experience tends to be more pronounced when a girl is eleven or twelve than when she is, say, eighteen. The dating of older boys may make her, in the eyes of her still immature female cohorts, an object of envy and jealousy. . . . Parents, too, may lack understanding and empathy: "What's going on between Scott and you?" "He's five years older than you are!" "Why can't you be like Jan, your best friend?" "She stays home and never causes her parents to worry." A generational conflict thus ensues. (p. 165)

Frisk, Tenhunen, Widholm, and Hortling (1966) confirmed in their patient group that heterosexual interests was one of the main sources of parental conflicts associated with early maturation.

Empirical studies have shown higher incidence of social and personal problems (Frisk, Tenhunen, Widholm, & Hortling, 1966; Simmons, Blyth, & McKinney, 1983; Simmons, Carlton-Ford, & Blyth, 1987) school-related problems (Andersson, Dunér, & Magnusson, 1980; Davies, 1977; Simmons, Blyth, & McKinney, 1983) and family problems (Crocket & Petersen, 1987; Frisk, Tenhunen, Widholm, & Hortling, 1966; Garwood & Allen, 1979; Hill, Holmbeck, Marlow, Green, & Lynch, 1985; Jones & Mussen, 1958; Savin-Williams & Small, 1986; Smith & Powell, 1956; Steinberg, 1985; Stone & Barker, 1939) among the early-developing girls relative to late-maturing girls. Although there are studies confirming the role of timing of sexual maturation for adolescent pregnancy, substantial relationships between timing of reproductive maturity and adolescent pregnancy have yet to be demonstrated (Gottschalk, Titchener, Piker, & Stewart, 1964; Kiernan, 1977; Presser, 1978).

Paranthetically, limited parental support of the social behavior concurrent with physical development among girls, in contrast with the generally strong support for social development among parents of boys, might comprise one of the main watersheds that cause early maturation to have primarily positive consequences for boys, whereas the results are more mixed for girls (cf. Hill & Lynch, 1983).

The issue of uneven or asynchronic development in an individual is an important, but a difficult one to tackle, both theoretically and empirically. Post Hoc, a reasonable interpretation of data indicating problem behaviors and coping inefficiency among early-developing girls was that these girls were not emotionally ready for such a rapid social development. However, the issue has not yet been subjected to prospective studies. Attempts have been made to develop instruments measuring "emotional age" (Pressey & Pressey, 1933), but the impact on developmental research has been limited, mainly due to the problem of how to conceive of "growth" in the affect domain. Clinicians recognize the problems involved in defining one type of emotional reaction as more mature than another. Excellent reviews and discussions on this subject have been given by Eichorn (1975) and Ausubel (1954).

Many researchers have speculated about the possibilities of uneven development in individuals. However, as mentioned, the physically and socially mature but an emotionally immature girl is still a developmental configuration that has no empirical confirmation. The data at hand are indirect evidences of such occurrences. From studies showing higher incidence of social problems, school achievement, and school problems among the early-matured girls compared to late- matured, for example, using between-individual comparisons, within-individual asynchronies cannot be established.

Limited Preparation. A case for uneven development, due to limited preparations for pubertal onset, has been advanced by Peskin (Peskin, 1967, 1973; Peskin & Livson, 1972). From his ego psychology viewpoint of adolescent development the latency years are an asexual period in which the person adapts to his or her environment with curiosity, exploring it in a practical and concrete way, gaining ego resources to be used in the next phase of the pubertal drive. Peskin has argued that to mature early physically is to cut short the latency period. The phase between infantile sexuality and adult genital sexuality will be shorter for the individual who enters puberty early and longer for the late-matured. The ego mechanisms developed during the latency period, to be used in coping with the sexual drive of puberty, will hence be less adequately developed in the early-matured individual. Consequently, the person will be less prepared for the emerging developmental tasks of adolescence.

In support of this argument, one might expect the early developers to have a more problematic pubertal entrance at or immediately around pubertal onset and to "show more intolerance of pubertal impulse, more fear of losing control than would the late maturers" (Peskin & Livson, 1972, p. 350). This hypothesis was tested separately for males and females in the Guidance Study subject sample by Peskin (1967, 1973). Equating the subjects for the age of pubertal onset (independent of the chronological age of the subjects), Peskin confirmed for males that prepubertally few differences were manifested between early- and late-developed males whereas, at the time of the pubertal onset, the early developers differed from the late-developed males in a variety of respects, such as showing lower activity level, less exploratory behavior, less socially initiating behavior, lower intellectual curiosity, less sociability and so forth. Overall, the early developers were characterized as more socially assertive, conforming and emotionally rigid, whereas the late maturers showed more activity, expressiveness, and emotional spontaneity. The results suggested that the early developers' fear of the inability to control impulses had consequences for their social adjustment and was reflected in greater social conforming than among late-developed males. A similar comparison of the early and the late-developed females (Peskin, 1973) equalized for the onset of puberty (2 years prior to the menarche) showed that the early developers expressed more stressful reactions (higher ratings for temper tantrums and whining) and more socially introverted behavior (lower ratings for social exhibitionism, higher for shyness). They also had lower ratings for irritability and higher ratings for dream recall and introversion, something that was attributed to "a turning receptively toward inner feelings." Contrary to what was expected, as many differences between the groups were evident prior to puberty onset as after. A similar type of research, but in this case linking changes in family relations and changes in intrapsychic functioning to (before, at, after) the point in time when puberty arrives among females has been presented by Hill and Holmbeck (1987); Hill, Holmbeck, Marlow, Green, and Lynch (1985); and MacFarlane, Allen, and Honzik (1954).

Stressful Change. A special case of the uneven developmental hypothesis concerns girls' mental preparedness for the menarche and their factual knowledge of physical change processes in themselves.

How girls interpret and experience the change in bodily configuration, the menarche, and the feelings that are associated with the rapid physical growth, depends on their prior knowledge and preparation for pubertal change (Brooks-Gunn & Ruble, 1982a, 1982b; Shainess, 1961). Positive or negative feelings associated with pubertal change may have lasting consequences for personal and social adjustment during adolescence as well as later on (Shainess, 1961). Early timing of maturation might be especially stressful and anxiety provoking. The early-matured girl will have less time to integrate the experience emotionally, and must solve to a greater extent by herself how to use tampons, develop menstrual hygiene, and adjust to how her physical activity and moods are linked to the menstrual cycle (cf. Deutsch, 1944). Koff, Rierdan, and Sheingold (1982) reported that early-developed girls among interviewed college girls regarded their menarche more negatively, had less prior information, and were subjectively less prepared than were later-matured females. This is consistent with findings reported by others (Woods, Dery, & Most, 1982). Furthermore, some data indicate that girls with early menarche experience more menstrual discomfort than girls who have their first menstruation later on (Furu, 1976; Ruble & Brooks-Gunn, 1982).

The external conditions surrounding the menarche in a narrow sense, and physical maturation in a wider, also might be more anxiety provoking for the early developer (Whisnant, Brett, & Zegans, 1979). The early developer will have fewer friends to confide in and with whom to discuss the potential meaning of the change. A consequence of this may be that the girl feels ashamed and tries to conceal the menarcheal fact from the close interpersonal environment (Ritvo, 1977). Petersen (1985) summarized some results from a study of early-developing girls and commented that some of these girls tended to conceal their advanced pubertal development by dressing in particular manners, and that almost every second early-developed girl did not report the correct time that she had had her first menstruation but waited until a later point in time. Some even concealed the menarche from their mothers.

The "Deviancy" Hypothesis

What has been termed the *deviancy* hypothesis, or the experience of not developing at the same pace as everybody else, has frequently been discussed in the physical growth literature as having special implications for psychological adjustment and interpersonal relations. Reaching puberty at a time when none of one's same-age peers, some of the peers, and most of the peers have reached this developmental point is thought to mean very different things for an individual. Accordingly, any attempt to understand the psychological reaction of a girl to her physical change must acknowledge the developmental level of the peer group at the time the person enters his or her physical growth acceleration (cf. Brook-Gunn & Ruble, 1982b; Faust, 1960: Petersen, 1987).

The exposed situation for the early-developed girls in early adolescence coexists with the period in which peer conformity reaches its maximum in adolescence (Constanzo & Shaw, 1966). Therefore, to differ in appearance from others at this point is likely to be a strongly distressing experience (Petersen & Spiga, 1982; Stolz & Stolz, 1944). Being physically different from others could have the consequence of defining onself as odd (Higham, 1980), leading to estrangement (Weatherley, 1964) and increasing self-consciousness of one's appearance. The experience of falling out of range with peers has been said to impair personality development and to lead to social isolation or to a search for alternative nonconventional networks that better satisfy the person's needs.

Of the total range of girls, those who mature earliest and those who mature latest enter their puberty as "minorities." When the early developer has reached puberty, she will differ physically from her immature same-age peers, whereas when the late developer attains puberty she will belong to an immature group distinguished from the vast majority of girls who have come a long way toward adult physical status.

The early-matured girls will appear physically conspicuous in the eyes of the peer group. They will stand out from others in height and in weight, differing not only from the same-sex peers but being comparatively more matured physically than opposite sex peers (Bayley & Tuddenham, 1944; Jones, 1949). Greater height and weight do not carry any immediate social advantages for the early developers (Faust, 1960) but might operate as a social stigma (Jones & Mussen, 1958). An analogous exposed situation is true for the late-matured girls during late adolescence. Faust (1960) noted that for these girls "the defense against immature physical status in junior high school seems to be more of a withdrawal and an attempt to be inconspicuous in the group" (p. 180).

Probably the best empirical illustration of the isolated situation for the early- and late-matured girls at respective end points of the pubertal interval is provided in the study of social prestige by Faust (1960). In her study, early-maturing girls in early adolescence were a minority of the total female population and were non-normative for their age group; at this time they were the least popular girls. At later adolescence, they were more "in phase" with their peers, and at this age they were ascribed the most prestigious qualities by their peers. It was now the developmentally deviant late-matured girls who were rated lowest in social prestige.

GENERAL AND DIFFERENTIAL IMPACT OF PUBERTY

Although empirical data may attest to effects on behavior at some point in adolescence of the age for puberty entry, it is not always clear how to conceptualize this impact. Diverse models can be formulated, and failures to distinguish between them have consequences for interpretations and explanations.

At least two models can be formulated (Magnusson, Stattin, & Allen, 1986). The first model refers to a *general* maturational impact. It suggests that timing of maturation has a general effect on behavior when girls attain some point or phase in their puberty sequence. When they reach this point, they might, to take an example, acquire more mature forms of functioning (socially, emotionally, etc.). Hence, the fact that we can observe individual differences in behavior being associated with timing of maturation at some point in time in adolescence is due to the circumstances that some girls have already passed this point in their puberty sequence whereas others still have not come as far. Because all girls ultimately will pass the pubertal point, and subsequently change in a uniform way, we should not expect any persistent or long-term consequences. If physical maturation is general over individuals, it implies that the analysis of early- versus late-matured subjects will be analogous to investigating how puberty itself intervenes in the life of females. Note that advocates of psychodynamic theory have not been so much concerned with the issue of the impact of maturing early versus maturing late as with the psychosexual consequences of attaining puberty per se. Puberty has a general, and similar, impact independent of whether or not the girl is early or late matured.

One very obvious example is menstrual hygiene. Whether the girl is an early or late developer, she will at some point in time or another be confronted with this issue. Another example is the personal and social meaning commonly attached to menarche. Psychodynamic advocates, in particular, have claimed that the menarche constitutes a turning point for the organization of subsequent life experiences, for defining oneself as woman and adult (Hart & Sarnoff, 1971; Kestenberg, 1961). Again, this change in girls' perceptions of themselves should be equally applicable whenever girls attain menarche.

This first interpretation predicts that the relationship of pubertal development to behavior is mainly a temporary and population-sequential phenomenon. Pubertal development has an impact on behavior, but the variations among girls in pubertal timing will only affect the point in time when these behaviors are activated. From this interpretation we would assume that the late-developed girls would catch up with the early developers with the passage of time, and that the relationship between maturational timing and psychosocial functioning would attenuate over time.

Another interpretation, a *differential* maturational impact hypothesis, is also possible. In opposition to the model of a generalized impact this model does not assume that the total range of females will be affected by puberty in the same way. Rather, the impact of biological maturation will be different for different subjects.

Some life experiences will only pertain to the earliest matured girls. A case in point is the social support available among same-age peers for sharing confidences and problems associated with the menarche and subsequent menstruations. What is unique for the early-developed girls and does not apply to the later-matured girls is that the early developers will have none or very few friends in their circle of same-age friends with whom they can share information, discuss, and compare common experiences associated with the menarche.

The earliest- and the latest-developed girls will also have something in common that does not apply to the majority of girls, namely, that when they enter puberty they will automatically be out of range with their peers physically. The early-matured girl enters puberty when the vast majority of her same-age peers are still physically immature, and the late-matured girl enters puberty in the context of a majority of mature peers.

The differential effects model is the one that is of particular interest from the point of view of the researcher looking for long-term effects of maturational timing. If, indeed, some girls, due to their pubertal timing, have distinct experiences and will perceive different reactions from their interpersonal environment than is the case for other girls, distinctly different forms of personality and social functioning may evolve among girls both contemporaneously and in a most lasting way. When discussions on the long-term implications of puberty timing have been brought to the fore, it is most often this type of pubertal impact that has been considered because the model assumes that a subgroup of girls will have in common unique or more frequent experiences of a certain kind and that these experiences may subsequently influence the social and emotional development into adult life.

CONCLUDING COMMENTS

In the preceeding section we dealt with two models for pubertal impact. According to the first model, some behaviors manifest a typical developmental pattern over time in adolescence. For these behaviors, physical maturation has an impact when a girl enters puberty (or reaches the menarcheal point), but variations among girls in maturity timing will only determine the point in time at which the behavior sequence is activated. At the end of adolescence all girls have changed their behavior in the same direction and to the same extent. No persistent effects on psychosocial functioning can be expected.

More persistent impact of being early or late developed can be expected from the second model. This model assumes that girls with different timing of maturity will develop and perceive themselves in a distinctly different way from other girls or develop a distinct behavior repertoire, so that different types of personality and psychosocial functioning relative to other girls may emanate. Either this impact is restricted to a *shorter period* of time or it has consequences for behavior both in the *short-term and the long-term perspective*. An example of a short-term differential effect might be that adolescent problem behavior would be more connected with early maturing than with late or that family conflicts related to typical adolescent problems would be overrepresented among the early developers, but, that, after the adolescent years no differences in social adjustment or with respect to family relations between early- or late-developed females could be traced.

Any differences between girls that are demonstrated using a common pre-postmenarchel design at a particular point in time in the pubertal interval cannot directly be interpreted as a function of the operation of either specific type of puberty impact model we have described. For example, if 12-year-old postmenarchael girls are found to display a more feminine self-concept than a comparison group of 12-year-old prepubertal girls, both types of maturity impact is possible. It might be the case that the attainment of menarche generally brings about this self-concept change. If so, and if we were to retest the same sample of girls when they were 14 years old, all girls who at this age are postpubertal might show evidence of adopting this feminine self-conception to the same extent as the former 12-year-old postmenarchael girls, whereas the now 14-year-old premenarchael girls would present the same lack of this self-concept as the former 12-year-old premenarchael girls.

Or the puberty impact might be *differential*. The change in physical appearance toward a more matured woman's body, encountered at a time in early adolescence when this change is not common for girls, may exert a disproportionally greater influence on the earliest-developing girls, making them adopt a fixed stereotype of the traditional feminine identity. When tested at age 12, the earliest-matured girls would differ from the premenarchael girls in such a feminine self-conception. Retested at the age of 14, the same differences would appear, with the earliest matured again being the subgroup of girls who have adopted such a feminine identity, even compared with the other postmenarcheal girls.

From these examples, it should be noted that ascribing the impact of puberty to any one of the two pubertal impact types cannot be made definitely until the developmental progressions are known in their totality over the age span covering puberty. Needless to say, in order to gain a knowledge of the eventual long-lasting consequences of early versus late maturing, this requires following the subjects further into adult life. Any single cross-sectional study conducted in the pubertal years can only hint at the most plasible path. Only through a longitudinal design, following the same group of girls over the adolescent years into adulthood, can more firmer developmental progressions be established.

The distinction between general and differential impact of puberty is an important distinction for several reasons. In addition to its implications for long-range influences it draws attention to the fallacy of drawing inferences from qualities normally associated with pubertal development to differences in some aspect between early and late developers. The finding, for example, that a feminine figure is a valued attribute for heterosexual interactions does not necessarily permit the conclusion that the feminine figure acquired earlier by the early-developed girl gives her the social prestige and the opportunities for establishment of heterosexual contacts that normally accrue to more physically mature girls. There might exist areas in which this is more true. Nevertheless, direct analogies between processes operating on the population level and processes operating for subgroups of individuals should not be drawn automatically.

As an epilogue, one conclusion that can be drawn from the previous review of research hypotheses and empirical findings on puberty impact in girls is that whatever may be the developmental paths linked to pubertal timing, they do not converge in a simple manner. At the present stage of knowledge the judgment of whether early or late physical maturation is beneficial or disadvantageous for girls should not be made at all. One and the same developmental progression might be judged as positive or negative depending on where (and when) we focus our attention. For example, on the positive side of early physical maturation might be the privileges and the support and encouragement given by the immediate interpersonal environment for engaging in adult-type activities. One the negative side are the conflicts with parents if such mature behavior is channelized into early heterosexual relations. Hence, it might be advantageous for some girls, in some domain, at some period of time, but for the present time we know too little to draw any general, or for that matter, any long-term conclusions. That there seem to be different kinds of transitional mechanisms involved—personal, emotional, social, interpersonal—that these effects change as a function of chronological age, and that they probably are connected with the overall socialization process and the attributes of the persons prior to the initiation of the pubertal sequence, makes it fairly obvious that a comprehensive model for the influence of pubertal development in girls requires taking an interactional perspective on development. Simple correlations or other association measures between psychosocial variables and pubertal development must be complemented with attempts to go beyond these relationships, attempts that address the question of the conditions that make certain developmental progressions more likely than others. To complicate the interactional network of changing factors over time even more, models purportedly trying to explain the relation between pubertal change and psychosocial processes must recognize the factors—within the person and in her environment—that mediate these relationships (Petersen & Taylor, 1980).

Petersen (1979) has stressed the point that puberty is not to be conceived of as an event but as a process working over time. Research on the role of timing of puberty thus will require formulating process models, mapping the different types of influences that prevail at different times in adolescence, and examining the associations between these psychosocial constellations at different ages. Only by studying how girls (early- and late-matured) adjust to the changing social and interpersonal circumstances that prevail at different points in time in the adolescent period can we arrive at a total picture of how the pubertal factor operates, impairs, or enhances progression.

The present study is one attempt in the direction toward an interactive model in research on puberty impact. We investigate the relation between biological maturity and psychosocial development in various areas, both in the short and in the long run. Attempts are made to determine the conditions under which relations between the timing of physical maturation and psychosocial functioning can be expected to occur or not to occur.

2
The General Approach and the Basic Model

INTRODUCTION

Chapter 1 described some of the obstacles that any systematic analysis of the role of pubertal timing for psychosocial development must face. They concerned, among other things, the changing external conditions surrounding early or late pubertal development over time, the different types of individual experiences developed from very early age that might affect coping with transition tasks in adolescence and one's pubertal maturation, and the complex interplay of biological, psychological, and social factors in adolescence, which render the role of pubertal development difficult to isolate. No single framework can encompass the diverse findings in previous investigations within a simple developmental formula. Therefore, any comprehensive study of the role of pubertal timing in female development has to address the issue from different perspectives.

The first chapter reviewed three types of general frames of references that have been formulated explicitly or implicitly in earlier literature in order to examine the significance of pubertal timing in female development and its possible effects on behavior. The present research integrates to some extent these three approaches. As is outlined in more detail here, the present study addresses the consequences of variations in maturational timing from the unitary development, the asynchronic development, and the "out-of-range" viewpoints. The research reported is based on an explicit interactional formulation. It attempts to map the interplay of biological, psychological, and sociocultural influences in order to disentangle the role of pubertal maturation, and it proposes a framework within which to analyze the significance of pubertal timing for social behaviors in adolescence.

A model that delineates likely changes in various aspects over the adolescent period, as a function of entering puberty early or late, is presented in this chapter. The model outlines the major areas of person and social processes that can be expected to be affected by variations in the timing of pubertal maturation, and the mediating factors—within the person and in the environment—that are connected with timing of maturation and behavior. The chapter concludes with a set of propositions that are tested empirically in subsequent chapters.

Before introducing the general model, we provide a setting for understanding the impact of timing of maturity on the general conditions under which an individual experiences adolescence. Topics addressed are both conceptual and methodological.

The first topic concerns methodology. It deals with the distinction between pubertal status and pubertal timing. It is argued that, from a developmental perspective, the study of how behaviors, feelings, interests, and attitudes are related to differences among girls in physical growth is most meaningfully accomplished using a pubertal-timing differentiation rather than a pre-post-menarcheal dichotomy.

The second topic concerns the visibility of physical growth in females. When we form impressions of others, the basic information is what meets the eye. Attributes that are immediately present for sensory perception also constitute major building blocks for people's inferences of traits and attitudes of others. The information that the person supplies to the immediate environment by his or her physical appearance is a sort of "social" information to which social expectancies and attitudes are linked (Tobin-Richards, Boxer, & Petersen, 1983). From studies conducted from several decades ago (e.g., Jersild, 1952) up to the present time, when teenagers are asked to mention in what aspects they would change themselves, physical appearance has been found, from study to study, to be an unusually salient feature of the total make up of the person. It is a person-bound quality that the majority mention more often than lifestyle, intellectual capacity, and one's social repertoire. Strong stereotypes exist as to what is the ideal body shape and look for boys and girls. Empirical data suggest that physical appearance is of more concern for a girl's personal adjustment in adolescence, and for the way other people react to her, than it is for boys. It is argued here that physical growth has a high degree of visibility and that the overt manifestations of growth accompanying puberty entrance determine self-perceptions as well as others' perceptions of a girl.

The third topic concerns social comparison. In agreement with other research work in this area, it is emphasized that it is not the fact of merely being early- or late-matured that determines differences in girls' conceptions of themselves and feelings associated with pubertal timing. It is the girl's developmental level as it is related to the developmental level of her same-age peers. To be in-phase or out-of-phase with the mainstream of same-age peers affects self-definitions and behaviors.

The fourth topic concerns issues related to sociocultural influences. Physical growth has no meaning in itself. Its consequences for a girl must be evaluated against the sociocultural norms, values, and clichés that are prevalent in society, particularly as they are connected with the reactions from the primary socializing agents. Furthermore, from the developmental perspective, the impact of early versus late maturity cannot be seen independent of the character of those social behaviors of girls in general that are "typical" of the particular age level under investigation. The consequences of being early-matured or late-matured will vary over time as a function of the prevailing social patterns at different ages. The match of age-graded social behavior with the timing of pubertal development is proposed here to constitute a framework from which to understand maturational timing in female development.

PUBERTAL TIMING AND PUBERTAL STATUS

The first conceptual distinction to be made is the one between pubertal *timing* and pubertal *status* (cf. Petersen & Crocket, 1985). Both these concepts refer to a differentiation among girls on a chronological age continuum.

Pubertal status can be said to constitute the intersection of the menarcheal event (or another growth measure) and a defined point in time. Those girls who have not yet had their first menstruation at a particular point in time are assigned the label *premenarcheal*, whereas those who have experienced their menarche are denoted *postmenarcheal* girls. Naturally, we might think if pre–post something else. One maturational criterion that might be useful is whether or not the girls have reached adult status as determined by the ossification of the bone (Tanner, Whitehouse, Marshall, Healy, & Goldstein, 1975). In the Oakland Growth Study, among the first longitudinal studies on the social and psychological implications of the timing of puberty, skeletal age was the measure used to differentiate between early- and late-matured girls. The reason this measure has not been used more often is that the menarcheal event is thought to have a unique significance in female development. It has been said to constitute "the single most important event of puberty" (Koff, Rierdan, & Silverstone, 1978, p. 635), signifying and symbolizing feminine identity and maturity, attributes that will constitute the core of the personality in subsequent development toward adult status.

Research utilizing the pre–postmenarcheal differentiation have primarily comprised cross-sectional studies, with a few exceptions (Koff et al., 1978; Ruble & Brooks-Gunn, 1982; Simmons, Blyth, & McKinney, 1983). Issues that have been dealt with concern girls' attitudes toward the menarche before and after, how the event is integrated into the concept of oneself as a woman, how it is emotionally integrated, its relation to self-concept and psychopathology, and other aspects and circumstances surrounding the first menstruation. In essence, the questions raised concern the psychological and social significance of the status of having reached menarche.

We might address the question of whether entering puberty earlier or later than what is normative has implications for the girls, concurrently and for future life. Stated in this manner the question at hand pertains to the *pubertal timing* in girls. What matters here is not whether girls have passed their menarche at a specified age, but if the mere fact that some girls are earlier than others, some are normal, and some are late, developmentally speaking, affects their way of thinking, feeling, and acting. For investigators employing the pubertal-timing differentiation of girls, the information that some girls are very early, some moderately early, some normal, some late, and some very late, is of vital concern.

The researcher who uses pubertal-timing information for differentiating girls might, but need not, be interested in the impact of the menarcheal event. More often the information about the first menstruation is used as one criterion among others of developmental maturity in a wider sense. If different pubertal maturation indices show high interrelations, than one could pick any one of them, and approximately the same rank order of girls on a maturity scale would appear. If the menarcheal age is the most easily accessible and economic measure, this would obviously be the one to use. But other developmental criteria would do equally well.

Pubertal status and pubertal timing are not mutually exclusive measures of individual differences in physical growth. They are somewhat different ways to look at the same phenomenon. Although pubertal status, as determined by the menarche, is an absolute event that might be measured independently for an individual, pubertal timing refers to a comparison of a girl with her agemates. Under special circumstances the status and the timing measures might be derived from the same data. For example, having continuously collected information about the pubertal status of girls from late childhood into adult life, a more fine-graded subdivision of the girls from those who had their menarche very early to those who had their first menstruation very late could be compiled. What then are the reasons for upholding the distinction, and does it matter whether we use one or the other of the two individual differences measures?

The pubertal status measure is a dichotomized measure, defined by the occurrence or nonoccurrence of the menarche. As such it is a rather static differentiation. The utilization of a measurement of physical growth based on the presence and the absence of the first menstruation at only a certain point in time has methodological shortcomings with its limited differentation of subjects. Apart from psychometeric reasons there are other more important circumstances limiting the utility of the pre–post menarcheal design. The pre–postmenarcheal dichotomy has been used in particular to draw conclusions on the psychological implications of passing the menarcheal point. However, and this is not always recognized, when girls are subdivided in a cross-sectional design into those with established menarche and those who have not yet passed the first menstruation, differences in some respects between these two groups cannot automatically be interpreted as being a function of the menarcheal experience. On the one side, among those who have started to menstruate there are the girls who had their menarche very early, in some cases a couple of years before the time when the subdivision was made. There are also those who started to menstruate just shortly before the subdivision point. Obviously, the time elapsed before the point in time for the dichotomization has psychological significance. The earliest-developed girls will have had longer time to integrate the menarcheal experience into their personality than will the later-developed postmenarcheal girls. Significant differences in some criterion variable between pre- and postmenarcheal girls might be due to the localization of the effects to the very early-matured postmenarcheal girls. Thus, the physical time over which such self-perceptual processes have operated might have had a decisive influence on the emergence of differences in comparison measures between the pre- and the postmenarcheal girls (i.e., there is a confounding of pre–postmenarcheal status and physical time). The crude subdivision of girls into pre- and postmenarcheal subjects misses this important time-dependent information.

To conclude, a proportion of the variance that the physical status dichotomy shares with some dependent variable will contain a portion of variance that can be attributed to physical time itself. This part of the variance will be largely unknown. Some means for reducing uncertainty are available with a longitudinal design. In this case, the comparison at two points in time of the girls who were premenarcheal at the first test occasion and postmenarcheal at the second with girls who were in the same status at both measuring points (premenarcheal at both occasions or postmenarcheal at both occasions) will yield a more relevant comparison basis for investigating change. Furthermore, this design offers the possibility of using the girls who changed pubertal status from the first occasion to the next as their own control group. In the end it is probably these types of longitudinal designs that will enable us to determine the role of the menarcheal event in female development.

Girls do not suddenly reach a maturity state and remain on this level, as is implied in the pre–post design. Rather, there is a gradual development ranging from entering puberty very early in time to very late. From this broader developmental perspective using the actual time when the menarche occurs as a criterion for general pubertal maturation is probably a more fruitful way to deal with long-term relationships. Effects concentrated to small portions of the research sample will often go unnoticed with the pre–postmenarcheal design. In addition to taking account of the full range of maturity, the use of a pubertal timing differentiation of girls carries with it opportunities to detect such nonlinear relationships as, for example, the concentration of an effect to the very early-matured or to the very late-developed girls. The pubertal-timing differentiation of girls has been used consistently in the present research.

VISIBILITY

The somatic changes of puberty are endocrinologically regulated. The growth in stature from early age is due to several hormones, in particular the secretion of growth hormone—somatotropin—produced by the pituitary gland. Adrenal androgens play an important role for the accelerated growth in puberty. The combined action of follicle stimulating hormone (FSH) and luteinizing hormone (LH) from the anterior pituitary gland initiate production of oestradiol and progesterone in the ovaries. These changes are mainly responsible for the development of genital hair and genital organs, for the composition of the subcutaneous fat, for sexual maturity, and for breast development.

Before the teenage girl notices any visible sign of puberty, hormonal activity has already prepared her for the forthcoming development of the secondary sex characteristics and for reproductive maturity (see Katchadourian, 1977). The onset of sexual maturation depends on the activity of a powerful hormonal system commonly referred to as the hypothalamic-pituitary gonadotropin-gonadal axis. The main coordinator of this hormonal activity is the hypothalamus. It operates through the pituitary gland, situated just below the brain. The pituitary gland secretes hormones into the blood stream with the endocrine glands (the thyroid, the adrenal, and the mammary glands and the ovaries) as the primary target glands, or, as in the case of the growth hormone, directly on body tissue. Although production of the sex hormones can be found already in early childhood, at the age of 8 or 9 the neurons of the hypothalamus increase sharply the secretion of luteinizing hormone–releasing hormone (LH–RH), which, via the anterior pituitary gland, stimulates the secretion of LH and FSH. These hormones act on the gonads. They are mainly responsible for the development of the ovarian follicles, and they stimulate the ovarian follicles to secrete sex hormones. Female oestrogens contribute to the development of the breasts. After the menarche, oestrogens are produced primarily during the follicular phase of the menstrual cycle (i.e., the first 14 days of a cycle of 28 days). Progesterone is formed by the corpora lutea during the latter part of the cycle.

The hypothalamic-pituitary gonadotropin-gonadal system described, with the critical factor for the reaching of reproductive maturity being the stimulation of FSH on the ovarian production of oestrogen, is not a one-way process (Katchadourian, 1977). The feedback effect of sexual steroid hormones on the hypothalamus and the pituitary forms a delicate system, the details of which are not fully understood even today. Hence, the precise interaction between the CNS and the peripheral sexual glands is still the subject of many investigations, with relevance for pharmaceutical contraception and relief of perimenopausal complaints.

Although much of the hormonal activity prior to the menarche goes unnoticed, its consequences for the manifest physical changes accompanying puberty will be tremendous. The overt changes in feminine physique from late childhood into adolescence, governed by the hormonal and the endocrine systems, make up a pubertal sequence over time—breast budding, pubic hair, height and weight acceleration, menarche, growth of axillary hair and the approach of adult-type proportions, breasts, and hair (Tanner, 1962, 1978). The first signs of the maturation of the secondary sex characteristics for most girls concerns the breasts. It is discerned as elevations of the area around the areola, "the bud stage," and increased areolar size and enlargement of the breasts. It appears typically around the age of 11 with a range from 8–9 to 13 years. The girl also acquires more rounded hips. The genital organ itself undergoes change, with growth of the uterus and vagina, and increased size of the labia and clitoris. The function of the epithelium in the vagina and the urethra is markedly dependent on oestrogen supply. Skin becomes more oily. Pubic hair appears, first with sparse growth along the labia. The growth spurt takes place when there still is more than a year before the menarche occurs. The age at peak weight velocity closely follows the height spurt. Before the age of 13 the majority of girls have had their first menstruation, which is to be followed by a period of irregular menstruations and anovulatory menstrual cycles until the ovulatory menstruation appears as full sexual reproductive maturity is achieved. After the menarcheal event the breast tissue develops further and acquires the adult-type form and pubic hair and axillary hair become more adult in form.

For a given sample of individuals, it is commonly found that individuals tend to maintain their rank orders across various indices of pubertal maturation. At the same time, although puberty initiates a number of bodily changes in a typical sequence over time, this sequence is not altogether fixed. Its duration and order may vary from individual to individual. In some individuals the time of physical transition may take just a couple of years, whereas for most it spans over twice as a long period, and although the typical start of development of the breasts occurs before pubic hair growth, the reversed order is found in a sizeable proportion of females.

There are strong individual variations with regard to the timing of pubertal development. Just a cursory glance at an age class of teenage girls shows great variations in pubertal maturation with a range of 5 to 6 years between the earliest- and the latest-developed girl. In fact, an early-developed girl in a class might have her first menstruation around 9–10 years of age whereas the latest-matured girl might not attain reproductive maturity before the age of 17. The "normal" expected age for menarche ranges from 11 to 15 years, and when it occurs before age 9 it is customarily referred to as *precocious puberty*.

The same strong individual differences with respect to when in time girls have their first menstruation occur for height and for weight. Imagine having the opportunity to observe the same group of children grow in stature from late childhood and throughout the adolescent years. At the earliest age there will be moderate differences in average height between the sexes and moderate variations within the respective groups of boys and girls. Around the age of 10 we notice an acceleration in height among some of the girls, and from the age of 11 to 13 we will actually find girls on the average to be taller than boys. We will find the age at peak height velocity among girls to occur just around the age of 12. The height spurt in boys will be observed at a later point in time, in fact about 2 years later. However, this acceleration among the boys will not escape us. Whereas girls will be a couple of centimeters taller than boys between 11 and 13 years of age, the superiority in stature for boys at the age of 16 will be almost 10 cm.

Detailed information on the time of appearance of the growth spurt was presented by Lindgren (1975). Data was obtained longitudinally for a group of Swedish subjects. At the age of 11 years almost every fourth girl had reached her maximal peak height velocity point. At this age this had happened for less than 1% of the boys. At the age of 13, more than 9 of 10 girls had passed their peak height velocity, whereas this had happened for just slightly more than 1 out of 4 boys. Whereas almost all girls had passed the peak point at the age of 14.6 years, there were still close to 3 out of 10 boys who had not.

Not only are there shifting growth patterns in height stature occurring *between* the sexes in the adolescent years, more important from our point of view is the rather great difference in height stature *within* the groups of girls and boys. Among girls these differences appear most pronounced between 10 and 13 years of age, and among boys between 12 and 16 years. For example, the differences between the 10% tallest and the 10% smallest girls at age 12 was 17.4 centimeters in Lindgren's material. After almost all girls had passed the menarche at the age of 16, the difference had decreased to 13.4 centimeters. At 10 years of age, before the growth spurt had started in most boys, the differences in height stature between the top 10 and the bottom 10% of boys was 15.4 cm. When it was at its maximum of 14 years of age, the difference was 22.8 cm.

The developmental tendencies that can be observed for height are about the same for weight. The weight acceleration starts earlier in girls than in boys and from the age of 11 to 14 years girls are in general somewhat heavier, about 1 kg difference. By the age of 16, however, boys are more than 10 kg heavier than girls. Marked individual variations can be observed from the age of 12 for girls and from the age of 13–14 for boys. Lindgren (1975) reported a weight difference of 13.2 kg between the top 10 and the bottom 10% girls at age 10. This difference increased to 19.9 kg at the age of 13. For 10-year-old boys the difference between the top and the bottom 10% was 12.7 kg, whereas it was double that size, 25.3 kg, at the age of 15. Over the age span 10 to 16 years, boys more than doubled their weight, and the average girl was about 75% heavier at the age of 16 than she was at 10 years of age.

The data just presented shows considerable sex differences as well as individual differences during the age span covered, and they show that the developmental differences within and between the sexes are closely connected with pubertal timing. During a period in early adolescence the early-matured girls will stand out from the majority of girls. They will be considerably taller and heavier than the rest of their same-sex peers. If a comparison is made with the boys in these early years of adolescence, an even more marked difference appears: "while she is physiologically a year or two out of step with the girls in her class, she is three or four years out of step with the boys—a vast and terrifying degree of development distance" (Jones, 1949, p. 77).

The situation will be similar but not as pronounced for the late-matured girls at the end of the adolescent interval. They will differ from their same-sex peers at this period, just the same as the early-matured girls stood out from the majority in early adolescence. However, the differences in general maturity will not be as large when compared with boys. In fact, their pubertal timing will be close to the timing for the average boy. So, if one compared perceptually the maturity process for the early- and the late-matured girls with "on-time" development as the norm, the early-matured girls will fare worse, being out of range both with same-sex and opposite-sex peers. Among boys, the late-developed boy will differ physically from his same-sex peers during a period in late adolescence, but even more so if compared with the timing of physical growth for the average girl.

It was these perceptual differences that led Bayley and Tuddenham (1944) to conclude that early-matured girls and late-matured boys would be particular risk groups in the adolescent period: "As these children attended a co-educational school in which grade placement was largely by chronological age, the two groups who would stand out in the school-room would be the large, early maturing girls and the small, late-maturing boys" (p. 53). In the clinical situation early-developed girls and late-developed boys have been reported to be overrepresented subgroups among teenage patients (Frisk, Tenhunen, Widholm, & Hortling, 1966).

One of the premises for the present investigation is the circumstance that pubertal maturation in females has a high degree of visibility. The point we want to emphasize is that the physical development in girls is something that is readily visible to the perceiver—to the girl herself, to her same-sex and opposite-sex peers, to her parents, teachers, and others (cf. Blyth, Simmons, & Zakin, 1985; Brooks-Gunn, Warren, Rosso, & Gargiulo, 1987; Brooks-Gunn, Warren, Samelson, & Fox, 1986; Duke, Litt, & Gross, 1980). People are generally quite accurate in arranging a group of girls in the order of how far toward adult physical status they have already developed. This consensus in opinion also includes the girls themselves. A girl can rather accurately determine which of the same-sex peers in her class are early- or late-developed for their age as well as readily subordinate herself on this early–late continuum (Brooks-Gunn & Petersen, 1984; Tobin-Richards, Boxer, & Petersen, 1983). The high visibility of pubertal development is likely to have strong consequences for a girl's self-concept and for other people's reactions to her. Perception of the girl's developmental stage will determine what interests and behaviors to expect from her and will influence other people's attitudes and behaviors relative to the girl. With the overt manifestation of physical development in the girl also enter social and cultural expectations, such that psychosocial and psychosexual development is expected to keep pace with the level of physical development. In their discussion about precocious and delayed puberty, Money and Clopper (1974) argued that "one's almost automatic response is to expect a person to behave consistently with the chronological age impression inferred from the physique age" (p. 175).

SOCIAL COMPARISON

Many contemporary notions about how individual differences among girls in pubertal maturation are transformed into individual differences in self-conception and behavior are based on the idea that physical growth acquires its psychological meaning against the background of what is the normative growth rate among same-age peers: the deviancy hypothesis. Consequently, the issue of early versus late pubertal maturation becomes an issue of *social comparison* (Money & Clopper, 1974; Rierdan & Koff, 1985; Tobin-Richards, Boxer, & Petersen, 1983; Weatherley, 1964). The girl compares her physical growth status against the mainstream of same-age peers, and it is this comparison, by herself and by her peers, that determines her perceived maturity:

> Not only do children observe these changes in themselves, but they observe these changes in their peers. . . . Given the inevitable social comparisons, it is not surprising that adolescents may mark puberty in terms of the age span that covers their peer group's development rather than their own. In fact, perceptions of pubertal timing may be more important for psychological functioning than actual pubertal status. (Brooks-Gunn & Petersen, 1984, p. 185)

The viewpoint that the impact of early versus late pubertal maturation should be conceptualized in terms of the level of maturation in the same-age peer group was the basic argument that Faust (1960) advanced to explain the change in social prestige for girls 12 to 15 years of age. In her study the prepubertal girls were the ones who in the sixth grade received the highest scores on social prestige, whereas it was the early-matured girls who were so favored at Grade 7 through 9. Faust's interpretation of these findings was that in the earliest grade early-matured girls were a small minority and not "in phase" with the developmental level of peers. At this age early maturation was not a prestige-lending quality. In the upper grades, being ahead of other girls, but now not so physically different from the majority because a greater proportion of girls had entered puberty, early maturing was connected with higher social status. In contrast, the late-matured girls were now in a distinct minority, and these girls also received the least favorable prestige scores by their peers.

Faust's study has had wide impact on how we conceptualize physical growth today. The investigation makes plain that answers to the question of whether developing early or late relative to peers is advantageous or disadvantageous, cannot be answered with a simple yes or no. Even for the same girl the relative advantages and disadvantages will vary as a function of chronological age.

Conceptualized as an issue of social comparison, the critical period for the early maturer is the early teens, whereas it is the late adolescent years for the late developer. The time at which the problems arise for these two subgroups of girls, then, is different because the physical growth status of the majority of peers at this time will be different; but the underlying motive is the same—being different from the mainstream.

The minority position does not necessarily have the same implications for the early as for the late-developed girls. Arguments have been advanced that late-maturing girls have an easier time because they do not stand out from the majority of girls as do the tall, early-matured girls. Jones (1949) advocated that it is the early-developed girls who will bear the most heavy burden and that late maturation, in fact, might be beneficial for the late maturers. The disadvantageous situation for the early developers Jones described as:

physically conspicuous . . . embarrasingly tall and heavy . . . decreased skill in physical activities involving running and jumping . . . the males of her own age are unreceptive . . . if she moves into an older peer group she may fall under parental restrictions, and, in any event, may lack the social maturity necessary to make a good adjustment among others of greater experience. (pp. 77–78)

In sharp contrast, the late-developing girls grows to

conform closely to our American standards of beauty and figure . . . the parents and the girl herself have a longer time in which to get used to the new interests, new impulses, and new requirements . . . the two-year lag in maturity patterns of boys as compared with girls is reduced or eliminated among those girls who mature late, and their interests in mixed social activities, when they emerge, are more immediately satisfied. (Jones, 1949, p. 79)

There are empirical data to support the idea that late maturation is not as beneficial as was suggested by Jones (1949). Typical problems of delayed puberty have been reported to be connected with doubts regarding one's own femininity, emotional disturbances, problems with regard to opposite-sex contacts (Adams, 1972; Higham, 1980; Rice, 1978), and social adjustment problems (Abernethy, 1925). Delayed breast development in particular might have negative implications for heterosexual relations (cf. Kelly & Menking, 1979). For girls with delayed puberty, Money and Clopper (1974) reported a typical problem involved their being treated as younger than they were, a situation that eventually resulted in social isolation and in adaptive problems of merging gender identity with physical development. In his discussion of the prominence of body image in adolescence, Schonfeld (1971) commented on the late developer's situation in the following way: "Adolescents with physical handicaps in the area of development are second-class citizens in their group and are often cruelly ostracized, ignored by the opposite sex, or treated with contempt. In the race for favor with the opposite sex, teenagers are usually merciless in taking advantage of any shortcomings of rivals (p. 318), and Higham (1980) stated that "girls with a flat chest and thin torso, legs, and arms, are unable to compete in the courtship rehearsals of teenagers" (p. 484). In Jones and Mussen's (1958) investigation of extreme early- and extreme late-developed 17-year-old girls, with the measure of pubertal development being based on skeleton development, the late-developed girls showed more negative self-concepts. Apter, Galatzer, Beth-Halachmi, & Laron (1981), in a study of subjects with delayed puberty, found that different aspects of self-image were related to the height of subjects, such that worse self-image was associated with retarded height. Self-image, however, was not related to delayed puberty in itself.

AGE-GRADED INFLUENCES

The psychosocial functioning of adolescents is commonly interpreted, directly or indirectly, against the background of the developmental tasks that the teenagers are or are about to encounter and the actual socializing conditions that prevail at different points in time in adolescence. Social norms, roles, and social expectations are intimately connected with chronological age (Foner, 1975; Jessor & Jessor, 1975; Riley, Johnson, & Foner, 1972). People behave differently at different ages, and seem quite conscious of whether they are "early," "on time," or "late" with respect to major social life events in the life cycle (Neugarten & Datan, 1973). The assumption of a unitary psychosocial development suggests that to each chronological age point there is connected a "social clock" (Neugarten & Datan, 1973) or certain expected developmental tasks and associated social behavior. The point of view that links psychosocial adaptation to the expected age-graded behavior niche refers to influences on human development that Baltes, Reese, and Lipsitt (1980) have labeled as normative age-graded.

Some of the age-related social roles and attached behaviors are formalized in society, expressed as prescribed minimum ages. If the reader were a Swedish citizen, this formalization would constitute a social timetable into adulthood. One starts school when one is 7 and stays there 9 obligatory years. Together with most of one's 16-year-old peers one goes on to the gymnasium and stays there an additional 2 or 3 years. Then, if the possibilities are available, one progresses to the university by the age of 18 or 19 at the earliest. A person is treated as a "minor" by society until the age of 15. After 15, the situation changes rapidly. Fifteen years is the age of criminal responsibility. At the age of 15 one is also given the formal right to one's own sexual life. Moreover, the individual, for the first time, is allowed to have employment, to see action movies and thrillers, to enter some discotheques and dances, and to drive a moped. Family allowances that have been sent monthly to the adolescent's parents stop at the age of 16. At the age of 18 one is treated as an adult citizen by the law and is allowed to vote in general elections, enter adult dances, and see "adult movies." At the same age, one may also take driving lessons in order to apply for a driver's license.

In summary, one is treated like a child with a minor's rights up to the age of 15, when the individual is given certain legal rights and when certain responsibilities are demanded by the society. At this age the doors to public social entertainment swing open. The formal transition to adult status occurs at age 18.

Most social behaviors are not of this formal type, in terms of being legally endorsed or otherwise regulated. They are of a more informal nature. To take an example from our own research, in a free-response task we asked the girls in our research group, when they were between 14 and 15 years, what benefits, if any, they saw in becoming older. Almost unanimously they looked forward to it. It was very obvious that the girls at this age thought being older meant having more legal rights and, generally, more freedom from restrictions. Being older meant handling one's own affairs, having more freedom to do as one liked, having more to say about daily routines, and being accepted as an independent person in control of one's own fate. Some girls wrote that they wanted to be older because their circle of friends were older, and they wanted to be on equal terms with them. A substantial proportion of the girls mentioned that if they were older, they would be spared their parents' admonitions and be accepted as they were. Finally, a few girls answered that to be older, for them, meant being allowed to have their own children.

There are few reasons to believe that the impact of early versus late pubertal maturation can be seen disengaged from the age-prescribed and the age-normative social behaviors in society. The frequent findings in the physical growth literature that maturation timing is connected with certain areas at one point in time but with other areas at other points in time, and that there are changing relationships with maturity timing for particular variables with the passage of time, strengthens the argument that physical growth cannot be viewed in a stable environment.

In fact, we feel that too little attention has been paid in research to the consequences of being early- versus late-matured with respect to the match between age-specific social expectancies, roles, and norms, on the one hand, and variations among girls in pubertal maturity on the other. Whether early or late physical development is advantageous or disadvantageous is dependent on how society has arranged the social activities at different ages, what behaviors and transition tasks are typical at these ages, what characterizes parent–child relations and peer relations at these ages, as well as what conditions are prevailing for boy–girl contacts.

Chronological age, connected with age-appropriate social behavior patterns, has sometimes been polarized against pubertal age in the literature. The question has been stated: "What means more for the emergence of new social behaviors in adolescence—chronological age or biological age?" (Dornbusch et al., 1981). Such research questions give the impression that we are talking about two independent operants, although these two ways of mapping development, in fact, do not form a state of opposition. On the contrary, they are, of course, closely linked and mutually dependent. Many of the new social behaviors and the developmental tasks that girls face during adolescence are probably quite closely linked to chronological age. At a particular age point, however, it is an important empirical issue to determine how and to what extent the factor of physical growth affects behavior. The question of what means more—chronological age or biological age—has to be subordinated to the general issue of how changes in behavior in adolescence come about against the background of what Neugarten and Datan (1973) label the "social clock" of the environment and the "biological clock" of the girls.

We view the match or the fit between social time and biological time as offering a most fruitful conceptual frame within which to analyze the impact of early versus late pubertal maturation. The linking of maturity timing to social life patterns at different ages provides the frame for understanding why differences between early and late physically matured girls appear and outlines which differences are likely to occur at these ages.

THE PRESENT MODEL

If we seriously attempt to analyze the impact of pubertal timing against the background of the age-normative psychosocial conditions that prevail at different points in time in adolescence we arrive at some conclusions that seldom have been explicated in research on physical growth in females.

Timing of Maturation and New Social Behavior

When examining the consequences of the timing of physical growth, the usual period of investigation has been the early teenage years, around 12 and 13 years. The outlook that these years are the point in time when the major psychosocial effects due to individual differences in the timing of physical growth are to be expected is quite reasonable from the point of view that it is here that the maximal differences in overt physical growth characteristics are found between early and late developers. For example, between 10 and 13 years the greatest variations among girls in height stature will be obtained, and around 12 and 13 years the same will be true of weight. Around this age the menarcheal age variable is most closely connected with other growth parameters such as peak height velocity, breast size, and pubic hair (Petersen, 1979). Later, these other growth indices become increasingly more independent of when the menarche occurred. Hence, if the impact on psychosocial adjustment of maturity timing is closely connected with individual differences in manifest growth characteristics, the early teens would consequently constitute the natural period for study.

There are other reasons that have also been offered. Major shifts in intrapsychic organization and self-concept have been linked to the menarcheal event (Blos, 1962; Hart & Sarnoff, 1971; Kestenberg, 1961). The menarche is thought to be associated with sex-role identity and identity as a woman and adult. Age at menarche has also been comparatively extensively investigated with regard to how it relates to change in body image, to interpersonal relations, and to social and emotional adjustment (Davidson & Gottlieb, 1955; Garwood & Allen, 1979; Goldberg, Blumberg, & Kriger, 1982; Koff,

Rierdan, & Silverstone, 1978; Rierdan & Koff, 1980a, 1980b; Smith & Powell, 1956; Zakin, Blyth, & Simmons, 1984). These studies have been conducted on samples of girls, 12 and 13 years of age, and they have mostly been of the pre–postmenarcheal design. Because girls normally attain menarche around this time, and it is the time for the optimal cutting point for differentiating between girls who have passed and girls who have not yet attained menarche, it seems quite reasonable that investigations have been concentrated to these early teenage years. However, although previous studies have correctly identified an age period of maximal differentiation as to physical growth characteristics, they have failed to look for delayed results of timing differences.

In contrast to earlier expectations concerning the appropriate time for differential results, the point of departure in the present study can be summarized as follows: Individual variations in pubertal maturation have comparatively little impact on behavior until independence and autonomy, intensive and intimate peer contacts, and the establishment of heterosexual contacts constitute central features in the daily life of adolescent girls. This proposition regarding when the behavioral consequences of early versus late pubertal maturation are likely to appear in the adolescent interval reflects a view that differs from what is implicit in many earlier studies in the field. Rather than saying that the consequences appear when the physical differences among growing girls are at their maximum, it suggests that the most differentiating time will be when differences among girls with respect to adopting "teenage-typical" social behaviors and major transition tasks of adolescence are at their maximum.

In psychosocial terms, the period of adolescence can be grossly characterized as a progression from other- to more self-directed action. In the early adolescent years, parents primarily exercise a type of authority that requires undirected respect for and conformity to their standards. Over time, adolescents demand from their parents and are granted more freedom to handle things on their own and to establish a more self-reliant way of living. The disengagement from the dependence on the family normally accelerates up in mid-adolescence. The change over time to more self-directed action has implications for the view of the impact of puberty. We cannot assume that physical changes trigger, release, or elicit subsequent changes in behavior in a mechanistic way independent of the character of the social environment. If conditions generally are more restrictive in families of females in their early teens, with regard to regulations at home like time to be in at evenings, bedtimes, spending money, choices of clothes, and so on, the early developer has few possibilities to transform her more physically matured status into more independent forms of activity in the early teens. More options will appear later on when parents generally are prepared to give their daughters more freedom to handle their daily living on their own. At that time the early-matured girl may derive more advantage from her advanced physical development than will her later-developed peers and may create more opportunities to channel her matured status into processes of independence and autonomy.

The change over time in limit-setting in families also has consequences for the expression of an early physically matured girl's adult image of herself and her personal identity. Undoubtedly, there will be changes in self-perception of one's maturity status induced by pubertal maturation in a wide sense, and, more narrowly, by the menarche itself. But the prospect of transforming changes in self-concept into overt behavior will vary with the responsivity of the close environment. A progressive consolidation of these maturity attributes can be expected with the passage of time, such that more limited channels to behavior will exist in early teens, but new types of social life options will emerge later in adolescence.

Generally speaking, to the extent that maturity timing in females is associated with major transition tasks encountered in adolescence—the loosening of ties to parents, developing independence and autonomy, establishment of a self-chosen way of living and personal moral code, identification as a woman and adult, and the establishment of a sexual identity and heterosexual relations—then early adolescence is too early a time to look for relationships between variations in pubertal maturation and transition issues. Such transition behaviors and shifts in social network characteristically begin to become normative among teenagers in mid-adolescence: around 14 to 15 years of age.

In psychoanalytic theory, what characterizes the mid-adolescent period is the emancipation from parents and the evolving of nuanced object relations (cf. Blos, 1962). Friendships that in late childhood were primarily built in unisexual groups at this time shift to peer interaction in heterosexual groups. Participation in crowds is a mid-adolescent phenomenon, and clique formation increases in perceived importance from early adolescence to the midadolescent years (Crocket, Losoff, & Petersen, 1984). The mid-adolescent years are also the period in which contacts with the opposite sex begin to become established and in which mixed-sex activities become frequent. Steady dating appears around 14 years of age (de Anda, 1983; Rice, 1978). In mid-adolescence the pattern of the older boy who dates a younger girl becomes normative (Blyth, Hill, & Thiel, 1982).

In the early teenage years a majority of teenagers conform to their parents' standards of behavior. Over time, a greater range of contacts with other persons occur, and an increasing proportion of teenagers adhere to their peers' rather than to their parents' opinions in situations eventually leading to breaches of norms. In one study in our longitudinal program, Henricson (1973) showed that about half the subjects in the research group at middle adolescence had opinions on norms that were fairly consistent with those of adults. The other half questioned the authority of adults. The study showed that these teenagers compromised between their parents' and their peers' standards. Furthermore, when peers were the dominant reference group, norm breaking was more tolerated and norm violations more frequent (Magnusson, Dunér, & Zetterblom, 1975). Hartup (1970) has claimed that it is during the mid-adolescent period that one can truly speak about an opposition between a parental- and a peer-endorsed course of action with respect to norm issues. Conflicts between adolescents and their parents also tend to be greatest around 14–15 years (Ellis-Schwabe & Thornburg, 1986).

Adolescent problem behavior shows a typical sequence during adolescence, rising particularly sharply at around 13–14 years of age, reaching a peak at 15–16 years, and decreasing in late adolescence (Douvan & Adelson, 1966; Jessor & Jessor, 1977; Offer, 1969). The close connection between entering mid-adolescence and engaging in criminal activities has been noted repeatedly (Hirschi & Gottfredson, 1983). Nowhere in the life span is the acceleration into crime so high as around 14 to 15 years. Furthermore, as determined by the rate of change in prevalences, a sharp deceleration occurs around 17 to 18 years. Farrington (1986) reported the peak age of offenses for English males and females in 1983 as 15 and 14 years, respectively, and, in 1982, as 17 years for American males and 16 years for American females. According to Farrington, the most parsimonious interpretation is the one offered by many developmental theorists: disentanglement from the bonds of the home and entering into the circle of peer influence.

That indeed the relation between pubertal timing and adjustment problems tends to accentuate in the midadolescent period has empirical support in the pubertal development literature. For example, Frisk (1968), investigating psychiatric and medical problems in an outpatient clinical sample, reported that the physically early-developed girls' problems were not brought to the fore until around 14–15 years of age. Andersson, Dunér, and Magnusson (1980) reported that school problems among the early-developed girls were more accentuated at age 15–16 than at 13 years of age.

In legal terms, mid-adolescence also constitutes the particular transition period from "minor" to "major" status. As mentioned earlier, before the age of 15 the parents in Swedish families have the ultimate responsibility for the upbringing conditions of their offspring. At this age teenagers are now permitted to take full-time or part-time jobs, and many new types of public leisure-time activities that were not available in the past now are within reach.

The association of individual variations in pubertal timing with mid-adolescent behavior may seem paradoxical in view of the fact that various indicators of physical growth at that time largely show low relationships with maturity timing, and the major shifts in intrapsychic organization associated with menarche commonly are thought to occur earlier. However, there is no reason to believe that the identification of oneself as socially and psychologically mature stops short of the menarche or of, for that matter, the broader pubertal maturation process encompassing this event. Rather, there is a progressive consolidation of these adult- and female-defining features over time and a gradual integration of these features with the rest of the personality.

This progressive consolidation is perhaps most obvious with respect to heterosexual interests. Early physical development means little for the early-matured 12-year-old girl's success in heterosexual contacts when most of the same-age boys are still lagging behind her physically by 3 or 4 years. Moreover, at this age chronologically older boys perceive her as a little girl with few sexual stimulus attributes. At a later point in time, the early-matured girl, who is now 14–15 years old, has had a longer time than her same-age later-developed peers to prepare herself psychologically for heterosexual contacts, and she has had a longer period to develop security in her sexual identity. Thus, she may feel more secure in the company of older boys than will her later-developed peers. Rather than eliciting certain types of feminine or adult-type activities in early teens, the feminine and adult orientation is gradually integrated, and at a later time, when these are matched by environmental responsivity, they will find natural outlets in behavior.

Basic Assumptions

The general hypothesis that underlies the present investigation, a hypothesis already elaborated in the 1920s, is the following: Many of the social behaviors, manners, customs, interests, and attitudes that teenage girls encounter as a function of entering adolescence are synchronized with physical growth.

A certain amount of conjoined progression in different areas is assumed to be characteristic for adolescent development. To some extent, particularly as it concerns recognition of the girl's maturity status and the subsequent demands and expectations of grown-up behavior from the important reference groups for girls—parents and peers—it can be presumed that physical growth is an impetus for development in the social domain (cf. Ausubel, 1954; Tanner, 1962). Much of other people's reactions to and expectations of the female adolescent's social and sexual behavior is a function of visual information: how "old" she appears in the eyes of the perceiver.

Consequently, at a given point in time, individual variations in adopting of "typical" teenage behaviors and more adult-type behaviors can to a certain extent be understood from the point of view of differential biological maturation. Individual differences in social behaviors in part indicate that persons enter the normal process of transition into new social patterns at different times. For example, girls in the mid-adolescent years pass through a time of liberation from parents and a time of change in norms, roles, and reference groups, but early-maturing girls pass through this phase at an earlier time than do their late-maturing peers.

This first working hypothesis is similar to the hypothesis that Stone and Barker (1937, 1939) formulated to test in the Oakland Growth Study. Their purpose can be summarized simply as: The girls who mature early are more apt at an earlier point in time to adopt behaviors, interests, and attitudes that characteristically occur more frequently among chronologically older girls than are the physically late-matured girls. Thus, at a particular point in time in the adolescent years, a greater number of the early-matured than of the late-matured are more likely to display behaviors, interests, and attitudes that are more normally occurring among older girls.

The impact of pubertal maturation on behavior must also be linked with the age-normative social behaviors in the adolescent period. In the preceding sections we argued that a match or fit between social time and biological time offers the most useful conceptual frame within which to analyze the impact of maturational timing on psychosocial adjustment. As described earlier, the present view suggests that the most differentiation should be sought when the differences among girls, with respect to adopting "teenage-typical" social behaviors and completing major transition tasks of adolescence, are at their maximum. We then come to the second assumption on which the present investigation is based, namely: Individual differences in the timing of pubertal maturation have comparatively little impact on new social behavior until a certain degree of environmental responsiveness to a girl's pubertal maturation is available—most notably in the mid-adolescent years.

A General Process Model

So far, our expectations with regard to the new social behavior that girls establish in mid-adolescence suggest an association with biological maturity in girls, such that social behaviors that commonly begin in mid-adolescence and behaviors that are of a more "mature" type will be adopted earlier by early physically matured girls than by late-matured. However, empirical evidence of relationships between the timing of pubertal development and acquisition of new social behaviors in mid-adolescence does not imply a direct relationship. Therefore, it is important to formulate and test specific hypotheses regarding the conditions under which the relationship is likely to be manifested.

In one of the first studies in the literature to deal with the role of biological maturation in development, Stone and Barker (1937, 1939) explicitly stated that their endeavor was exploratory, aimed at generating hypotheses and data from different domains. Even with the final results in hand, that postmenarcheal 12- to 15-year-old girls favored more "mature" responses in the domains of "heterosexual interests and attitudes," "adornment and display of person," and "imaginative and secretive activities" to a greater extent than did premenarcheal girls of the same ages, they admitted that these differences between the pre- and the postmenarcheal girls could be attributed to a number of factors, from a direct effect of the physiological system and/or indirectly through social transmittors:

> Regrettably, the methods employed in this study are not instructive as to what casual factors brought about the differences in responses by our premenarcheal and post menarcheal groups. . . . It is generally known that certain physical changes associated with the menarche are to be accounted for in terms of specific tissue responsiveness to hormones of the anterior lobe of the pituitary and to hormones from the gonads. Perhaps certain behavioral changes as manifested in interests and attitudes also can be accredited to hormones from one or both of these glands. In this connection, however, one must not overlook or underestimate the importance of extrinsic factors, particularly those of an ideational type which undoubtedly would play a more potent role in man than in lower animals lacking man's cultural background. . . . Finally, there may be differences in the social stimulation directed by adults to postmenarcheal and premenarcheal girls. (Stone & Barker, 1939, p. 60)

This state of affairs, a manifold of possible progressions for how differences in pubertal timing is related to differences in actual behavior of girls, and limited testings of such models, still prevails in present day research. Attempts have been made in specific domains to detail the path from pubertal maturation to behavior (cf. Simmons, Carlton-Ford, & Blyth, 1987). However, no more general attempt has yet been tried.

Probably the most comprehensive and insightful discussion of this issue was presented by Petersen and Taylor (1980). By way of introduction they noted the virtual absence of systematic theoretical frameworks in this field, a consequence, as they conceptualized it, of the tremendous difficulties in specifying the relationships among the operators occurring in the biological, the psychological, and the social systems. Empirically, researchers have focused their attention on some parts of this complex interactive system at the expense of others, resulting in scattered and often inconsistent findings.

Petersen and Taylor differentiated between two types of models for psychological adaptation to puberty: (a) a direct, unmediated effects model, and (b) a mediated effects model. The first type of model is close to the suggestions by psychodynamically oriented theorists that there is a direct link between the sexual drive and affective problems in adolescence (Kestenberg, 1967a, 1967b, 1968). Petersen and Taylor recognized the role of the gonadotropins for moodiness as an example of a direct connection between physiology and behavior. With respect to this first model, they did not find a single study in the literature purportedly designed to test the existence of a direct hormonal influence on psychological functioning. Thus, direct influences of biological changes on behavior still await empirical confirmations.

The mediated effects model has been more common in the field. It proposes that there are intervening variables that are responsible for the changes in the psychological functioning observed to be related to changes in the biological system. Such intervening variables might refer to processes working *within the person* (sex-role identity, adult image, integration of other people's reactions to one's growth status, etc.) or to factors operating *in the environment* (reactions and expectations of people in circles close to the person, sociocultural values of maturity, etc.). Petersen and Taylor proposed a path going from the biological system (genetic potential, endocrine changes, secondary sex characteristics development, and time of onset of puberty) through the sociocultural system (attractiveness standards, peer and parental responses, and sterotypies of early and late maturers), to the individual system (personal responses, body image, self-image, self-esteem, and gender identity).

In concurrence with Petersen and Taylor we do not think it likely that the direct link between bodily or hormonal changes in girls and social behavior is the principal relationship in development. Rather, the effects of the timing of pubertal maturation are most likely to be mediated by an immediate environment that changes as a consequence of changes in the girl.

The fact that maturation takes place physically and personally, as well as socially, requires taking a comprehensive look at the role of the physical growth component in general development. A receptive and prepared individual, an interpersonal environment that is favorable to change, and some form of opportunity structure—these are regarded here as the essential elements in maturation. Transformed to the issue of the behavioral impact of differential biological maturation, all three components—the *intraindividual*, the *interpersonal environment* and the *setting conditions*—are needed in order to understand and explain how early pubertal maturation is channeled into an earlier transition into more mature life patterns.

A general model can be proposed, delineating changes in four systems as a function of entering puberty early or late: (a) the biological, (b) the psychological, (c) the interpersonal, and (d) the behavioral. The model, and the operating features within each one of the four systems, are described in Fig. 2.1. The model is presented as a path diagram. The age dimension is introduced in terms of social response in the interpersonal system and self-perception in the psychological system.

A comment should be made with respect to the paths presented in Fig. 2.1. The model proposed should not be considered as a reductionistic biological model with individual differences in social transition behaviors being primarily determined by differential physical growth in development. What the model outlines is that, to the extent physical growth is associated with transition issues in adolescence, its role for behavior is more likely to be indirect than direct.

According to the model, early pubertal development is thought to have the primary consequence of instilling changes in the psychological (self-perception and attitudes) and in the interpersonal system. A reciprocal influ-

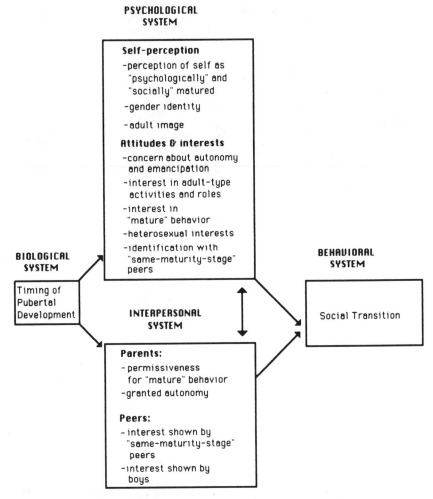

FIG. 2.1. A model for the influence of pubertal maturation on social transition behaviors in mid-adolescence.

ence is thought to operate between the psychological and the interpersonal system. A girl's experience of her psychosocial maturation is as much a function of how the environment responds to her developing body, as the social support in her environment is dependent upon the girl's willingness and readiness to change in the direction of more mature forms of behavior. Petersen and Taylor have likewise discussed the psychological adaptation to puberty as a simultaneous or "dual responsiveness" to both internal (subjective meaning of maturity, body image, gender identity, etc.) and external influences (sociocultural values, family and peer influences).

No direct effect of the biological system on the behavioral system is thought to take place. Instead, the impact of biological maturation is assumed to operate through the psychological and the interpersonal systems. Within these, *self-perceived maturity* and the *characteristics of the peer network* are the main mediating factors proposed to be operating for the influence of biological maturation on behavior. These influences primarily contribute to an earlier social transition among early-developed girls relative to later developers.

The general model could in principle be applied to all girls. However, in the process of transition from childhood to adulthood a basic view is that girls who develop physically early are subjected to more abrupt changes in social behavior in adolescence, and, generally, will exhibit a more intensive rate of passing through the developmental tasks normally associated with the adolescent life phase. Therefore, the pathways described in Fig. 2.1 could be rewritten for the early developer in the following way. Relative to her late-maturing peers, the girl who develops early is presumed at a particular point in time in mid-adolescence to:

Self-perception

- have a stronger perception of herself as "psychologically" and "socially" mature
- show stronger gender identity
- show stronger adult image

Attitudes and interests

- show more concern with liberation from dependency on parents, with autonomy and emancipation
- show more interests in adult-type activities and roles
- show more interests in "matured" behaviors
- show greater heterosexual interests
- prefer association and identification with girls of her own age who are at her own "maturity stage"

Parents

- be met with greater permissiveness for "mature" behaviors
- be granted more autonomy

Peers

- be shown more interests by same-age peers at the same "maturity stage"
- be shown more interest by boys, particularly older boys

Behavior

- have an earlier psychosocial transition, including approaching typical "teen issues" and more matured forms of social behaviors earlier in time

A more detailed outline of the mediating factors in Fig. 2.1 is given in the following section.

Self-Conceived Maturity

A girl's awareness of her own maturational status is considered as something of a prerequisite for more rapid acquisition of new social behavior among early-developed girls than among later-developed girls in the teenage years. Perception of oneself as more matured than others is the primary *intraindividual* component in the process of acquiring more matured behaviors, and it defines the preparedness for changing toward more matured forms of social behavior. Awareness of one's maturational stance has its roots in the girl's perception and in the emotions associated with the change in bodily shape and functions, in the perceived level of her pubertal maturation relative to other same-age girls, and in the reactions to her growth status from parents and other adults, from peers, and from members of the opposite sex.

The physical deviance from the normative in growth characteristics in early adolescence has been addressed in the literature essentially as a negative experience for the early developer. It sets her in a minority position relative to the majority of still-undeveloped peers, causes distress, annoyance and even estrangement, and might eventually result in a negative self-concept. When we consider the impact of physical differentiation from the present perspective, however, another interpretation is feasible.

The obvious difference in growth characteristics, relative to those of other girls in early adolescence, and the difference in sexual maturation will have their primary effects of increasing the early-developed girls' awareness of the factors of physical and sexual maturation, sex-role identity, and behaviors characteristically attached to maturation. Besides, the earlier sexual maturation of these girls will have a prestige value the more imminent to regular opposite-sex contacts the girls come in their development. Therefore, the awareness of one's greater maturity relative to peers, instigated by physical processes in early adolescence, will intensify later processes of autonomy and independence, normal social transition tasks, and emerging opposite-sex contacts, among the early-developed girls. These girls will be sensitized to behaviors commensurate with their pubertal maturation level and biological sex role.

Support Systems

Eichorn (1963) has made the general point that "much of human behavior is a function, not of structure per se, but of the environmental supports, outlets, expectations, and reactions relative to that structure" (p. 52). The attitudes, expectations, and reactions, stemming from early physical growth, which are directed from significant others and from the broader society, is here regarded as the essential *interpersonal* component determining the course of development for the early-developing girl. In daily life these attitudes, expectations, and reactions mainly involve the two primary socialization agents—parents and peers.

Parents. The expectations on grown-up behaviors on part of the girl directed from parents, their support but also their demands for adult-like manners, is one of the sources that provide the girl the opportunity to transform her maturational status into actual behavior. It is argued here that girls' pubertal maturation and their parents' permissiveness and expectations regarding appropriate social behavior go hand in hand, such that the early matured girl generally will encounter a parental attitude more favorable of mature social behavior than will the late-matured girl. Thus, it is assumed that parents base their support and demands for social transition behavior in their daughters to a great extent on the girls' pubertal maturation features.

Higher permissiveness on the part of their parents for more matured social behavior among early-developed girls does not ensure a conflict-free parent-daughter relation. As is reviewed later in this book, the empirical evidence strongly speaks against such a conclusion. Rather, what could be expected is stronger parental ambivalence. Fairly convincing findings on parent–daughter relations have accumulated, which show more strained parental relations among early- than among late-matured girls, particularly as they relate to the issue of heterosexual contacts. Although adult behavior includes sexual activity, that type of adult behavior in the daughter is likely to be discouraged by many parents. A girl with early reproductive capabilities faces a different set of problems from those faced by a woman, albeit both have the ability to reproduce. Paradoxically, then, we are in a position both to expect higher permissiveness for advanced social behavior and autonomy strivings among the parents of the early-developed girls at the same time as we have reason to expect greater parental conflicts, related precisely to these transition issues, among the early-developed girls compared to the late-developed.

Peers. Outside the family, the peer network is the socialization agent with a primary potential to instigate new life patterns and to provide the adolescent opportunities to transform maturational status into actual behavior. This role of the peer group generally in the adolescent period has been expressed by many. Ausubel (1954) concluded the following:

> In urban societies, the culture is not organized to provide children with many important opportunities for extrinsic status. Mature roles must be learned by a course other than gradual participation in family or communal responsibilities. . . . In his peer group he is given a chance to obtain the mature role-playing experience from which adult society excludes him and which his parents are unable to furnish. (p. 191)

Similarly, Hill and Lynch (1983) commented that

> during adolescence it is not only parents but peers of the same and opposite gender and other adults who define roles in gender-differential terms. In adolescence, "sex roles" are much, much more than a matter of displaying behaviors transmitted from parents through imitation and direct tuition. One does not learn "dating" from parents nor are universalistic norms of the sort encountered in schools typical of family interaction. (p. 205)

An extensive body of literature suggests that peer relationships and opposite-sex contacts are a particularly important feature of the life situation for adolescent females (cf. Rosenberg & Simmons, 1975). Although this view is widely accepted, little is known about the character of female peer groups (Savin-Williams, 1980) or the manner in which specific peer types exert particular influences on social behavior in adolescence (Campbell, 1980). In the present research, the issue of social transition among females is assumed to be closely linked to friendship selection and peer support of new behavior patterns. The comparatively higher interpersonal orientation among teenage girls, relative to adolescent boys, is thought to make girls particularly sensitive in their attitudes and behavior to the attitudes and the behavior patterns among their peers and their boyfriends (Rosenberg & Simmons, 1975). Therefore, social behaviors characteristically beginning in adolescence among females must largely be understood in the context of the patterns of social behavior existing in the peer ecology during the midadolescent years.

From the present point of view, the understanding of biological maturation's role in the process of acquiring new social behaviors must be approached primarily from the *kind* of peers the girls encounter in daily life and the social behavior among these peers. To the extent that peer influences are a key instigating factor for the impact of early biological maturation on new social behavior, then one cannot expect the types of peers and their social activities to be similar for the early- and the late-matured girls. As more fully described below, our basic hypothesis is that girls in the mid-adolescent years tend to associate with peers who match their physical development, making for differential opportunities to experience new social behavior in the peer culture.

The attributes that have been found to constitute the basis for friendships among adolescent girls in the peer literature are demographic characteristics first and foremost: gender, age, school grade, and neighborhood area (Blyth, Hill, & Thiel, 1982; Coleman, 1961; Duck, 1975; Dunphy, 1963; Hartup, 1976; Kandel, 1978; Kandel & Lesser, 1972). The teenage girl's best friend is often a girl in the same class or another girl at the same grade level in the same school, who is living in the same neighborhood. Availability in the physical environment encourages a certain type of friendship relations.

Friendship formation with other school pupils on the same age level is what should be expected. Coleman (1961) concluded that "among pairs of friends, the one item that the two members had in common far more often than any other—including religion, father's occupation, father's education, common leisure interests, grades in school, and others—was class in school" (p. 76). As children enter school they are locked into an age-segregated environment in which experiences are forced to be shared daily with same-age peers. Society's organization of its social institutions in this way shapes a particular form of interpersonal networks, that of persons with similar chronological age. Not surprisingly, there is less age homogeneity in out-of-school peer groups than in school peer groups (Montemayor & Van Komen, 1980). There are other, psychological reasons as well for expecting peer associations based on age homogeneity. Friends of the same age share the same historical background, similar life experiences, similar value systems, and they are expected by their parents to associate with agemates (Hess, 1981). A smoother adjustment to others' behaviors and a better recognition of others' motives also seem to occur for same-age subjects rather than for interaction with peers at other age levels (Hartup, 1976). Therefore, not unexpectedly, friendship formation based on age similarity has long been recognized as a chief factor (cf. Jenkins, 1931; Van Dyne, 1940).

That friendship primarily occurs between girls of the same age and the same school environment has led researchers to believe that same-age–same-school friends cover the important peer reference group. Kandel and Lesser (1972) drew the conclusion that "since all but a small minority of friendships among both American and Danish adolescents are with schoolmates, a description of friendships and peer influences within the school serves adequately as a description of all adolescent friendship groups" (p. 171). The frequent use of class-based measures in sociometric investigations to determine the social prestige of adolescent boys and girls illustrates the same belief: Peer interaction within the class is a valid microcosm of peer interaction generally among teenagers.

However, empirical results from our own research indicate that the same-age–same-school friends of adolescent girls do not adequately cover the total or necessarily the most important circle of their friends. A general hypothesis suggesting that girls in the mid-adolescent years tend to associate with peers who match their physical developmental stage was posited by Magnusson, Stattin, and Allen (1985, 1986). They proposed that for girls who develop physically at the normative rate, the friends within the same-sex–same-school age sphere constitute the "expected circle" of friends whereas for the physically early- and late- developing girls, a somewhat different peer constellation is likely to be discovered. From the matching viewpoint, this hypothesis challenges the age-graded friendship formation viewpoint and assumes a greater likelihood that the peer network of both the early- and late-developed girls cuts across the age borders.

From the perspective that girls seek out and are sought out by peers who are congruent with their biological stage of maturity, the early-developed girls are assumed to make friends with others who match their advanced physical growth status, suggesting an association and identification with chronologically older adolescents. These peers do not necessarily refer to school pupils. Some of them might in fact have quit school and begun to work. In effect, a substantial part of the circle of friends of the early developed girls will be outside the "expected habitat" of the average teenage girl, differing both in age and in social attributes. In earlier literature, such association with chronologically older peers on the part of earlier-matured girls has been reported by Faust (1960) and, for females with precocious puberty, by Money and Walker (1971).

The choice of friends at the same maturational level as oneself also should be the case for engaging with peers of one's own age. For the early-developed girls it is hypothesized that they more often than other girls engage with other same-age peers who, like themselves, are early-developing. Incidently, in a study of attitudes and emotions associated with the first menstruation, Brooks-Gunn and Ruble (1982a) reported that the postmenarcheal 12-year-old girls had more friends who, like themselves, had started to menstruate than had the premenarcheal girls of the same age, supporting a position on association with agemates based on a maturational stance. Jones (1948) also reported pubertal maturation to be a significant factor for friendship formation of adolescent females.

Finally, a coexistence between sexual maturation and psychosexual development is expected. The earlier reproductive maturity and the accompanying sexual interests and urges emanating from an earlier sexual maturation, together with the interest devoted from males to the early-developed girls, are assumed to make early-maturing girls establish stable relations with members of the opposite sex in mid-adolescence to a greater extent than do the other, physically-later developed girls. What should be observed in girls' contacts with the opposite sex is that steady dating predominantly refers to steady relationships with boys who are chronologically older than the girls themselves

(Berzonsky, 1981; Blyth, Hill, & Thiel, 1982; Dunphy, 1963). Schofield (1965) reported that the girls who had sexual intercourse before the age of 17 almost exclusively had chronologically older partners.

Three hypotheses, then, are suggested regarding the early-developed girl:

1. She seeks, more often than later-developed girls, friends who are chronologically older than herself and friends who are already engaged in full-time employment.
2. Among her own age class of girls she seeks as friends other early-matured girls.
3. She will have more established relationships and more advanced heterosexual experiences with boys than will her late-maturing peers at the time in mid-adolescence when stable heterosexual relations become common among females.

In the following, the label *nonconventional peers* refers to peers who are chronologically older than the girls themselves, peers who are employed, and members of the opposite sex. It is assumed that by representing socially more advanced behavior and more mature life styles than do same-age, same-class, and same-sex friends, association with the three types of nonconventional peers will have similar behavioral implications for adolescent females.

A reversed situation obtains for the late-developed girls. The general assumption of friendship formation based on a match with physical growth characteristics suggests for the late developers that they, more often than the early-developed girls, will associate with peers who are younger than themselves, more often associate with same-age peers who are late-matured like themselves, and have fewer heterosexual relations and experiences than do other, earlier-matured girls. To the extent that boyfriend relations are established, these will involve more same-age than older boys. Such a conclusion was also drawn by Berzonsky (1981), who commented that

> while a fast maturer may find herself to be out of her element when dating older boys, the slow maturer finds herself to be in step, dating boys, perhaps, with whom she has grown up and come to know. Male classmates who develop normally or somewhat precociously would be maturing at approximately the same pace as the slightly late-maturing girls. These circumstances suggest that late-maturing girls should experience minimal transition difficulties. (p. 166)

OPPORTUNITY AND SUSCEPTIBILITY TO NEW SOCIAL BEHAVIOR

The argument for a match between level of physical and sexual maturation in girls and the characteristics of their peer network forms the basis for expecting *differential opportunity* for encountering new social behavior, between early- and late-developed girls. To the extent that the peers of the early-developed girls are chronologically older than the peers of late-developed girls, the early-developed girl will have more opportunities to encounter new and more advanced habits and leisure-time activities and to encounter attitudes, expectations, and demands of a peer value system that positively reinforce her own behavior in that direction. Day-to-day exposure to more mature social lifestyles in a more advanced circle of peers, whose selection reflects the girls' own readiness to engage in more mature behaviors, thus defines the *setting condition* for expecting earlier transition into more mature lifestyles among the early-developed girls.

Not only will the peer network for acquiring new social behaviors be broader for the early-developing girl, it will also have greater impact on her behavior than will the same peer influences when they affect the later developer. With early-developed girls being more "prepared" or "ready" to change in the direction of more sophisticated social life patterns than are late-developed girls, the influence on their behavior of engaging with older peers and with boys will be more powerful on the early- than on the late-developed girls. In this connection, we might talk about *differential susceptibility* among girls to adopt more advanced social behavior.

The distinction between differential opportunity and differential susceptibility is a central one. Part of the difference in behavior between early- and late-developed females we assume to be attributable to differences in opportunity to encounter more advanced lifestyles among certain peer groups. Often, most notably in criminological theories (cf. Sutherland & Cressey, 1970), one calculates with only the opportunity component and thereby assumes a direct relationship between contacts with the characteristic behaviors and motivations of particular peer groups, on the one hand, and the subject's own behavior, on the other: New behavior is directly learned in interaction with these peers. This is also our claim on differential opportunity. More of the early- than the late-developed girls will be found in nonconventional peer groups, and, subsequently, more of the early- than the late-developers will be influenced by these types of peers.

However, the opportunity structure is not the whole story. If the peer contact were the critical factor per se, then we would expect that all girls who associated with nonconventional types of peers, irrespective of whether they were early- or late-developed, would change their behavior and attitudes toward the lifestyle of the nonconventional peers to the same extent. We do not believe this to be the case. By virtue of their earlier pubertal maturation, greater awareness, willingness, and preparedness for change will be expressed by the early-developed girls for behaviors that are indicative of a more mature status, lifestyles that are more adult-like, and interests and attitudes that characteristically belong to older age strate of teenagers. Thus, the impact of these peers with a more mature life orientation will be stronger on an early-developed girl who engages with nonconventional peers than on a late-developed girl who associates with the same type of peers.

SOCIAL AND EMOTIONAL ADJUSTMENT

An expected earlier social transition is just one facet of the developmental outcome of early pubertal development. Another is the increased risk for social and emotional problems that such early transition creates for the early-developed girl. The hypothesis that early-maturing girls pass through a more stressful transition into adult life than do their late-maturing counterparts is an issue that is investigated in depth through the course of this book.

It is assumed here that individual differences among females with regard to social transition behaviors in mid-adolescence are, to a large extent, a function of the influence of *nonconventional peers*—older peers, working peers, and older boyfriends. The more nonconventional peers in one's circle of friends the earlier in time more mature behavior takes place and the higher the risk for social adjustment problems. Through her biological resemblance to older age groups of teenagers, the early-developed girl is considered to be the most receptive to the impact of this nonconventional peer influence, for better or for worse.

In her more frequent associations with older peers, employed peers, and with older boyfriends, the early-developed girl is likely to encounter situations and experiences that are more unfamiliar to the girl with conventional friends only. She will encounter the more tolerant attitudes toward norm violation that characterize older groups of teenagers, and through association with them she will more often be likely to meet and participate in situations that may lead to rule breaking. The fact that the nonconventional peers with whom the early-developed girl associates in mid-adolescence are passing through their most intensive norm-breaking period in adolescence accentuates the risk for involvement in such activities.

Many of these experiences will probably be considered as deviant in terms of the norms of the conventional society, more tolerated among older teenagers but, being non-normative and not age-appropriate for the mid-adolescent girl, seen as a sign of social maladjustment on the part of the early-developed girls.

A further complication for the early-developed girl is the fact that she has to cope with and adjust to lifestyles among two age strata of peers: same-age and chronologically older peers. The marginal position of adjusting both to the norm climate and social life typical for agemates and to the social life more typical for older age strata of peers might make her vulnerable for developing emotional problems. Such cross pressures not only make it difficult for an early-maturing girl to be comfortable in her self-perceptions but also magnify her tenuous place in both same-age and older reference groups. Overall, early-developed girls can be assumed to be subjected to more difficult problems than those that confront later-matured girls. The accelerated tempo at which social transition occurs among the early developers, the greater demands on "mature" behavior from the reference groups, and the cross pressure of evaluations, attitudes, and norms from same-age and older peers means that these girls live under complicated and stressful environmental conditions in adolescence. Consequently, we hypothesize the presence of more emotional problems among early than among late developers, presumably connected with the former's peer contacts. The two hypotheses regarding the susceptibility of early-developed girls to problem behavior and emotional problems are more fully outlined in subsequent chapters.

In contrast, late-developed girls are assumed to be less apt than the early- or even the average-developing girls to acquire sophisticated social life patterns in midadolescence and to engage in norm-violating behaviors. In the midadolescent years they will be less sought out as friends among older peer groups and be met with more indifferent attitudes among older boys. Their less mature status has the consequence of more frequent association with chronologically younger peers, whose childish social life functions as a protective factor for engaging in more socially deviant activities. Less pressure from their peers for engaging in advanced social behavior, in comparison with the more advanced peer network of the early-developed females, means less risk for cross-pressure influences and emotional difficulties among the late-developed females.

SUMMARY

The present model explicitly connects the physical growth component with particular psychological processes working within the individual and particular factors operating in the environment. It involves specific hypotheses regarding the conditions under which the pubertal impact is likely to be manifested in behavior and under what conditions its emergence will be less likely. Thus, it is open for direct empirical verification and disconfirmation.

The major perspective of the model proposed here is the postulate that social development co-occurs with pubertal maturation in girls, such that early biological maturation corresponds to an earlier display of new social behaviors and transition tasks. Basically, our argument is that the earlier display of transition and "matured" forms of behaviors among early-maturing girls occurs in the pursuit of adjusting to the lifestyle and social behavior of the particular peer reference group with which they engage.

Underlying the argument is a model for peer influences on behavior, which can be described in Fig. 2.2.

SUMMARY

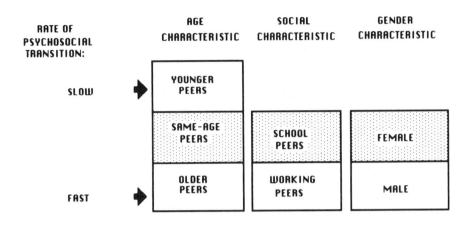

FIG. 2.2. A model for peer influences on the rate of psychosocial transition in adolescence.

According to Fig. 2.2, the expected circle of friends of teenage girls are—in terms of age, social, and gender characteristics—same-age peers, school peers, and same-sex peers. To the extent that a girl's circle of peers contains friends exclusively in these categories—a strictly conventional peer network—it is expected that her social transition from child to adult status largely follows the age-normative and the age-appropriate pattern. Her transition into adult lifestyles and her adopting of typical adolescent behaviors—beginning to drink, initiation of heterosexual relations and sexuality, the striving and gaining of independence from parents, and so forth—follows the pace that is normative for her age generation of peers.

The rate at which psychosocial transition occurs might be slowed down or accelerated for those girls who associate with peers outside the conventional peer network. To have younger peers in one's circle of friends is assumed to be linked with a slower rate of adopting more mature behavior forms. The opposite is the case for associating with older peers and working peers, as well as with males (boyfriends who are generally older than the girl herself). Such a peer network is assumed to accelerate psychosocial transitions into more advanced behaviors and adult-like lifestyles.

Association with nonconventional peers is proposed to be linked with girls' pubertal maturation. Girls who are more advanced pubertally are more likely to have peers who, like themselves, are more mature. They will more often than other girls be found to engage with older peers and with peers who have quit school and have begun to work. An earlier sexual development for these girls is also posited to be linked with more intimate contacts with members of the opposite sex. In this context of more "advanced" peers an accelerated psychosocial transition is assumed to occur. Later-matured girls, by contrast, are more likely to seek the company of chronologically younger peers, and, consequently, their psychosocial transition is expected to occur at a slower pace.

To summarize, the assumptions presented in this chapter can be illustrated as a chain of consecutive propositions:

1. First is the proposition that individual differences in the timing of pubertal maturation are thought to have consequences for the self-system, resulting in differences in perceiving oneself as "socially" and "psychologically" mature.

2. Second is the argument that social behavior must be understood in the context of the patterns of social behavior existing in the interpersonal ecology of teenage girls during the midadolescent years, drawing attention to the peer network in these years.

3. Third, to the extent that peer influences are the key mediating factor for the impact of biological maturation on new social behavior, the acquisition of such behavior among the early-matured girls must be found outside the frame of the ordinary peer interactions of the normal teenage girl (i.e., outside same-sex–same-age–same-school friends).

4. Fourth is the proposed match between the characteristics of these nonconventional peers and the level of physical (sexual) maturation in girls, which implies a differential opportunity to encounter new social behavior between early- and late-developed girls. The early-developed girls will seek and be more readily accepted as peers by older adolescents, by teenagers who are already working full time, and among boys. Subsequently, more of the early-developed girls than of the late-developed will encounter the more advanced social behavior characterizing these nonconventional peer groups.

5. The fifth argument concerns the peer impact on behavior. It is argued that the impact of nonconventional peers on the girls' behavior must be conceptualized as a joint function of characteristics of the girl and the type of peers under consideration. Due to differential susceptibility regarding the adoption of the advanced social behavior of nonconventional peers, with early-developed girls being more willing to do so than late-developed girls, the peer influence will be more powerful on the former group.

6. Besides an earlier psychosocial transition among early-developed girls, association with older peers who have more liberal norm systems makes these

SUMMARY

girls vulnerable for engaging in norm violations. In addition, the cross pressure of older peers and agemates on "appropriate" behavior might create an emotionally problematic life situation for the early developers. More norm violation and more emotional problems among early-maturing girls than among later-developed girls are expected.

Girls' friendship formation is a good example of reciprocal influences in development. From the point of view of the developing girl, friendship formation is dependent on how well peers possess those qualities in which the girl is interested. Thus, the defining of oneself as more (or less) mature, psychologically, socially, and heterosexually, relative to other same-age peers, brings with it both association with peers whose behavior repertoire coincides with this definition and a differential readiness to participate in advanced social behavior.

So far a general biosocial model for female development during adolescence has been presented and the relevant theoretical viewpoints and empirical results from earlier research reviewed and discussed. In the following chapters this general, interactional model and the assumptions derived from the model are the objects of empirical tests in a series of studies. These studies are based on data from a longitudinal research program that is briefly presented in chapter 3.

3
Data

A LONGITUDINAL PROJECT

Under the title "Individual Development and Adjustment" a longitudinal research program was initiated by the second author in the middle of the 1960s at the Department of Psychology, University of Stockholm. The first data collection took place when the subjects were 10 years of age, and the project is still in progress. The subject population consisted of all the school children in a mid-Swedish community of about 100,000 inhabitants who were receiving normal schooling in Grade 3 of the compulsory school in 1965. They were 1,027 pupils, 517 boys and 510 girls. Data collections have been performed for the pupils successively throughout the school system and thereafter through postal questionnaires, interviews, and medical examinations.

The general aim of the project is to examine how person and situation factors—independently and in interplay with each other—operate and influence the course of personal and social development from childhood into adulthood. Main subprojects within the program are directed toward the study of the developmental paths to alcoholism, criminality, and mental illness (for information about the program, its main aim, research strategies, and data collections, the reader is referred to Magnusson, 1988; Magnusson & Dunér, 1981; Magnusson, Dunér, & Zetterblom, 1975).

An interactionist perspective on development (Magnusson, 1988; Magnusson & Allen, 1983a) has guided research within the program. Recognizing that the individual from birth through life is involved in a constant reciprocal and dynamic interaction with his or her environment, research questions have been focused on how the individual, through his or her perceptions, thoughts, emotions, physiological processes, and behavior, functions and develops as a totality in relation to his or her environment. A challenge for research has been the attempt to pinpoint how the subsystems of perceptions, thoughts, physiology, and action develop in conjunction with one another.

Data collections have included psychological and biological factors on the person side and physical and social factors on the environmental. Measures of mediating psychological variables (intelligence, creativity, norms, values, etc.), social network factors, and manifest behavior have been supplemented by measures of various biological functions of interest, with an emphasis on hormonal functioning. At the present stage of the research program, data are available for various biological, psychological, and micro- and macrosocial variables for a representative sample of males and females from the age of 10 to 30.

SUBJECTS

At each grade level at school, all those present in the school system were included, irrespective of whether or not they participated in preceding data collections. This feature of the research design, that the subjects at each age covered comprised a complete school grade cohort, is noteworthy. All types of classes within the ordinary school system in the community at respective points in time for data collections were represented, including normal classes and classes for deaf pupils, pupils with reading problems, and pupils with conduct disturbances. The schools were all public schools as there were no private schools in the community (and very few in the rest of Sweden). Almost all of the children at the appropriate age in the community belonged to the target population. In 1965 there was just under 1% of the children in the community who were not included in the project's target population (due to severe problems, e.g., severely retarded children, psychotic children, etc.). Studies within the project have shown the subject population to be representative of Swedish pupils in general in central cognitive, biological, and social variables (Bergman, 1973; Magnusson, Dunér, & Zetterblom, 1975; Stattin, Magnusson, & Reichel, in press). Therefore, the conditions for an effective investigation of the consequences of biological maturation can be regarded as favorable as the subjects involved represent a rather unbiased age class of children, encompassing almost the entire range of social background conditions for the children in the community in question.

The sample of females constituting the research group for the present investigation consists of the girls, born in 1955, for whom complete data for pubertal development were obtained when they were approximately 14 years of age (in Grade 8) in 1970.

A total of 590 girls were registered pupils in 1970, and data on pubertal development were obtained for those 509 (86.3%) girls who were present at school on the days that data were collected.[1] Most of these 509 girls were born in 1955. However, 2% were early school starters (born in 1954), and 8% were late school starters or pupils who had not moved in the ordinary manner up to the next class. In order to control for differences in chronological age between the girls, and also to control for the fact that some girls, for various reasons, started school early or had not moved up to higher grades in the expected manner, it was decided to restrict the research group only to those girls who were born in 1955, had started school at the expected year (1962) for children born in 1955, and had school-leaving certificates from Grade 8 in 1970.

Of the 509 girls who were present the day data on pubertal development was collected, 466 girls satisfied the three criteria for inclusion.

The analyses presented in this volume are based on two types of data: (a) for biological maturation, and (b) for various aspects of psychosocial and emotional development. Because biological maturation is a control factor in the analyses, the property of this type of data is analyzed more in detail.

DATA

Age at menarche serves as the basic measure of biological maturation throughout this book. This measure satisfied our purposes in different respects. It refers to a pubertal developmental event that has been shown to be an important developmental marker for growing girls, and that, as determined from previous research is both reliable and valid, as is demonstrated here as well.

In the process of transition from childhood to adulthood, it is our basic hypothesis that girls who develop physically early are subjected to more abrupt changes in social behavior in adolescence, and, generally, live under more complicated and stressful environmental conditions. Consequently, our focus of research is on the early side of the early–late physical maturation continuum. In terms of measurement, variations in early maturation are of more vital concern for discrimination than are variations in late maturation.

The analyses reported in the volume are based on data from a question on the girls' age at menarche in a self-report instrument, The Adjustment Screen-

---

[1] We want to correct the population figures presented earlier. The total population of eighth-grade girls comprises 590 girls, not 588 as has been stated in earlier publications.

An additional 8 girls in the community were registered pupils at Grade 8 and born in 1955. However, they never appeared at school and had no school-leaving certificate. One had run away from home, one was pregnant and stayed at home, and the rest had the school's permission to be out working. Four of these girls were registered for psychiatric problems and/or criminality.

ing Test, given at the average age of 14.10 years (14 years and 10 months).[2] In its original form in the inventory, girls were asked to state on a 6-point scale the age when they reached menarche. According to the answers to this question, 5 girls (1.1%) had had their menarche before 10 years of age, 44 (9.4%) between 10 and 11 years, 108 girls (23.2%) between 11 and 12 years, 186 girls (39.9%) between 12 and 13 years, 99 girls (21.2%) between 13 years and 14.10 years, and 24 girls (5.2%) had not had their first menstruation at the time the inventory was administered. The median age for the self-reported menarche was 12.86 years, which corresponds closely to national figures for the age cohort in question. For Swedish girls born in 1954–55, Lindgren (1976) reported an average age for the menarche of 12.98 years using a similar data collection method.

With the present focus on the early side of the maturity continuum, a four-category variable of menarcheal age was formed with a finer discrimination of early pubertal development. The distribution of girls for the categories of the pubertal timing scale is presented in Table 3.1. The table also presents the average chronological age at the time of the data collection for the four menarcheal groups. Analyses indicated that there were no significant differences as to chronological age among the four groups of females.

Reliability and Validity of Data

Self-reported age at menarche is considered to be a quite accurate measure of pubertal age, and it is the method that has been used most frequently in the literature. It is not without problems, however, including distortions involved in remembering the actual point in time, irregular first menstruations making it problematic for girls to establish when they attained menarche, and status issues: Girls might underreport the actual age at which it occurred (Bergsten-Brucefors, 1976; Greulich, 1944; Lindgren, 1976; Weatherley, 1964).

Available results indicate that females rather accurately remember the time of their menarche. Damon, Damon, Reed, and Valadian (1969) obtained a correlation of .78 between actually documented age at menarche and recalled menarcheal age after 19 years. It was reported that 77% of the adult women recalled the age of menarche within 1 year of its actual occurrence. Livson and McNeill (1962) obtained a similar high correlation of .75 after 17 years between recalled and actual menarcheal age. For women aged 17 to 53 years, with an average age of 34 years, Bean, Leeper, Wallace, Sherman, and Jagger (1979) reported that 90% of the sample reported their age at menarche within 1 year of its actual occurrence. A correlation as high as .60 between registered

---

[2] In the following references to chronological age, the points in time of the data collections in this project use a period notation, such that 15.10 years refers to 15 years and 10 months.

TABLE 3.1
Recalled Age at Menarche among Females Born in 1955 (n=466) and
Chronological Age at the Time of Measurement

Age at menarche	N	%	cum %	Chronological age
−11 yrs	49	10.5	10.5	14.10
11–12 yrs	108	23.2	33.7	14.10
12–13 yrs	186	39.9	73.6	14.10
13– yrs	123	26.4	100.0	14.11

menarche and recalled menarcheal age after 39 years was reported by Damon and Bajema (1974).

In a Swedish growth study on a nationwide sample of females, a correlation of .78 was obtained between girls' reports of their menarcheal age in a postal questionnaire at age 17, and the menarcheal age according to school nurses' notes collected during 6 years, age 10 to 17 (Lindgren, 1975). One could argue that because early maturation might imply higher status, girls would be apt to report their point in time for the menarche earlier than it actually occurred. There were no indications in this Swedish sample, however, that such was the case; nor were there any indications that the early-matured girls tended toward the "normal" range. Girls in the lowest quartile reported their first menstruation as 12.28 years or earlier. The same figure, according to the school nurses, was 12.31 years. Neither did the late-developed girls report their menarcheal age as occurring at an earlier point in time. The 25% lastest-developed girls stated the time for their first menstruation at 13.65 years or later, in comparison with 13.66 years or later according to information given by school nurses.

Despite the findings that females fairly accurately remember the time of their first menstruation and that they do not systematically tend to deviate from the actual point in time of the event, there is no guarantee that this recall method is free from all the types of unreliability that characterize retrospective reports. Obviously, some independent means of validating the measure is needed in order to be confident that it reflects what we are interested in, namely biological maturation.

As determined from its relation to other maturation measures reported in earlier research, menarcheal age seems to be a fairly good measure of physical maturity in a broad sense (Tanner, 1962). For example, the age at which breast buds appear is typically associated with menarcheal age on a level of correlation of around .80. In the Swedish growth study (Bergsten-Brucefors, 1974; Lindgren, 1976), menarcheal age showed moderate to high correlations with other criteria for physical development. There was a correlation of .69 with mean age at peak height velocity, .61 with the peak weight velocity point, .59 with secondary sex characteristics, Stage II, .65 with State III measures, .67 with Stage IV, and .54 with secondary sex characteristics, Stage

V. Coefficients in the magnitude of .70 to .85 have been reported for the relation of the age at menarche and peak height velocity, skeletal maturity, pubic hair, and breast development (Nicolson & Hanley, 1953; Shuttleworth, 1937).

Some data for validation were obtained in the present study for a smaller sample of subjects. Measures for skeleton maturation, height and weight for the females in this subsample are used here in order to validate the menarcheal age measure.

For economical reasons, such as that more time-consuming examinations could not be done within the normal school schedule and that some investigations were difficult to administer because they had to be performed outside of school and required considerable time, special equipment, outside expertise, and so on, data for the total research group were complemented with more intensive investigations of a subsample of subjects—*the medical sample*. This subsample was formed at Grade 6 and was planned to involve about 250 pupils who would be extensively investigated, particularly with respect to physiological and hormonal factors.

The medical sample in Grade 6 included nine complete classes, stratified by type of housing area. In Grade 6, at the average age of 12.9 years, measures of height and weight were collected for these subjects, as well as data for electrical activity of the brain (EEG measures), physical performance capacity (oxygen absorption during physical work: ergometer cycling), and secretion of catecholamines (noradrenaline and adrenaline excretion during stress and nonstress conditions).

In Grade 8, close in time (February–March 1970) to the date at which data for the menarcheal age was collected (October 1970), the medical sample took part in a medical examination at the community central hospital. Measures involving X-rays of the wrist were collected to determine the course of ossification.

Ten years thereafter, when the subjects in the medical sample were, on the average, 25.10 years of age, they were again contacted. Those who still lived in the community or in other big cities in Sweden were asked to participate in an investigation involving an interview, a test session, and a medical examination. Of the original medical sample, 230 subjects were located and contacted, and 173 subjects agreed to participate in the medical examination. The examination involved collecting measures of blood pressure, blood sample and bleeding time, samples of urine, data about fertility, medication, and the use of contraceptives, and data for height and weight.

Skeletal Age

Skeletal age is a highly objective method of determining physical maturity and is probably the most efficient measure of physical growth available at the present time. The bones in the body develops from conception in a fixed sequence; starting with a primary ossification center, developing epiphyses, and ending with epiphyseal fusion. High correlations between skeletal maturity and menarcheal age have been reported in previous literature. Simmons and Greulich (1943) reported a correlation between age at menarche and skeletal maturity according to Todd Standards of $-.86$ for girls 14 years of age. Coefficients on the same magnitude were presented by Nicolson and Hanley (1953).

Such high correlations could not be expected in our case because the optimal measurement conditions for the girls in the medical sample was not met. Skeletal age was determined with the TW2 method, developed by Tanner, Whitehouse, Marshall, Healy, and Goldstein (1975). The method assesses how far from adult ossification status the bones have come in their development. A single development score, in our case with the maximum 1,000, was calculated for 20 bones in the hand. At the time the ossification measure was taken, more than half of all girls had already reached adult bone age status. Of the 98 girls who participated, 50 girls were classified as having reached an adult ossification status.

Although the TW2 measure differentiated among the group of late-developed girls, data for menarcheal age differentiated in the group of early-maturing girls. To adjust for these circumstances, the skeletal maturity measure was dichotomized, thereby contrasting the group of girls who had reached about grown-up ossification status (50 girls) with the girls who had not (48 girls). Among the 73 girls who reached menarche before age 13, 47 girls or 64.4%, had nearly reached grown-up ossification status. Of the 25 later-matured females only 3 girls (12%) had done so. The phi coefficient, measuring the association between menarcheal age and skeletal maturity, was $-.46$ ($p < .001$), which is reasonably high in view of the distributions of the two maturational measures.

The percentage of girls in each of the four menarcheal groups who had reached adult ossification status at 14.3 years is presented in Fig. 3.1.

As can be seen in this figure, almost 9 out of 10 of the earliest-developed girls were classified as having reached grown-up ossification status, in comparison with about 1 out of 10 girls in the latest-developed group. The biseral correlation coefficient was $-.51$, which was significant at the .001 level, at least (gamma = .74).

Height Stature

Ninety-three girls had data on height stature at 12.9 years. Table 3.2 shows the height stature at this age for the females in the four menarcheal groups. Figure 3.2 shows the difference between adult height stature (25.10 years) and height at 12.9 years for the 60 females who had data on height stature both from early adolescence and early adulthood.

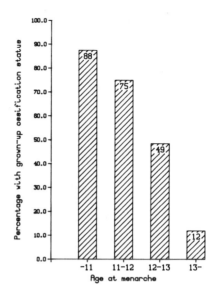

FIG. 3.1. The percentage of subjects in four menarcheal group who had reached adult ossification status at 14.3 years.

TABLE 3.2
Height Stature among Girls in the Biological Sample at 12.9 Years, by Menarcheal Age

Age at menarche	Height stature (cm) at 12.9 years		
	M	sd	N
−11 yrs	162.2	4.9	11
11–12 yrs	159.3	3.7	27
12–13 yrs	155.4	6.4	31
13–yrs	154.0	5.4	24
Total	157.0	5.9	93

As shown in Table 3.2, there were marked differences among the females in the four menarcheal groups with respect to height stature at 12.9 years. There was an average difference of about 8 cm between the most early- and the most late-developed groups of girls. In correlation terms, age at menarche and height stature at 12.9 years was highly related ($r = -.47$, $p < .001$).

Figure 3.2 shows how far from their adult height the girls in the four menarcheal groups had come at 12.9 years. The difference between the height stature at this age and adult height was around 6 to 7 cm among the early-matured girls, whose menarche had occurred before or at the age of 12, whereas it was almost double this size, 10 to 14 cm, among the later-developed girls. The correlation between adult height stature and menarcheal age was not significant ($r = -.15$, $n = 60$, ns).

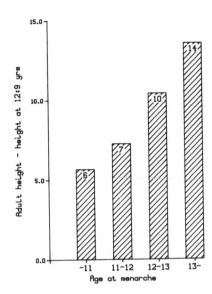

FIG. 3.2. Differences between adult height and height stature at 12.9 years for girls in four menarcheal groups (centimeter).

TABLE 3.3
Weight among Girls in the Biological Sample at 12.9 Years, by Menarcheal Age

	Weight (kg) at 12.9 years		
Age at menarche	M	sd	N
−11 yrs	53.5	10.1	11
11–12 yrs	50.1	7.4	27
12–13 yrs	44.2	8.8	31
13–yrs	40.8	5.1	24
Total	46.1	8.8	93

Weight

Weight data for the medical sample of girls at 12.9 years, broken down by menarcheal age, is shown in Table 3.3

Table 3.3 shows that there were considerable weight differences between early- and late-matured girls at 12.9 years. The earliest-matured group of girls was almost 13 kg heavier than the latest-developed group of girls at this age. The correlation between menarcheal age and weight was −.50, which was significant at the .001 level.

In Fig. 3.3 is presented the mean differences between the adult weight and the weight in early adolescence for girls in the four menarcheal groups. Data refer to the 57 females in the medical sample who had complete weight

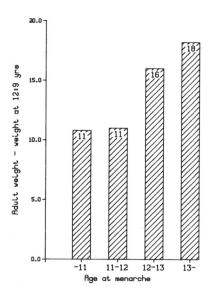

FIG. 3.3. Differences between adult weight and weight at 12.9 years for girls in four menarcheal groups (kilogram).

information on the measurement occasions as both 12.9 and 25.10 years. Three females who were pregnant at 25.10 years were excluded.

The difference between adult and adolescent weight was considerably less for the girls who reached their menarche before age 12 than for the girls who had their first menstruation at the age of 12 or later. There was a nonsignificant correlation of −.17 between weight at 25.10 years and menarcheal age.

Body Mass

What is the appropriate weight for a person with a given height? This is a question that refers to bodily proportions and to physical attraction. A measure for bodily proportions which simultaneously takes into account height and weight was developed by Quételet in Belgium in 1871, and it is a frequently employed body mass index today. Mathematically, it is defined as the weight in kilograms divided by the square root of the height, measured in meters. The normal interval for adult females is between 20 and 24.

The differences in bodily proportions among the girls in the four menarcheal groups are presented in Table 3.4.

The average girl at the age of 12.9 years had a BMI of 18.6 Those who had their menarche before 11 years attained an average body mass score of more than three units higher than those who menstruated after 13 years of age. The correlation between the BMI measure and menarcheal age was −.39 ($p < .001$). This association tells us that, relative to their height, the girls

TABLE 3.4
Body Mass Index (BMI) for Girls in the Biological Sample at 12.9 Years,
by Menarcheal Age

	Body mass index at 12.9 years		
Age at menarche	M	sd	N
−11 yrs	20.4	4.1	11
11–12 yrs	19.7	2.8	27
12–13 yrs	18.2	2.7	31
13–yrs	17.2	1.7	24
Total	18.6	2.7	93

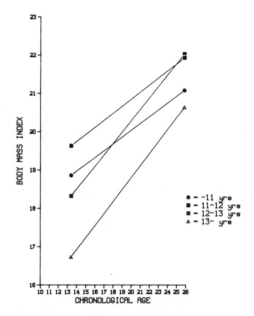

FIG. 3.4. Body mass index at 12.9 and 25.10 years of age for females in four menarcheal groups.

who menstruated before age 13 were more sturdily built than their later-developed peers at the age of 12.9 years.

The growth in bodily proportions from age 12.9 to 25.10 for the 57 females in the four menarcheal groups with complete data from both test occasions is detailed in Fig. 3.4.

At the age of 25.10 years, the total group had an average BMI of 21.6, which is at the lower part of the normal interval. At this age in early adulthood, few differences in body mass were observed among the menarcheal groups, and there was a nonsignificant correlation ($r = -.13$, ns) between BMI and menarcheal age.

To summarize, the results for the relationships of menarcheal age to height, weight, and body mass in Grade 6, and to skeleton maturation in Grade 8, were in the expected directions. The size of the coefficients are comparable to those reported for other samples (cf. Simmons, Blyth & McKinney, 1983; Zacharias, Rand, & Wurtman, 1976). As was expected from previous studies, the early-developed girls in the present subject sample were taller, weighed more, and were stouter at the age of 12.9 years than were their later-developed counterparts. The relationship between weight and menarcheal age was on the same magnitude as that between height and menarcheal age and stronger than the relationship between body mass and menarcheal age. Despite certain scaling problems, a clear-cut relationship between self-reported age at menarcheal age and the ossification index for skeleton maturation at the age of 14.3 years was obtained. At early adulthood, at the age of 25.10 years, there were still some indications that the early-developed girls were taller and heavier than the later-developed girls. However, no significant correlations with menarcheal age were obtained at this age. Altogether, these validation data attest to the usefulness of the menarcheal measure as a general measure of individual differences in pubertal maturation.

Other Comparison Criteria

Of particular interest, for the evaluation of the results of further analyses in this volume, is the relationship of menarcheal age to intelligence, socioeconomic background, and family size.

Intelligence. At the age of 14.10 years in Grade 8, an intelligence test, WIT III (Westrin, 1967), was given to the subjects in the research group. Intelligence data were obtained for 98% of the girls. WIT III contains four subtests measuring inductive ability, verbal comprehension, deduction, and spatial ability. A split-half reliability of 0.93 is reported in the manual. There was a correlation of 0.02 (ns) for the 456 subjects who had both menarcheal and intelligence data, demonstrating that age at menarche is virtually unrelated to general intelligence. Most earlier studies of pubertal timing and intelligence have also reported insignificant or weak negative coefficients: the earlier matured, the higher the IQ. In their review of the literature, Newcombe and Dubas (1987) reported a small but consistent advantage across studies for early-developed girls with regard to general intelligence.

Socioeconomic Status. In previous studies a weak negative correlation has frequently been found between menarcheal age and socioeconomic status, meaning that lower class girls attain menarche somewhat later than do upper class girls. This is usually explained by better health and nutrition standards in families of higher social class. However, there are empirical indications that the relationship between sociometric status and menarcheal age is disappearing (Diers, 1974; Tanner, 1962). Studies have been reported showing no significant relationships between menarcheal age and social class (Douglas, 1964; Poppleton, 1968; Zacharias, Rand, & Wurtman, 1976) and even weak positive relationships (Jenicek & Demirjian, 1974; Rona & Pereira, 1974). A lack of a relationship between menarcheal age and socioeconomic status has been reported for other comparable Swedish samples (Furu, 1976; Lindgren, 1976).

The present measure of socioeconomic status was based on the educational level of the parents. In Grade 9, the parents of the girls answered a postal questionnaire containing, among others, questions concerning their level of education. The answers were coded into seven categories: (a) unskilled laborer; (b) vocational training, vocational school, or equivalent; (c) lower secondary school, girls' school, folk high school, etc., (d) education at intermediate level; (e) higher secondary school; (f) advanced education above matriculation, but of shorter duration than university studies; and (g) university-level education.

The correlation between the menarcheal age of the girls and their mother's education was positive, low and insignificant: 0.04 ($n = 368$, ns). The same correlation calculated for the one of the parents with the highest achieved education was 0.08, which was not significant at the 5% level ($n = 374$). Although menarcheal age shows some relationship to parents' education, the strength of the association is quite marginal and is assumed not to affect any of the reported relationships between menarcheal age and social adjustment variables.

Family Size. Age at menarche had previously been found to be positively related to family size, suggesting that smaller families are more common among early- than among late-developed subjects (Douglas, 1964; Poppleton, 1968, Tanner, 1962). Douglas (1964) claimed that the relationship between physical and mental growth was accounted for by the fact that the early-developed subjects more often came from families with just one child. In these small families the parents had more time and opportunity to channel their interests and their aspirations onto the child; spending more time in discussion and interaction would subsequently heighten the child's verbal ability (Nisbet, 1953). When family size was partialled out, Poppleton (1968) reported that the significant association between menarcheal age and school achievement (English) was reduced to nonsignificance.

Age at menarche was also, for the present group of subjects, found to be connected with family size. However, a prominent feature of the data was that it was not the number of siblings per se that was responsible for the obtained relationship, but the number of sisters the girl had. This is illustrated in Table 3.5 for family structure data collected at 14.10 years. Overall, there was a marginal relationship (at the .10 level) between the number of siblings

TABLE 3.5
Menarcheal Age as Related to the Number of Siblings at 14.10 Years

Age at menarche	Number of sisters[a]			Number of brothers[b]			Number of siblings[c]		
	0	1	>1	0	1	>1	0	1	>1
−11 yrs	35.7	45.2	19.0	42.2	37.8	20.0	14.6	46.3	39.0
11–12 yrs	38.8	38.8	22.3	36.0	42.0	22.0	14.0	33.0	53.0
12–13 yrs	46.5	31.8	21.8	42.3	38.9	18.9	18.8	33.5	47.6
13– yrs	26.3	41.2	32.5	37.1	43.1	19.8	8.8	30.7	60.5
Total	38.2	37.3	24.5	39.4	40.6	20.0	14.6	33.9	51.5

[a] $X^2 (6, N=425) = 14.15, p=.03$
[b] $X^2 (6, N=425) = 1.68$, ns
[c] $X^2 (6, N=425) = 10.58$, ns

and age at menarche, and there was a nonsignificant relationship with the number of brothers. However, girls who menstruated later had more sisters than the girls who attained menarche earlier. In correlation terms there was a positive, significant correlation of 0.09 ($p = .03$) between menarcheal age and number of sisters. The correlation remained at the same level after controlling for parents education.

PSYCHOLOGICAL DATA

The data on psychosocial development in the present research, which will be related to the girls' biological maturation, are multifaceted. The comparison between early- and later-developed females will encompass a variety of types of measures for data collection: sociometric methods, student questionnaires, vocational questionnaires, interviews, parent questionnaires, register data, medical examinations, and objective tests.

Data covering these aspects of individual functioning are collected at the following points in time:

Grade 3 (age 10)
Grade 6 (age 13)
Grade 8 (age 15)
Grade 9 (age 16)
Age 25–26

Register information on criminal offences, collected from the police, covers the offence activity up to age 30 and for registered alcohol abuse, up to age

25. Register information on psychiatric disturbances during the teenage years covers Grade 7 through 9 (i.e., age 14 to 16).

A survey of the variables and methods used in the present research is given in Table 3.6 A more detailed description of the data is given in running test.

REPRESENTATIVENESS

Although the subject cohort almost completely covered the total number of 14- to 15-year-old subjects in town, who were supposed to enter Grade 8 in 1970, all females in Grade 8 could not be reached to ensure menstrual information. Some of them were absent the days collection of data were made. Of the girls born in 1955, 466 were present and 60 were absent when data on pubertal development were collected.

In order to gain more knowledge regarding the extent of personal and social problems and their effects on the generalizability of the data in terms of systematic drop out, it was decided to make certain comparisons between the absent girls and those who were present at school.

The two groups of girls did not differ significantly from each other for different measures of classroom behavior, educational aspirations, sociometric status, and self-reported school satisfaction 2 years earlier in Grade 6 in 1968. Neither did they differ in socioeconomic status. The groups of girls differed significantly only for one measure. The absent girls in Grade 8 in 1970 had lower intelligence ($p < .01$) in Grade 6 in 1968 than did the girls who were present.

It was further decided to examine relevant comparison variables, not only for data collected earlier in time, but also for measures that were collected after the collection of data for pubertal development. This type of drop-out analysis is seldom presented in the literature. The reason for conducting such an analysis was that school absenteeism, for Swedish pupils, to some extent has been found to preduct future personal and social problems, while, at the same time, its concurrent connection with such adjustment problems might be more limited (see also Rutter, Graham, Chadwick, & Yule, 1976). Therefore, comparisons were performed between the absent girls and those who were present at school for two important measures: documented evidence of psychiatric problems and of delinquency.

Not unexpectedly, the absent girls had more problems of a psychiatric type than the girls who were present at the day for the data collection. Of the 466 girls who were present, 55 girls or 11.8% were found during Grade 7 through 9 in the registers of one or more of the three existing professional networks in the town, which they might consult when having serious personal, social or academic problems (the Counseling Service of the Schools, the Child and Youth Psychiatric Services, and the local Social Welfare Authorities). Of the

TABLE 3.6
An Overview of Measures and Instruments

Measures	Instruments Used
Age 10 (grade 3)	
Peer ratings of social status Self-rated social status	Sociometric methods
School satisfaction Peer relations	Pupils' questionnaire
Sex-role identification Self-evaluation	Semantic differential test
Subject's school satisfaction Subject's burden of work at home Parents' interests in the subject's schooling Parent's aspirations for the subject's education Subject's adjustment problems	Parents' questionnaire
Classroom behavior: (aggressiveness, motor restlessness, concentration difficulties, timidity, disharmony) School motivation	Teacher ratings
Standardized achievement	Objective tests
Marks Hours absent from school	Information from school records
Age 13 (Grade 6)	
Peer ratings of social status Self-rated social status	Sociometric methods
School satisfaction Peer relations	Pupils' questionnaire
Educational aspirations	Vocational questionnaire
Sex-role identification Self-evaluation	Semantic differential test
Subject's school satisfaction Subject's burden of work at home Parents' interest in the subject's schooling Parents' aspirations for the subject's education Family constellation Subject's adjustment problems	Parents' questionnaire

(Continued)

TABLE 3.6
(Continued)

Measures	Instruments Used
Age 13 (grade 6)	
Classroom behavior: (aggressiveness, motor restlessness, concentration difficulties, timidity, disharmony)	Teacher ratings
School motivation	
Standardized achievement	Objective tests
Length stature	Medical examination
Weight	
Body mass	
Marks	Information from school records
Hours absent from school	
General intellectual capacity	Standardized Swedish intelligence test
Age 15 (Grade 8)	
Perceived maturity	Pupils' questionnaire
Perceived difference from peers	
Body image	
Drug use	
Alcohol drinking	
Psychosomatic reactions	
Parent relations	
Topics of disagreement with parents	
Peer relations	
Parent vs. peer orientation	
Age at menarche	Adjustment Screening Test
Weight problems	
Self-evaluation	
Peer relations	
Frequency of staying out in evenings	
Opposite-sex relations	
Sexual intercourse	
Going steady	
Attitudes toward sexuality	
Personal problems	
Mother relations	
Father relations	
Parent relations	
Teacher relations	
Emotionality: (psychosomatic reactions, anxiousness, depression, hyperactive behavior)	

TABLE 3.6

Measures	Instruments Used
Age 15 (grade 8)	
Chartacteristics of the peer network: (number of friends, age characterization, social characterization, amount of time spent with peers)	Peer Relations questionnaire
School satisfaction	Vocational questionnaire
Preference for same maturity stage peers Same-sex ratings of social status Opposite-sex ratings of social status	Sociometric methods
Friendship-group membership Norm violations Alcohol drinking Parents' evaluations and sanctioning of norm breaking Peers' evaluations and sanctioning of norm breaking Parents' perceived expectations regarding norm breaking Peer reference group Wanting own children	Norm inventory
Marks Hours absent from school	Information from school records
Standardized achievement	Objective tests
General intellectual capacity	Standardized Swedish intelligence test
Skeletal age	X-ray
Age 16 (Grade 9)	
School satisfaction Perceived maturity Perceived difference from peers Sexual intercourse Runaway behavior Drug use Alcohol drinking Psychosomatic reactions Parent relations Mother relations Father relations Topics of disagreement with parents Peer relations Parent vs. peer orientation	Pupils' questionnaire

(Continued)

TABLE 3.6
(Continued)

Measures	Instruments Used
Age 16 (grade 9)	
Parents' interests in the subject's schooling	Parents' questionnaire
Parents' aspirations for the subject's education	
Parents' level of education	
Subject's perceived school interest	
Standardized achievement	Objective tests
Age 25–26	
Length stature	Medical examination
Weight	
Body mass	
Family life	Postal questionnaire
Present work conditions and past work history	
Education	
Retrospective view of school	
Future educational plans and eventual obstacles	
Sense of control of own life	
Social support	
Alcohol drinking	
Alcohol drinking	Interview
Drug use	
Use of tranquilizers	
Smoking	
Parent relations	Tests
Personality (somatic anxiety, psychic anxiety, muscular tension, social desirability, impulsiveness, monotonic avoidance, detachment, psychastenia, socialization, indirect aggression, irritability, suspicion, guild, inhibition of aggression)	
Sex-role identity	
Life values	

60 absent girls, 15 girls (25 %) were found at one or several of the networks in the three upper grades in the comprehensive school. This comparison between those absent and those present, taken from register information, somewhat exaggerates the problem situation for the absent girls. One third of them had nothing more serious than a note on school transfer or a consultation with the school psychologist for school fatigue.

A search of criminal records was made for all Grade 8 girls. Data were collected from the local police, local welfare boards, and from the register of the National Police Board (Stattin, Magnusson, & Reichel, in press).

Complete information on registered criminal offences was obtained for all girls in the research group. Considering the total period up to 30 years of age, 40 girls (7.6% of the research group) had committed some offence. Of these 40 girls, 9 belonged to the absent group of females. Thus, although 6.7% of the girls for whom menarcheal information was obtained had criminal records, the figure (15%) was significantly higher ($p < .05$) among the absent girls.

The results presented with respect to characteristics of the participating and the nonparticipating females impose certain limitations on generalization. The nonparticipating females had lower intelligence in Grade 6, and they were more extreme with respect to future delinquency and psychiatric problems. However, the loss of subjects should not be exaggerated. Indeed, the vast majority of the future "problem girls" belonged to the research group. It included 78% of all females born in 1955 who had criminal records and 79% of the girls who were found at one or several of the three professional networks in adolescence. With our main task of determining the relationships among variables, rather than their prevalence at different ages, the loss of data for the 60 absent subjects is judged to have limited effects. At the same time it must be acknowledged that the manner in which biological maturation intervenes in the life of the nonparticipating females cannot be ascertained by our research.

The situation with respect to generalizability in the present case is probably more favorable than in most previous studies on the psychosocial consequences of early versus late maturation. It is not uncommon to find investigations that contain a sample loss of 30% to 40% initially, and it is more true than not to find a drop-out rate in longitudinal research exceeding this magnitude when the subjects have been followed up in adulthood. In our case, the feedback of results to the pupils and the information given by the project staff to the parents, to the parent organizations at school, to the school personnel, to mass media, and to social institutions secured cooperation. No family refused to let their daughter be tested at school. The sample loss over time was also small. For example, when the subjects were on the average 25.10 years old, the postal questionnaire that was sent out was answered by 89.5% of the females in the original research group. In conclusion, the situation with respect to generalizing the findings of the present research, can be said to be quite satisfactory.

Statistical Generalizations

In chapter 1 appropriate methodologies for data treatment in the study of developmental problems were analyzed and discussed. With reference to the multidetermined, stochastic nature of developmental processes and the existence of nonlinearity and interactions among operating factors it was argued that the common linear structural models for data treatment are not generally applicable to our problems. However, the methodology applied here for data treatment leads to a manifold of significance testings. An obvious consequence of this fact is that we have to consider the problem of mass-significance and interpret single measures at any given level of significance with caution. The emphasis should be placed on the consistency of results across variables, across groups of individuals, and across age levels.

4
Psychological Adaptation and Self-Concept

INTRODUCTION

The concept of "self" is customarily meant to refer to an individual's knowledge, ideas, and evaluation of him or herself. It can be conceptualized as possessing both a domain-specific content and an overall evaluative aspect (Burns, 1979). The content aspect of the self-concept is referred to as *self-image* (self-regard). It stands for the subject's cognitive and affective beliefs and views about him or herself in different areas: status, roles, mental capacity, practical skills, ethnic and sexual identity, personality traits and so on. *Self-evaluation* or self-esteem is a superordinate concept, referring to individuals' overall approval and acceptance of themselves. In the present chapter we are concerned with how females psychologically adapt to alterations in their own bodies, as a function of entering puberty early or late, and whether or not, and how, these physical changes affect growing girls' self-concept, both with respect to self-image and self-evaluation.

The self-concept domain is one of the domains in the physical growth literature that researchers commonly have approached from the point of view of a *general* impact of pubertal development (see chapter 1). Such changes in self-concept are part of the *normal adolescent development* where pubertal entry is expected to influence the point in time when these changes in self-definitions are supposed to take place. This is perhaps most obvious with respect to the menarche. That differences in self-concept among teenage girls are observed at a particular point in adolescence is partly attributed to the fact that some girls have already had their menarche at this time, and have subsequently changed in their views of themselves, whereas other girls have not yet attained this developmental point. Over time, because all females sooner or later will have their first menstruation, the self-concept of the later-matured girls will match those of the early developers. This individual timing perspective has often meant utilizing a pre–postmenarcheal design at a certain point in time in adolescence in order to demonstrate the effect on self-conception of having passed versus not having attained menarche.

Although many attach great importance to the personal significance of physical maturation, there is, as of yet, little consensus as to how to conceptualize the meaning of bodily maturation, the extent of psychological adaptation it requires for the early versus the late developer, and how the changes in bodily proportions and appearance accompanying pubertal growth are connected with changes in self-concept. Different opinions have been expressed about the particular features of the linkage between self-concept and early versus late maturing. Therefore, in order to try to understand the possible implications for self-conception of variations in pubertal maturation we first must address the issue of the meaning ascribed to puberty generally, its salience, the particular demands and options it yields for girls, and their actual experiences in connection with pubertal development.

BODILY DEVELOPMENT AND PSYCHOLOGICAL ADAPTATION

Much has been written about the connection between bodily development in adolescence, the adjustment it requires on part of the subject, and its connection with self-concept. In agreement with many clinicians' opinions, with numerous empirical surveys in this area, and with others who have dealt with adolescents' problems, we do not underestimate the role that physical appearance plays for the personal well-being of teenage girls. Much is focused on physical appearance in adolescence (Crocket, Losoff, & Petersen, 1984; Diers, 1974; Lenerz, Kucher, East, Lerner, & Lerner, 1987), and issues connected with body image tend to be of increasing concern from late childhood and onwards among females (Collins & Propert, 1983; Rosenberg & Simmons, 1975). Teenage girls are concerned with how they appear in the eyes of others. Virtually all past and present studies in this area show that teenage girls are quite self-conscious and critical of their physical appearance. Aspects related to physical attractiveness are often found to outnumber problems in other areas. The prominent role of attractiveness and the striving to reach the unattainable ideal are believed to underlie girls' worries, concerning and overcriticism of their perceived physical appearance (Rosenbaum, 1979).

The physical impression that young people make on others plays a prominent role for the kind of reactions they encounter. The attributes that make for success as well as rejection among peers of the same-sex and of the opposite-sex, to a large extent, relate to appearance and behavior (Allen & Eicher, 1973; Crocket, Losoff, & Petersen, 1984; Petersen, Schulenberg, Abramowitz, Offer, & Jarcho, 1984; Cavior & Dokecki, 1973; Place, 1975; Snyder, 1972). From the perception of subjects, attractive persons are attributed with other positive traits, such as intelligence, sociability, competence, and so forth to a higher extent than are less attractive subjects (Dion, Berscheid, & Walster, 1972). Physically attractive persons also describe themselves as more happy, less neurotic, and higher in self-esteem than do less attractive persons (Mathes & Kahn, 1975). Lenerz, Kucher, East, Lerner, and Lerner (1987) reported that physically attractive early adolescents had more positive peer relations, were rated by teachers as higher in scholastic competence, social acceptance, and athletic competence, and had fewer conduct disturbances and behavior problems, as reported by themselves and their parents, than did less attractive adolescents. Simmons, Blyth, and McKinney (1983) reported that positive peer relations, physical attractiveness, and high self-esteem tended to go together in early adolescence. Correlations at the magnitude of .4-.6 between measures of satisfaction with one's physical features and general self-esteem have been reported, the relationships' being somewhat higher for females than for males (Pomerantz, 1979).

The role of physical appearance is presumably of more concern in the teenage years than in other life phases. Except for the very earliest period in the life of the individual, there are no other phases in the life cycle that brings so much change in bodily structure and form over so short space of time as during puberty. The radical changes in various growth parameters across these years heightens the teenagers' awareness, curiosity, and consciousness of their appearance, but also their feelings of inferiority and shame. Puberty has, not surprisingly, been characterized as a life phase of body narcissism (Adams, 1969).

Piaget (1950, 1969), Bruner (1966), and Werner (1948) have posited a general developmental trend in cognition from concreteness and egocentrism to abstraction and decentrism. At the age at which physical growth is most intense, cognitive organization is still governed, to some extent, by overt, perceptual features of persons and event. The role of perceptually salient cues for cognitive activity in these years is revealed in research on person perception, role-taking, concepts of causality, friendship formation, and moral development (Duck, 1975; Livesley & Bromley, 1973; Gollin, 1958; Kugelmass & Breznitz, 1968; Peevers & Secord, 1973). Therefore, it is no surprise that the direct, physical impression that adolescents make on one another determines much of their concepts of others and their view of themselves.

The time during which these physical changes occur co-exist with the time when peer pressures to conform are greatest. Such peer influences bring imperative demands to have the right look, to wear the right clothes and to project the right style. Some data indicate that such peer pressures more strongly affect females than males, particularly with respect to style and opposite-sex contacts (Brown, 1982; Dwyer & Mayer, 1968-1969; Rosenberg

& Simmons, 1975). Brown (1982) found that to be active socially, conforming to styles, having boyfriends, and smoking cigarettes in high school and college, were experienced as involving strong peer pressures among females, whereas having sexual intercourse was connected with high peer pressures among males. More consistent relationships between engaging in different behaviors and experiencing peer pressures to do so were found among females.

A particular reason for the prominent role of physical appearance in adolescence is its role for heterosexual relations (Dwyer & Mayer, 1968-1969). How successful the girl is in her contacts with the opposite sex also determines much of her social prestige among her same-sex peers (Schofield, 1981) and affects her self-esteem (Bardwick, 1971; Douvan, 1970; Mathes & Kahn, 1975).

Altogether, the reasons for the prominence of physical appearance in adolescence are manifold, ranging from overriding cultural demands, through interpersonal influences, to the unique experience of one's own physical development in the individual teenager. The great role of interpersonal relations for well-being in females, the prominence of physical attraction in females for opposite-sex contacts, and the closer link between physical attractiveness and self-esteem in females, relative to males, have all been assumed to sensitize females to the changes in their physical appearance (Lerner & Karabenick, 1974).

Underlying much discussion on self-concept in the physical growth literature is an implicit linkage between physical appearance and general self-esteem, making the changes in physical features induced by puberty likely to affect overall self-evaluation. Second, current research identifies the girls with early pubertal development as particular candidates for such self-definition changes, at least in the early part of the adolescent period. An expected *increase* in level of self-evaluation with pubertal development can be said to be founded upon two assumptions. First, to the extent pubertal development as a cultural and social developmental marker is positively evaluated by the society, the individual who develops earlier than others may draw benefits from her maturational stance personally and interpersonally. Second, the female who passes this phase of development early ascribes to herself this positively valenced attribute. For example, given the validity of the proposition that adolescent girls are sensitive to aspects of themselves as social objects and that self-evaluation is closely connected with popularity among peers (cf. Walker & Green, 1986), then, to the extent physical maturation bring about physical features that have these popularity-reinforcing properties, maturation might positively influence the self-evaluation of early developed females. If, however, reaching sexual maturation has a culturally negative meaning, or if one's early maturation is connected with negative experiences in daily life, there are fewer reasons to expect that maturation would be beneficial for the self-evaluation of the early developer.

"Growing up" is generally thought to hold positive connotations for the young teenager. (Blyth, Simmons, & Zakin, 1985). A mature body signals to others adult-like status and sexual and reproductive maturity. Clausen (1975) made the following comment:

> Just as there exist preferences for types of physique, it appears that there exist preferences for the timing at which various maturational stages are reached, though here the generalization rests less on direct data than on inference. In general, younger children try to emulate older ones, and especially as puberty is approached, children wish to be accorded more freedom from parental supervision and to be treated more nearly like adults. . . . Peers who appear to be older and more mature than their years are frequently the objects of admiration and envy. We would expect, therefore, that early maturing carries some of the same positive connotations as does an athletic build. (p. 26)

Gender identity and the behaviors commonly ascribed to the female identity are thought to become salient with the articulation of feminine features that follow physical and sexual maturation (Petersen, 1980). Changes in secondary sex characteristics particularly focus the issue of sexual maturation (Rierdan & Koff, 1980b). That popularity among the opposite sex is largely dependent on how far in her development toward adult physical status the girl has come is another factor which argue in favor of the early-matured girl with regard to positive changes in self-evaluation, at least in the early part of adolescence (Simmons, Carlton-Ford, & Blyth, 1987). To acquire a woman's bodily figure might be reassuring for feminine identity and be a source of pride for the early developer.

If these factors were the only ones, the early developer might benefit from the positive and personal meaning ascribed to changed maturity status, and, accordingly, develop a more positive self-evaluation. However, a few critical observations moderate this conclusion. First, girls' body image has been found to be less related to self-evaluation than is their self-image in other areas, such as peer relationships, coping ability, and educational aspirations (Petersen, Schulenberg, Abramowitz, Offer, & Jarcho, 1984). Second, clinical observations suggest that adolescents with negative body images often are the same persons who had negative body images before puberty (Schonfeld, 1963, 1971), Thus, if there is a connection between body image and overall self-evaluation, this connection was already present before entrance into puberty. In fact, available data do not unanimously suggest any marked age-related changes in self-evaluation across age in adolescence (Dusek & Flaherty, 1981; Simmons & Rosenberg, 1975; Wylie, 1979).

On the other hand, that individuals' self-evaluation appears to be rather stable over time in adolescence does not preclude age-related changes with respect to other self-perceptive features or with respect to self-consciousness

(Dusek & Flaherty, 1981). In the 1968 Baltimore study of the stability over time in self-consciousness, self-image, and self-esteem among one and the same group of girls studied from Grade 3 through Grade 10, the most radical change over time concerned the girls' self-consciousness, which rose particularly sharply around the age of 12 (Simmons & Rosenberg, 1975). Third, as was described in the first chapter, we should not automatically ascribe processes operating for the general teenage population to those operating for early- or late-developed girls. The finding that a more mature bodily figure is related to success in opposite-sex relations for teenage girls generally does not necessarily apply to the early-developed girls in early adolescence whose close environment largely consists of undeveloped males.

Furthermore, we should not lose sight of the fact that physical maturation may bring uncertainty and feelings of loss of control. Whatever the final outcome in bodily dimensions and general look will be, adolescents may feel that they are the object of growth forces that are largely out of their voluntary control. From the loss of control point of view, physical changes involved in pubertal development have been posited to increase girls' self-consciousness and give them a more unstable self-picture (Douvan 1970; Peskin, 1973; Simmons, Blyth, & McKinney, 1983). At a time when physical structures and functions develop rapidly, self-consciousness has been found to cause awkwardness, particularly for the subjects who develop physically at a non-normative rate (Mussen, Conger, & Kagan, 1963). Height and weight, two growth parameters that differentiate the early-developed girls from the later-developed girls during early adolescence, are growth features that are not particularly favored as feminine in our culture. Deviations from others with regard to these body features are reported as particularly distressing to adolescent females (Dwyer & Mayer, 1968-1969; Stolz & Stolz, 1944). The increase in body fat from early puberty is especially of concern to early-developed girls (Dornbusch et al., 1984). There is a certain communality of opinion in the research literature that the greater increase in body fat from early adolescence, among the early-developed girls compared to the later-developed, is one of the concomitants of pubertal development that brings particularly negative experiences for these girls. Increased weight does not fit with the cultural idea of a lean figure, and it has no prestige value in the peer group.

Thus far, as it concerns pubertal maturation in a general sense, it seems as if early physical development brings both negative and positive consequences. Therefore, an overall hypothesis of a positive or a negative change in self-evaluation connected with the passing of puberty early would probably be inappropriate.

When discussing the possibility for self-concept changes as a function of entering puberty, the menarche, rather than physical maturation generally, is thought by many to be the critical event capable of instigating lasting self-definitions (Deutsch, 1944; Kestenberg, 1961, 1969; Koff, Rierdan, & Silverstone, 1978). Abruptly, inevitably and irreversibly transporting the person toward adult status, the first menstruation has been considered a disruptive factor in female development, one with sufficient power to reorganize both self-image and self-evaluation. The issue then becomes focused on determining whether menarche has positive or negative connotations in development.

MENARCHE AND PSYCHOLOGICAL ADAPTATION

The first menstruation, indeed, seems to be a unique marker of maturation. Females, often several decades after it happened, can readily remember when it occurred, how they felt at the time, and what support they got from their immediate social environment (Golub, 1983). The menarche has been assigned a special personal significance as an event in development. Koff, Rierdan, and Silverstone (1978) strongly associated intrapsychic changes to the first menstruation:

> Although the actual feminizing changes of puberty thus begin long before, and continue long after, menarche, it seems that the onset of menstruation—a sharply defined biological event—is the particular time at which the psychological and biological changes occurring throughout adolescence are organized and integrated. (p. 636)

The role of the first menstruation as the definite marker of sexual maturity, with the feminine biological role attached to it, was also emphasized by Deutsch (1944).

Kestenberg (1961, 1967a, 1967b), in particular, has stressed the positive aspect of the menarche. Kestenberg argued—without any empirical findings to support the proposition—that the menarche and the menstruations that follow will cause trouble, physical discomfort, pain, and fear, but the suffering will have a fixated point of reference in the menarche, a change from vague fantasies in prepuberty to something concrete and firm by the time of the onset of the first mense:

> The onset of menstruation makes it possible for the girl to differentiate reality from fantasy. What she knew and what she anticipated can now be compared with how it happened to her. The sharpness of experience, the regularity of it, the well-defined way of taking care of it, the sameness of the experience as compared with her own anticipation of it and the experience of others—all these provide relief. It helps the girls to structuralize her inner and outer

experiences, to regain her ability to communicate and to perceive in an organized fashion. (1961, p. 34)

Further, the girl tolerates the continuous discomfort and pain because it is part of being a female.

Koff, Rierdan, and Jacobson (1981) reported that most postmenarcheal seventh- and eighth-grade girls, who responded to a sentence-completion test of a fictitious girl having had her menarche, stated that the first menstruation was a sign of "maturity" and "she seemed to have grown up and looked more like an adult." Among postmenarcheal, 12- to 15-year-old girls interviewed by Whisnant and Zegans (1975), some had started to reflect on their future, feeling closer to their mothers and thinking about their own children. One out of four girls in a study conducted by Clarke and Ruble (1978) agreed that "menstruation is something to be happy about." Ruble and Brooks-Gunn (1982) reported that the most frequent positive feeling about the menarche was that it was the sign of maturity, a view expressed by 72% of the fifth- and sixth-grade postmenarcheal girls in their longitudinal sample. One third mentioned the possibility of now being able to have own children, and every fourth girl mentioned being a woman as a positive experience.

Additional knowledge of the cultural meaning and the personal and emotional significance of the menarcheal event is gained by information from investigations of premenarcheal girls. They are the group of girls soon to pass this phase of pubertal development. These girls might, in comparison with postmenarcheal girls of the same age, be more attentive to the positive aspects and perceive the personal significance of the event, whereas, at the same time, their attitudes toward menarche are comparatively less inflated by physical symptoms and discomforts (cf. Rierdan & Koff, 1980a; Stubbs, 1982).

The available literature suggests that premenarcheal girls generally have positive connotations associated with menarche and that they look forward to this event. However, they prefer to have their first menstruation at the same time as most of their same-age peers. The premenarcheal girls in Koff, Rierdan, and Jacobson's (1981) study expressed positive anticipations, connecting menarche with "growing up" and "woman." A greater proportion of the premenarcheal girls, compared with the actual experience of the postmenarcheal, anticipated a positive change in body image, and they expected to tell their mother and their friends about what had happened. Similarly the premenarcheal girls interviewed by Whisnant and Zegans (1975) expected to announce the first menstruation to their friends and their mothers. They thought they would act differently and would be granted more freedom by their parents.

If we were to judge by these data exclusively, which refer to the personal significance of the menarcheal experience and its social significance, they would suggest that the postmenarcheal girls in the early part of adolescence would be in a more beneficial position relative to the premenarcheal. On the other hand, these positive aspects are just one side of the coin.

The adjustment to the changes in everyday behavior following the menarche might be unproportionally difficult for the earliest developer. If she is unprepared for the first menstruation, she might not see the full implication of this developmental marker; she might be unsecure as to significant others' reactions, and because deviations from what is standard in the peer group are assumed to be stressful for the teenager, early menstruation at a time when most of one's peers are premenarcheal might cause distress. The literature shows, indeed, that girls who mature early are more unprepared for the menarche, and this has been reported to have consequences for their menstrual experiences later in adolescence (Brooks-Gunn & Ruble, 1982b; Koff, Rierdan, & Sheingold, 1982).

The menarche has a diffuse societal status. The anthropological literature shows that menstruation is typically connected with guilt, shame, embarrassment, uncleanness, and other negative affects. Williams (1983) reported that the majority of fourth- to sixth-grade females agreed with the statement that "Menstrual blood is old blood that the body doesn't need anymore," and one out of eight girls agreed that "Menstrual blood has mysterious powers." Issues connected with menstruation still belong to a domain that, according to women today, should not be spoken about in public or with males, not even being a suitable topic for discussions at home (Milow, 1983). Probably the majority of adolescent females and also grown-up women hesitate to discuss issues connected with menstruation publicly (McKeever, 1984). However, compared with preceding generations of teenagers, the attitudes toward the menarche have become more positive and open today (Maidman, 1984; Stubbs, 1982), and knowledge about what to expect has increased considerably (Brooks-Gunn & Ruble, 1983).

Although the onset of the first menstruation is openly welcomed in many societies and is the overt indicator that the girls have definitely passed the border from childhood to adulthood, no such formal *rite de passage* exists in the modern western culture (Mead, 1952; Paige & Paige, 1981). Whisnant, Brett, and Zegans (1975) made a detailed analysis of commercial education materials available to young girls. They concluded that such information ignored the positive affective message to adolescent girls. Instead of emphasizing that the menarche implied a new role and a new status within the society, the message transmitted how to deal with menstruation hygienically. It was the modern rational society's answer and aid to the physical problems and discomfort that accompany a physiological change:

> The prevalent view that menstruation is like a sickness is conveyed through advertisement for "women's medications" and "feminine products." Preparation for menarche involves assembling a suitable array of products, which are conve-

niently packaged and available by mail order. Menarche is then portrayed as a hygienic rather than a maturational crisis. (Whisnant & Zegans, 1975, p. 813)

The view that menarche is an indifferent cultural experience in which "the menarcheal girl is left without the socially established support" (Logan, Calder, & Cohen, 1980) has led to discussions about how best to educate girls about the physical, social, and sexual aspect of the menarche (McKeever, 1984), as well as about what customs should be instigated in families to create a positive atmosphere at the time girls begin to menstruate.

In early psychoanalytical writings, the menarche was described as a negative and disruptive event. Deutsch (1944) regarded the first menstruation in a female's life as traumatic, evoking anxiety, distress and feelings of concealment: "All observations suggest that, whether or not the girl is given intellectual knowledge, even when she has the best possible information about the biological aspects of this process, and despite its wish-fulfilling character, the first menstruation is usually experienced as a trauma" (p. 157). Girls also commonly have been found to be secretive of their experience of the first menstruation just after it appears (Brooks-Gunn & Ruble, 1983) and to consider it as an unpleasant and anxiety-provoking event (Koff, Rierdan, & Jacobson, 1981; Whisnant & Zegans, 1975). Interviewed in retrospect about their menarche, adult women tended to remember it as negative (Bardwick, 1971; see also review by Greif & Ulman, 1982).

One of the earliest empirical investigations on reactions to the onset of the first menstruation was reported by Conklin (1933). Over half of all girls reported indifferent reactions and only every third girl reported positive emotional reactions (curious, interested, delighted or proud). Koff, Rierdan, and Jacobson (1981) found many evidences of negative attitudes towards the onset of the first menstruation among the postmenarcheal girls. Girls mentioned feeling "sick," "strange," "weird," and "scared," stated that it imposed limitations to their activities, and mentioned attempts at hiding it from their peers. Whisnant and Zegans (1975) similarly reported that the postmenarcheal girls were rather secretive of mentioning that they had their first menstruation, were fatalistic about the menstruation—"there's nothing you can do about it, anyway. You're stuck with it"—and expressed anxiety in connection with the onset of the first menstruation.

Shainess (1961, 1962) made a questionnaire investigation of the experience of menarche among a clinical sample of adult women. Most of these women reported having anticipated the onset of menstruation with fear, and a strikingly low proportion of them reported that their mothers showed a positive response to their menarche. An interesting finding was that the woman who showed no evidence of premenstrual tension reported that they were well-prepared for the first menstruation. However, in contrast, Woods,

Dery, and Most (1982) found little evidence of a connection between experiences connected with first menstruation and premenstrual symptoms and menstrual attitudes among 18- to 35-year-old women. Both studies utilized retrospective reports. The issue of menarcheal experiences and adult menstrual attitudes and symptoms awaits longitudinal data to be finally settled.

The most extensive study of girls' reactions to menarche hitherto was done by Ruble and Brooks-Gunn (1982). They collected information on this topic by means of a questionnaire given to a large sample of girls at three grade levels, 5–6, 7–8, and 11–12, along with interviews to a longitudinal group of fifth- to sixth-grade girls. Of the postmenarcheal girls in the cross-sectional sample, almost every second girl reported hassle as something negative about menstruation, 30% mentioned physical discomfort, 1 out of 5 girls reported behavioral limitations and emotional changes, and most of the girls (79%) mentioned some form of worry in connection with the menstruation. It should be noted that early-developed postmenarcheal girls expressed more negative feelings connected with the menarche than did later-developed, and they also reported more menstrual symptoms and more medication taken.

The premenarcheal seventh- and eighth-grade girls in Ruble and Brooks-Gunn's material (Brooks-Gunn & Ruble, 1982a, 1983; Ruble & Brooks-Gunn, 1982) rated menstruation less negative than the postmenarcheal. The premenarcheal also said that they would be more open about the event than the postmenarcheal actually had been. However, the premenarcheal girls estimated the expected physical discomforts connected with the menarche as more serious than the postmenarcheal girls actually experienced. Overall, these data indicated that premenarcheal girls tended to be more extreme, both positively and negatively, with respect to the menarche than were postmenarcheal girls. The greater severity of menstrual pain expressed by the premenarcheal girls was interpreted in terms of the cultural beliefs of a menstruation-pain interconnection (Brooks-Gunn, 1987; Brooks-Gunn & Ruble, 1982b). In an earlier study by Clarke and Ruble (1978), the postmenarcheal girls expressed greater dislike of menstruation and thought it had more adverse impact on moods and schoolwork than did the premenarcheal. The less negative attitudes expressed by the premenarcheal girls was in this study interpreted as lack of direct experience of physical discomforts on the part of these girls, or, alternatively, that these girls mainly had positive anticipations as to the forthcoming maturational sign.

Overall, one conclusion that can be drawn from studies investigating the significance of menarche in the lives of growing females is that girls' reactions tend to be quite diverse. The first menstruation seems to have both positive and negative valences attached to it and is connected with both positive and negative feelings in the same individual. This seems also to hold in a cross-national perspective (Logan, 1980). Using a projective technique, Petersen

(1983) found that 33% of seventh-grade girls' reactions to the menarche was negative, 49% reported positive feelings, and 5% showed ambivalent feelings.

The results from studies by Woods, Dery, and Most (1982) and Brooks-Gunn and Ruble (1982a) are also suggestive in these respects. The majority of women, aged 18 to 35, in Woods, Dery, and Most's study reported having been happy, proud, or excited, at the same time as most also reported having felt upset, embarrassed, and scared. In agreement with the findings reported in the early study by Conklin (1933), the postmenarcheal girls' reactions to the menarche was found by Brooks-Gunn and Ruble to be quite varied, with most girls reporting both positive and negative feelings. In fact, only 18% of the girls reported predominantly negative attitudes and 19% predominantly positive. What is perhaps more interesting is the fact that most girls in their study reported that the menarche was not a particularly emotionally laden event: "On the average, girls reported feeling a little bit negative (e.g., upset) and positive (e.g., excited)" (p. 1560). The possibility cannot be ruled out that girls denied their feelings, but at face value the potentiality of the menarche to change central parts of the self-concept is questioned by these findings (cf. Henton, 1961). Even among girls with precocious puberty, the matter-of-fact attitude toward the first menstruation tends to be the common attitude (Money & Walker, 1971).

The indifferent attitude toward the first menstruation, of something to be tolerated and lived with, has been conceptualized as a distress reaction to unfulfilled positive expectancies. We cite here in length Helene Deutsch's (1944) comments on girls' afterreaction to menarche:

> The young girl hopes that with the onset of menstruation her role with regard to her environment will change, that she herself will experience something new and momentous. Above all, she hopes to be recognized as grown-up and to acquire new rights. "Grown-upness" for her means freedom from her own inability to achieve anything and, above all, from the restrictions and renunciations that she has to suffer as a child and that are imposed upon her by the grown-ups, especially her mother. However, menstruation, that important sign of maturity, does not bring out any advantageous change. Young girls who have reacted to the first menstruation with depressions often openly admit that they were previously informed about the facts and yet experienced a painful feeling of being surprised. The surprising element was the sense of disappointment that may be expressed thus: "Here is the longed-for, tremendous event, yet nothing has changed around me or inside me." (p. 156)

In conclusion, in the light of the evidence that pubertal development generally and menarche specifically bring a diverse set of demands and options of both positive and negative kinds, an overall hypothesis concerning change in self-evaluation with passing puberty, cannot be made.

THE IMPACT OF PUBERTAL DEVELOPMENT ON SELF-CONCEPT

How strong, then, is the actual empirical evidence for an association between pubertal development and self-concept changes? In the following we review the literature on this subject, paying attention to both the content and the evaluative aspect of the self-concept.

Self-Image

Positive evidence for a change in self-image as a function of pubertal development was presented by Koff, Rierdan, and co-workers. In several studies, using both cross-sectional and short-term longitudinal strategies they investigated the sex-role identification of pre- and postmenarcheal girls employing the Draw-a-person test (Koff, Rierdan, & Silverstone, 1978; Rierdan & Koff, 1980a, 1980b). Giving 12- and 13-year-old girls the opportunity to draw pictures of males and females, they found that the postmenarcheal girls more often drew a female first and tended to produce sexually more differentiated drawings than did premenarcheal girls, thus providing evidence of greater feminine body identification and sexual differentiation after the first menstruation.

In a cross-sectional analysis, Rierdan and Koff (1980b) tested the idea that the awareness of one's physical maturation would be most articulated in early adolescence when the most dramatic changes in sexual maturation are at hand. They found higher sexual differentiation scores (explicit drawings of breasts) and a stronger tendency to draw females first among the postmenarcheal girls who just had attained menarche than among the postmenarcheal girls who had passed the event a longer time ago. They also reported higher awareness of sexual development markers among younger adolescents than among older, independent of menarcheal status (Rierdan & Koff, 1980a). These results were interpreted to support a hypothesis of sensitivity to female-defining features immediately associated with the menarche.

In another study (Koff, Rierdan, & Silverstone, 1978), one and the same group of 12-year-old girls were tested twice, with an interval of 6 months between the occasions. There were significant differences between the pre- and postmenarcheal girls with respect to drawing sexually differentiated pictures and drawing a female first on both test occasions. More important was that the group of girls who changed from pre- to postmenarcheal status from the first to the second occasion had a mean sexual differentiation score that was close to the mean for the stable premenarcheal girls on the first occasion and close to the mean for the stable postmenarcheal girls on the second test occasion. A higher proportion of these changing girls also drew a female first on the second test occasion compared with the first. According to the investigators, the results overall suggested that the menarche provided a confirmation of the feminine role and constituted a discrete point in time for the reorganization of body image and sexual identification. Undoubtedly, this study is one piece of the positive evidence we have at the present time for the impact of menarche on the self-image in girls. The design of the study overcomes many of the problems involving control of all relevant variables in cross-sectional investigations comparing premenarcheal and postmenarcheal girls.

Greater concern about heterosexual issues and a stronger heterosexual orientation among early-developed girls than among later-developed is another common finding reported in the physical growth literature. In one of the early investigations, Stone and Barker (1939) reported that postmenarcheal girls scored higher than the premenarcheal girls for issues connected with "heterosexual interests and attitudes," and "adornment and display of person"; areas that are commonly associated with a female identity. A stronger heterosexual orientation among early-developed girls, in comparison with later-developed, was also reported in the Milwaukee longitudinal study by Simmons, Blyth, and McKinney (1983). The differences between early- and late-developed girls that appeared at Grades 6 to 10, concerned perceiving oneself as popular among boys, being concerned about opposite-sex popularity and dating, and evaluating opposite-sex popularity higher than competence. Garwood and Allen (1979) reported more problems related to heterosexual contact among postmenarcheal 13-year-old females than among premenarcheal females of the same age. In an investigation of pubertal development on sex-role related interests, among other things comparing girls with regard to responses to pictures of infants, Goldberg, Blumberg, and Kriger (1982) found that postmenarcheal 12-year-old girls showed higher preferences of this sort, an effect that was independent of experience with infants.

An area that has produced strong and consistent results across studies is research on girls' body image. Early-matured girls have been found to have a more unfavorable attitude toward their physical appearance than do late-matured girls (Brooks-Gunn, 1987). Simmons, Blyth, and McKinney (1983) reported data for the longitudinal sample of the Milwaukee study. Comparing early, average, and late developers, they found the early-matured girls to be less satisfied with their weight (Grade 10), height (Grades 7 and 9), and figure (Grade 10; but more satisfied at Grade 6). The early-matured girls were also more concerned about their height (Grades 7 and 9), weight (Grades 9 and

10), and figure (Grades 6, 9, and 10). All these comparisons were performed with control for the actual weight and height of the girls. The study showed that weight was the critical variable behind the more unfavorable body image of the early developers. In a later study of the same sample, Blyth, Simmons, and Zakin (1985) noted that early developers in the sixth and the seventh grades reported less satisfaction with their weight than did other girls. Lower satisfaction with their weight among early maturers was also reported by Gross and Duke (1980), who employed data from the National Health Examination Survey for about 3,000 girls, 12 to 17 years of age (see also Duncan, Ritter, Dornbusch, Gross, & Carlsmith, 1985). The role for body image of weight was again confirmed in a short-term longitudinal study covering Grades 6 to 8 reported by Crocket and Petersen (1987). In Grades 7 and 8, the more physically developed girls were less satisfied with their weight than were the later-developed girls. Similar results were reported by Zakin, Blyth, and Simmons (1984), who asked pre- and postmenarcheal 12 to 13-year-old girls to rate their satisfaction with their weight, height, figure, and looks. Only the weight variable was significantly related to pubertal status, with the postmenarcheal girls reporting being less satisfied.

Tobin-Richards, Boxer, and Petersen (1983) concluded from an investigation of the body image of pre- and postmenarcheal seventh-grade girls that in contrast with boys, for whom facial hair was the physical growth characteristic most closely connected with self-perception of attractiveness and overall body image, weight was the physical growth variable among females that was significantly related to body image. In Tobin-Richard et al.'s study, there were no significant linear relationships between pubertal development (measured by various self-reported growth indices) and perceived attractiveness or body image. However, a curvilinear trend was found, such that girls who perceived themselves as either earlier-matured or later-matured than same-age peers had a more negative body image and perceived themselves less attractive than girls who considered their pubertal development to be "on time." In agreement with previous studies, it was particularly the girls who considered themselves as early-matured who presented this more negative body image. Similar results of a nonlinear relationship between pubertal development and satisfaction with weight was reported by Brooks-Gunn and Warren (1985b).

That the weight factor appears to differentiate early and late developers is of interest in view of the fact that the literature on body satisfaction generally shows teenagers, particularly adolescent girls, to be quite concerned about their weight. Although just a minority of adolescents need, for medical reasons, to lose weight, often it is a majority of girls that from time to time try to control their weight (Dwyer & Mayer, 1968-1969). An informative study on body satisfaction was reported by Clifford (1971). Overall, both boys and girls expressed most criticism for physical features connected with physical growth. For females, least satisfaction was reported with respect to weight, looks, legs, waist, and hips: "It is primarily in these aspects of the body . . . that weight gains are manifested. If there is a concern with weight, these become the critical areas for both sexes" (p. 124). Females reported less satisfaction with their body parts than did boys.

Self-Evaluation

In the light of the ambiguousness meaning of and reactions to physical maturation and menarche that have been reported in the research literature, it is no surprise that there is a lack of convincing data regarding relationships between overall self-evaluation and pubertal development. Garwood and Allen (1979) compared pre- and postmenarcheal girls, 12- to 13-years of age, with respect to level of self-esteem using the Tennessee Self Concept Counseling Scale. There was a weak tendency for postmenarcheal girls to rate themselves higher in self-esteem, but neither the overall score nor any of the subscales differentiated significantly between the two groups of girls. A further subdivision of the postmenarcheal girls with respect to when the menarche occurred did not change the basic results. Indications of heightened self-awareness among postmenarcheal girls were reported by Davidson and Gottlieb (1955). The postmenarcheal girls received higher, but not significantly higher, scores on the M and FK scales of Rorschach than did the premenarcheal girls. The subjects were 26 girls, 11 to 13 years of age. No differences in self-evaluation between 12-year-old and pre- and postmenarcheal girls were reported by Zakin, Blyth, and Simmons (1984) for a modified version of the Rosenberg Self-Esteem Scale. However, when rated attractiveness of the girls was introduced as moderator variable, unattractive and average-attractive postmenarcheal girls scored higher on self-esteem than did the attractive postmenarcheal girls, whereas higher attractiveness was related to higher self-esteem among the premenarcheal girls. The authors explanation for this finding was that pubertal development boosts self-esteem of less attractive girls. Their study was cross-sectional, however, and one needs longitudinal data to support such a directed conclusion.

Neither Simmons, Blyth, and McKinney (1983) nor Blyth, Simmons, and Zakin (1985) found any general differences in self-evaluation between early-, average-, and late-matured girls in the Milwaukee study. In a study of the same longitudinal sample using LISREL methodology, Simmons, Carlton-Ford, and Blyth (1987) confirmed the absence of a direct relationship between pubertal timing and self-esteem for seventh-grade girls. However, they observed an indirect relationship. Early-matured girls were more likely to be involved in problem behavior at school, and this problem behavior dimension was negatively associated with self-esteem. For the factor dimension, "likes self," no differences between premenarcheal and postmenarcheal seventh- and eighth- grade females were reported by Brooks-Gunn and Ruble (1983).

Moreover, in a matched-control study of precocious puberty in girls (Ehrhardt et al., 1984) no differences in self-evaluation emerged.

In one of the few studies reporting significant differences in overall self-evaluation, Jones and Mussen (1958) investigated 17-year-old girls with the TAT. Level of maturity was ascertained from skeletal age. The late-developed girls ascribed to the main figures in their stories more negative qualities than did the early developers, indicating more negative self-evaluation among the late-matured girls. It should be remembered, however, that this study dealt with delayed menarche in late adolescence and as such did not concern itself with the issue for girls in adolescence generally nor with early versus late maturation in the normal range. Utilizing an ecological approach, whereby subjects indicated their level of self-esteem in response to a randomly scheduled "beeper" that they carried with them, Jaquish and Savin-Williams (1981) reported that breast and genitalia development at Grade 7 was positively related to self-esteem at Grade 8. On the other hand, no significant relationship was obtained between pubertal development indices at Grade 8 and self-esteem measures at the same grade. The results are difficult to interpret, in part because of limited sample size ($N = 21$), and in part because the measure of self-evaluation to a great extent dealt with aspects of moods and feelings in specific situations.

Conclusions

Recognizing the shortcoming of interpreting data on self-concept without adequate knowledge of the subjects' self-concepts prior to puberty, nor of their prior experiences, the question of whether pubertal development is accompanied by changes in self-concept can be summarized as follows.

Data have been accumulated that early maturing is linked with certain *self-image* changes. Earlier-developed females present a picture of stronger feminine identification and a more integrated sexual identity than do later-developed, and there is a connection between body image and pubertal development, with early developers having more unfavorable attitudes, especially concerning their weight, than do later-developed girls. Furthermore, when compared for their sex-role related interests at a given point in time during the adolescent period, earlier-developed girls seem to have stronger interests of this type than do later developed girls.

The more critical attitude toward their appearance among early developers, compared with the later-developed girls, is partly a function of their greater gain in body fat relative to other later-developed girls. Girls who mature early tend to be more heavily built than those who mature late. The gain in body fat is also a normal feature of female development (Dornbusch et al., 1984) and a question of timing of physical maturation. Other factors contribute to the early developers' greater concern for their weight. It is probably the case that such concern expresses the role of physical appearance in heterosexual relations. The more salient their relations to boys, the greater their self-consciousness or overcritical their attitude (cf. Hill & Lynch, 1983). The finding of decreased body image among females generally, with the passage of time from adolescence, presumably reflects this increased consciousness of the role of physical appearance for heterosexual success with the imminence of opposite-sex contacts. A further comment should be made with regard to weight and other possible influences. In their large-scale study of a national sample of 12- to 17-year-old American females, Dornbusch et al. (1984) reported a relation with desire for thinness and social class, with upper class females expressing a stronger desire, even after controlling for actual fatness. Within each social class, early developers expressed least satisfaction.

There is no particularly impressive evidence that variations in pubertal development among females are related to individual differences in self-evaluation. It is probably the case that the girls' ambivalent outlook regarding the meaning of menarche in their own lives, as well as their response to other aspects related to pubertal development, is connected with the finding in the literature of few differences in self-evaluation between pre-and postmenarcheal girls. As mentioned before, an underlying rationale for expecting positive changes in self-evaluation as a function of characteristics of the individual is that if an attribute that the person possesses is positively evaluated by others, and if the person adopts this value as part of the overall evaluation of herself, self-evaluation might increase. Given the uncertainty of the outcome of physical growth and the disequilibrium with regard to appearance that puberty entails, perhaps the most obvious outcome for the evaluation of the self is neither a general negative nor positive effect, but rather an unstable and fluctuating self-evaluation over time. The test of this hypothesis requires a longitudinal design. It has not yet been performed.

Such an analysis also should be complemented with an examination of the interaction of different psychological and social factors behind changes in self-evaluation. Empirical data suggest that the influence of puberty on self-evaluation operates along different paths (cf. Simmons, Carlton-Ford, & Blyth, 1987), where the impact of physical maturation is indirect and moderated by features of social behavior. A case in which demographic factors contributed to the outcome was the finding by Clausen (1975) that early developers among middle-class girls in the Oakland Growth Study showed higher self-confidence, whereas it was the late-developers among the working class girls who showed higher self-confidence.

To conclude, changes in self-image but not in self-evaluation have been demonstrated to be connected with pubertal development. There are limited possibilities of discerning whether or not observed differences in some aspects of the self-image are connected with aspects referring to physical maturation generally, or, alternatively, directly tied to the menarche. The pre–postmenarcheal division of same-age girls does not guarantee that it is the menarche in itself which is connected with individual differences in self-concept (see chapter 2). The comparison of changing girls over two test occasions (from pre- to postmenarcheal status) with stable pre- and postmenarcheal girls on both test occasions (cf. Koff, Rierdan, & Silverstone, 1978) focuses the menarche as the effective determinant. However, the definitive design is the one that, in addition, controls for initial differences in growth parameters like size, weight, and breast development at the first testing, and changes in these respects by the second test occasion.

THE PRESENT STUDY

The Self-Concept: Stability and Change

What conditions should be met in order for an individual's self-concept to undergo change? One theme that has some generality across different types of self theories is the assumption that an individual's self-evaluation tends to remain rather intact over time in the normal range of upbringing and is largely unaffected by transient experiences. Parents' evaluations, their love and affection versus neglect and criticism toward the child in early years tend to be an integrated part of children's evaluation of themselves, making the parental evaluation of the child his or her lasting evaluation, one that is resistant to change.

Epstein (1980) has asserted that a stable self-structure and a stable self-evaluation is an absolute necessity if we are not to experience life and our part in it as chaotic:

> Individuals will defend their personality theories of reality at almost all costs because they need a theory in order to function. . . . The maintenance of self-esteem to the child, and later to the adult, is equivalent in importance to the maintenance of a love relationship with the mother to the infant. . . . The overall favorableness of self-assessment identifies one of the basic postulates in a person's self-theory. As a higher order postulate, self-esteem is resistant to change. Should it change, it has widespread effects on the entire self-system. (pp. 106, 113)

It follows from this proposition that because the person seeks to keep consistency in his or her self-system, we normally can expect only gradual and successive changes. Profound changes related to basic self-evaluation are likely to appear only under more extraordinary circumstances. In effect, an individual's self-concept is a structure that, in itself, sets limits on change, at least in so far as central, evaluative aspects are concerned. In normal development, changes that can be observed occur gradually over time, and peripherally rather than centrally in the system (Epstein, 1980).

The menarche has been thought to be a change-provoking event capable of yielding the conflict between self-structure and reality that is necessary for a reorganization of the self-system and for more lasting changes in self-evaluation. However, accepting that the menarcheal event could be the "single most important event of puberty," changes in self-concept that may occur at the time of the menarche should not divert our attention from the changes in self-perceptions that are linked to physical growth in the broader perspective. For most girls, the menarche does not come abruptly. The first menstruation does not open up puberty, but occurs at the end of the pubertal sequence. It is expected and is part of the broader gradual physical change process starting with breast development, passing through height and weight acceleration, and ending with the achievement of full sexual reproductive capacity. Apparently, a model for self-conception changes as a function of physical maturation must take into account the gradual maturational process. In agreement with others who stress the broader, physical maturation process (cf. Stone & Baker, 1939), in the present model changes in self-concept are viewed as a gradual integration of adult- and female-identifying characteristics from the very first pubertal changes. The awareness of self-defining qualities as woman and adult is probably enhanced through the menarche. However, possible changes in self-image do not come to a halt by the time of menarche but continues in a process of increasing integration with other aspects of the girl's self-defining maturational attributes.

We have discussed at great length in the present chapter the meaning of pubertal maturation for females. Research has documented quite diverse reactions and probably quite individualistic reactions and emotions. The findings that the self-image of the girls was affected by the factor of pubertal timing, whereas self-evaluation was not, suggest that the conditions for change in self-image and change in self-evaluation follow different paths. For an individual's evaluation of herself, rooted in early internalization of significant others' feelings and attitudes toward the person and daily reinforced in the family context, it is perhaps indispensable that more extreme events must take place before the person changes the evaluations of herself. In contrast, a person's self-image, that part of the self-concept that represents his or her status, role, ethnic and sexual identity, appearance, and so on, can be said to be more a culture-bound aspect of the self-structure. Being more dependent on the cultural meaning of the particular attribute in question and on the changing roles that society requires at various life phases, such facets are more open for change over time. The person becomes selectively attentive to certain qualities within herself that pubertal maturation signifies, and the social meaning connected with physical development will be the individual's own view of herself.

THE PRESENT STUDY

The basic postulate in the present research is one emphasizing the gradual integration, with the passage of time, of a biological and a personal and social maturation. An earlier physical maturation is accompanied by a tendency in the individual to view herself as mature relative to agemates—whether it concerns the *personal*, the *social*, or the *reproductive* domain. This self-image sensitizes the person to features in the environment that are connected with more mature patterns of social life, facilitates an internalization of those attitudes and norms connected with a more mature status, and makes the individual attentive to those interpersonal contexts and associated behaviors that signify a more mature stance. Differently formulated, the age of pubertal development, reflected in girls' perceptions of themselves as mature, will be one of the factors that determines the rate at which the normal growing-up processes occur.

Much of the previous literature on the subjective perception of being early, on time, or late in terms of physical development has emphasized the negative sides of deviating from the normative in early teens. At a time when peer contacts occupy a great part of the free time of girls, when peer conformity is a new interpersonal dimension, and when adherence to ideals and social manners of the youth culture become central in the life of the teenage girl, the early developer will experience herself as belonging to a minority group distinct from the mainstream of peers. Her social status is not increased by her more mature physical appearance. In these early years of adolescence, almost all same-age boys are still underdeveloped, and the girl herself is too young to attract the interest of older boys.

Exclusive focusing on the negative aspects of being out-of-range with peers developmentally misses the point that, being crystallized in early adolescence, girls' self-conception as early, as average, or as late developers may become active determinants for much of their social behavior from this age onward. Takings its starting point in early adolescence, individual differences in physical maturation bring information both to the individual and to those around her regarding her maturational standing among her agemates. For the girl herself, definition of oneself as more mature than one's agemates sets the direction for which behaviors are appropriate, what peers are to be sought, and what experiences are to be expected. For the socializing agents it brings information on what to expect of the girl socially and personally, for example, what recognition, privileges, and demands are suitable for her maturational position.

Attempts have already been made in the literature to relate behaviors and attitudes of growing females to their perceived rather than to their actual developmental stance (Rierdan & Koff, 1985; Tobin-Richards, Boxer, & Petersen, 1983; Weatherley, 1964). For the present, our aim is to examine whether or not such a connection between a girl's physical timing of puberty and her perceived maturity exists in mid-adolescence and to estimate the strength of this association. In chapters to follow, a more detailed analysis of the role of perceived maturity as a mediating factor for the impact of pubertal development on behavior is made.

In the empirical section to follow, we have, in line with the separation of the content and the evaluative aspects of the self-concept, made separate accounts for self-image and self-evaluation.

Self-Image

Perceived Maturity. The first issue is whether or not there is a pubertal timing basis for a girl's self-defined maturity in mid-adolescence. This is the first critical test of our premises. A wide variety of factors may account for individual differences in perceived maturity among girls in mid-adolescence. Some girls might be more politically and socially oriented and for that reason feel themselves more mature, with better knowledge of contemporary social life and events. Other girls might have ethical standards, values, and interests, that run contrary to those of the peer group and more akin to those of adults. However, if physical growth is a quality around which girls organize their perceived maturity status, and if this self-conception is comparatively based, then we would expect the early-developed girls to feel more and the late-developed girls to feel less mature than same-age peers in general.

This assumption of an intimate link between actual pubertal timing and a girl's perceived maturity was tested when the girls were 14.5 years old. No mention of physical growth nor any questions on physical maturity were asked at the time when the question of self-perceived maturity was given. The question was given on a pupils' questionnaire that included questions concerning peer and parent relations primarily. The cross tabulation of the menarcheal grouping and the perceived maturity measure—answers to the question: "Do you feel more or less mature than your classmates?" is given in Table 4.1.

Among the early-matured girls, 42% thought of themselves as more mature

TABLE 4.1
Perceived Maturity among Early- and Later-Developed Girls at Age 14.5 Years

	Do you feel yourself more or less mature than your classmates?		
Age at menarche	More mature	About as mature	Less mature
−11 years	41.7	53.6	2.1
11–12 years	37.1	60.8	2.1
12–13 years	20.9	75.7	3.4
13– years	16.1	64.3	19.6

$\chi^2 (6, N = 434) = 52.08, p < .001$

than their classmates, and only 2% said that they felt less mature. The situation was quite different among the latest-developed girls. Only 16% considered themselves as more mature and nearly 20% thought of themselves as less mature.

The data presented in Table 4.1 clearly indicate that girls' conceptions of their maturity in mid-adolescence are connected with their actual physical developmental timing. The apparent differences among the menarcheal groups of girls in perceived maturity are rather striking. The findings cannot be attributed to having attained versus not having attained menarche. At the time the assessment of experienced maturity was taken, most girls had long since passed their menarche, and many of them had almost reached adult physical status.

On a subsequent question in the pupils' questionnaire at 14.5 years, the girls were asked "Do you feel different from your same-age peers?". Strong differences in this respect were obtained among the menarcheal groups of girls. As can be seen in Fig. 4.1, more than three times as many girls among the early-matured answered in the affirmative than among the latest-developed girls. The relationship between maturity level and perceived difference from peers was significant at the .001 level.

The girls who answered the question on perceived difference in the affirmative were subsequently asked to indicate in what respects they felt different from peers. A number of response alternatives were specified, including being more religious than other peers, having other leisure time activities, feeling more romantic, having another political view, another way to look at life, or

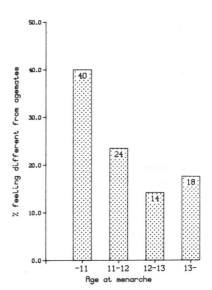

FIG. 4.1. The percentage of girls in four menarcheal groups who felt different from the agemates.

something else. The girls were asked to mark which one of these options which was closest to the way they believed they differed from other same-age peers. The answers to this question for girls in the four menarcheal groups is presented in Table 4.2.

The alternative that gathered most of the earliest-matured girls was "more romantic"; 46.2% of the girls in this maturity group declared that this was the alternative that was closest to the way they felt themselves different from other peers. No one specific content alternative gathered such a substantial number of responses among the girls in the other three maturity groups. Their answers were more varied over the different alternatives specified, and, in fact, the alternative "something else" subsumed the greatest number of girls among those with their first menstruation between 11 and 12 years, as well as among the girls having their menarche at 13 years or thereafter. The latter finding indicates that none of the alternatives given were relevant for many of these girls.

The results so far, pertaining to self-image at mid-adolescence, suggest that the girls who entered puberty early perceived themselves to be more mature than their same-age peers. Considerably more of them than of other girls declared that they felt different from their same-age peers. The most common reason for feeling different from peers was that these girls perceived themselves to be more romantic than other girls.

Body Image. We were interested to know the extent to which physical features were of concern to the girls, relative to other aspects, and whether there were any differences between girls with regard to the factor of pubertal development. Therefore, answers to an open-ended question, dealing with which of their self-attributes with which the sample of girls were *dissatisfied*, were content analyzed. The question was given in the Adjustment Screening

TABLE 4.2
Perceived Difference from Peers in Different Areas for Girls in Four Menarcheal Groups at 14.5 Years

	Age at menarche (years):			
	−11	11–12	12–13	13–
More religious	0.0	15.2	10.0	13.3
Other leisure-time activities	19.2	15.2	37.5	20.0
More romantic	46.2	21.2	22.5	26.7
Another political view	0.0	6.1	7.5	0.0
Another way to look at life	11.5	18.2	17.5	6.7
Something else	23.1	24.2	5.0	33.3
	100.0	100.1	100.0	100.0

$\chi^2 (3, N = 129) = 25.89, p < .05$

Test at 14.10 years. It read: "In what respect would you like to change yourself?". It was stated in the instruction that the girls should answer this question only if they perceived that they had some major attribute, or major area, with which they were dissatisfied. Every second girl admitted that they, indeed, wanted to change in some important respect. No differences in rate of answering among the four menarcheal groups could be detected. The specific answers, grouped into broader categories, for the girls in the four menarcheal groups are given in Table 4.3.

As can be seen in Table 4.3, mid-adolescent concerns among the girls were closely connected with their own body image. The domain that most of the girls in all four menarcheal groups picked was physical appearance. More than every third girl (35.3%) declared that she in some way wanted to change her appearance. Some girls wanted to be taller, some smaller, some thinner, and some wanted to add more weight.

A somewhat larger percentage of the early-matured girls complained that they weighed too much, compared to the later-developed girls. The contrast between the early- and late-matured girls in this respect is perhaps best illustrated with other data in the Adjustment Screening Test. In the inventory, a question with the following wording was given to all the girls: "Do you weigh too much according to your school nurse, school doctor, or another doctor's opinion?". Six percent of all the girls in the research group admitted that this was the case. As can be seen in Table 4.4, it was evident that the early-matured girls were overrepresented among the girls with weight problems. About 15% of the girls who had their menarche before 11 years stated that they were overweight in comparison with 3% of the latest-developed group of girls. These results agree with those reported with data

TABLE 4.3
Areas of Preferred Change for Girls in Four Menarcheal Groups at 14.10 Years

	Age at menarche (years):			
	−11 (n=27)	11–12 (n=56)	12–13 (n=93)	13– (n=59)
Change of physical appearance	37.0	32.1	31.2	44.1
Less temperamental	25.9	17.9	9.7	6.8
Nicer, more concern for others	7.4	17.9	9.7	10.2
Less shy, nervous	18.5	14.3	21.5	13.6
More self-confident	0.0	1.8	4.3	5.1
More school motivated	3.7	10.7	11.8	5.1
Change of specific behaviors	3.7	1.8	7.5	10.2
Other answers	3.7	3.6	4.3	5.1
	99.9	100.1	100.0	100.2
answer rate (%)	55.1	51.9	50.0	48.0

TABLE 4.4
Weight Problems among Girls in Four Menarcheal Groups at the Age of 14.10 Years

Age at menarche	Do you weigh too much according to your school nurse, your school doctor, or another doctor's opinion?			
	No	Somewhat	Yes	N
−11 years	63.0	21.7	15.2	46
11–12 years	78.6	14.6	6.8	103
12–13 years	84.7	10.4	4.9	183
13– years	92.4	4.2	3.4	118

$\chi^2 (9, N = 450) = 23.87, p < .01$

from the National Health Examination Survey by Dornbusch, Gross, Duncan, and Ritter (1987), among other studies.

The way the girls presented themselves to others, as judged by the two categories "physical appearance" and "shyness" in Table 4.3, accounted for more than half (52.8%) of all free descriptions given by the girls. The additional finding that academic achievement answers were more rare among the girls is in line with the general model of emerging sex differences in sex-role expectancies presented by Coleman (1961). Attributes such as physical appearance are more important than educational achievement and aspirations for girls, whereas the reverse is true of boys (cf. Crocket, Losoff, & Petersen, 1984). It should be observed that about the same proportion of girls in the four menarcheal groups mentioned features connected with appearance and shyness. This suggests that the interpersonal dimension in the self-concept among mid-adolescent girls is rather general and independent of maturational influences (although somewhat different in content between early- and late-developed girls).

Issues connected with social adjustment in Table 4.3, in most cases concerning adjustment at home (temperament, being nicer to parents, being more orderly) were also mentioned by a sizeable proportion of girls. Altogether, one out of every four girls (24.3%) mentioned conflicts, particularly with parents, due to her own temperament or because the girl perceived herself not to care for significant others' attitudes or wishes enough. The high frequency of interpersonal problems and family conflicts among the girls in this sample converges with earlier, comparable studies (Adams, 1964). It should be noted that answers in Table 4.3 covering social adjustment issues—less temperamental; nicer, more concern for others—were more often mentioned by the girls in the two early-matured groups (33.3% and 35.8%) than by the girls in the two later developed groups (19.4% and 17%). We return to the social adjustment issue in chapter 6. At this point it can be said that the obtained differences between earlier- and later-matured girls in this respect is in agreement with differences for other data on social adjustment at this age.

To sum up, from the girls' free descriptions given to the open question asking in what respects they wanted to change themselves, two themes were recognizable. The most profound related to the interpersonal area. More than every second girl mentioned physical appearance, shyness, and nervousness in contacts with others, particularly with peers, as the domain within which they wanted to see a change. How the girls presented themselves to others tended to be about equally important for girls in all four menarcheal groups, with weight, as expected, being of more concern for the early-developers. The second theme related to social adjustment—the desire to be less temperamental, to be nicer, and to be more orderly. Answers pertaining to this theme, with parents as the main target, were given more often by the early-matured girls than by the later-matured.

Sex-Role Identification. Before leaving this section on the self-image of early- and late-developed girls, we review some data on sex-role identification for the girls obtained before the mid-adolescent years. These data will pertain to the ages 10 and 13 years.

The semantic differential technique (Osgood, Suci, & Tannenbaum, 1957) is a common measure of assessing attitudes and values. It consists of rating a particular criterion concept on bipolar adjective scales, like "good–bad," "kind–unkind," and so forth. The semantic differential was used in Grades 3 and 6 (10.3 and 13.3 years) to determine the girls' attitudes toward themselves with respect to femininity orientation. For the femininity aspect, the concept "I—as I am as a rule" was rated on a 7-point scale ranging from "manly" to "womanly." An examination of the test–retest stability over 10 weeks indicated high stability for the scale both at Grade 3 and at Grade 6. At the former occasion 74% of the girls did not change more than one step over the 10-week period for the 7-point scale. The same proportion of girls who did not change more than one step in test–retest investigations in Grade 6 was 77%.

At the age of 10.3 years, indications of differences with respect to femininity orientation ($F = 2.39$, $df = 3, 283$, $p = .06$) were obtained, with the latest-matured girls reporting themselves feeling less womanly than the other girls (group 1: 6.4, group 2: 6.2, group 3: 6.3, group 4: 5.7), while no significant differences appeared at age 13.3 years ($F = 0.85$, $df = 3, 387$, ns). The finding that the latest-developed group of girls reported themselves feeling less womanly at the age of 10 is of particular interest. When the psychological situation for early- and late-matured girls in these early years of teens is discussed, the early-matured girls and the late-matured boys are often the special groups on which the focus is directed (Faust, 1960; Jones, 1949). However, against the background of the broader maturational process of the development of secondary sex characteristics, the lack of signs of the development of feminine body characteristics at this age might give rise to doubts regarding one's sexual identity and gender role among the late-matured girls. Further studies on gender-role identity in these early teenage years are needed to provide a more safe documentation of this finding.

Self-Evaluation

At the age of 14.10 years, a self-acceptance scale comprising 10 items was given to the girls among the other items making up the Adjustment Screening Questionnaire. The self-acceptance scale was made up of two subscales: general evaluation and preferred change. The first sub-scale of five items measured how the girls evaluated (from approval to disapproval) their perceived self. The items were designed with adolescent girls as the specific target population, and they referred to aspects of the physical self, the moral self, and self-worth. The preferred change scale measured the extent to which the person wanted to change herself from her present identity (being older, being younger, being some other person, being a boy). The total self-acceptance scale had an alpha reliability of .76. The following specific items made up the two subtests:

General Evaluation

1. Are you satisfied with yourself?
2. Are you pleased with how you look?
3. How often do you feel you are not good enough?
4. Do you often accuse yourself of things for which you are not to blame?
5. Do you feel ashamed of yourself?

Preferred Change

1. Would you like to be older than you are?
2. Would you like to be younger than you are?
3. How much would you like to change yourself?
4. Would you rather be a boy?
5. Would you like to be another person than the one you are?

As is shown in Table 4.5, for neither of the two subtests nor for the total self-acceptance scale were there any significant differences among the girls in the four menarcheal groups. A closer item-by-item analysis showed no differences among the menarcheal groups except for one item, "Do you often accuse yourself of things for which you are not to blame?". The early-matured girls accused themselves more often for such things than did the later-matured girls. This is not surprising in the light of results from analyses of data on

SUMMARY AND CONCLUSIONS

TABLE 4.5
Differences among Girls in Four Menarcheal Groups with respect
to Self-Acceptance at 14.10 Years

	F	p	post hoc
General Evaluation	1.59	ns	
Are you satisfied with yourself?	1.38	ns	
Are you pleased with how you look?	0.99	ns	
How often do you feel you are not good enough?	1.75	ns	
Do you often accuse yourself of things for which you are not to blame?	4.86	.01	1>3,4
Do you feel ashamed of yourself?	0.21	ns	
Preferred Change	0.24	ns	
Would you like to be older than you are?	1.38	ns	
Would you like to be younger than you are?	1.11	ns	
How much would you like to change yourself?	0.38	ns	
Would you rather be a boy?	0.07	ns	
Would you like to be another person than you are?	0.70	ns	
Total: Self-Acceptance	0.93	ns	

social adjustment reported at this age. These results are more closely described in chapters 5 and 6.

The results presented in Table 4.5 indicated that self-evaluations among girls at 14.10 years were not related to their pubertal timing. Neither were there any indications of differences in self-esteem between girls in the four menarcheal groups at earlier ages. At the age of 10 and 13, the girls' self-evaluations were measured with the semantic differential instrument. Seven scales were added together comprising an *evaluation* dimension (good–bad, interesting–dull, pleasant–unpleasant, clean–dirty, happy–sad, unfair–fair, and kind–unkind). These scales were rated for the concept "I—as I am as a rule."

No differences among the four menarcheal groups of girls could be traced for the evaluation of "own person" at age 10.3 years ($F = 0.10$, $df = 3,281$, ns) nor at 13.3 years ($F = 0.15$, $df = 3,385$, ns). Again, those results agree with previous literature showing that self-evaluation is a self-component that is generally not connected with pubertal timing in girls.

SUMMARY AND CONCLUSIONS

The model proposed, relating physical development to self-image among adolescent girls, was tested in the present chapter. It was hypothesized that the defining of oneself as mature relative to agemates would be connected with a girl's level of pubertal development. In agreement with the proposition of a gradual integration over time of pubertal maturation with a self-definition as mature, it was confirmed that differences in self-perceived maturity in mid-adolescence, at a point in time long after most girls had passed their first menstruation, were related to the pubertal timing of the girls. The early-developed girls generally regarded themselves as more mature than their same-age peers more often than did the late-developed girls at this point in time. They also perceived themselves, to a greater extent, as feeling different than their same-age peers, a feeling many attributed to being more romantic than other girls.

Another analysis of the self-image of the girls was made, comparing their sex-role identification in early adolescence. At the age of 10 years, the early developers regarded themselves as more mature than did the late developers. However, this difference between early- and late-developed girls was not evident 3 years later. More detailed analyses, covering sex-role-related interests and opposite-sex relations as they pertain to mid-adolescence, are given in the next chapter.

In agreement with many previous studies, few differences were found in mid-adolescence in overall self-evaluation between the early- and the later-physically developed girls. Significant differences among girls in the four menarcheal groups appeared only for one item that dealt with self-accusations for things for which the girl was not to blame. The early-developed girls reported more self-accusations than did the later-developed girls. A further analysis comparing the menarcheal groups of girls for self-evaluation aspects when they were 10 and 13 years of age yielded about the same results as were obtained in mid-adolescence, namely, that there existed no significant relationship between self-evaluation and entering puberty early or late.

Much has been written in the developmental literature on the close connection in the teenage years between self-image and physical appearance. In agreement with earlier studies, the girls in the present research group were found to be quite concerned with the impressions they physically and mentally made on others. Of the girls who perceived that they had some quality that they wanted to change, every third girl wanted to change some aspect of her appearance, and more than half of all girls mentioned their physical appearance and shyness in contacts with others as something with which they were dissatisfied. About as many of the early- as of the later-matured girls mentioned such features, thus supporting the propositions of many psychologists that the interpersonal component in the female identity is a dimension that cuts across individuals (Coleman, 1961; Marcia, 1980; Miller, 1979).

There were differences among the menarcheal groups of girls with regard to the type of physical attributes with which the girls were dissatisfied. The early-matured girls were more critical of their weight. Against the background of the results regarding the association between menarcheal age and weight

SUMMARY AND CONCLUSIONS 129

problems presented here and also reported by others, a number of issues are brought to the fore. The first one is the association between girls' conception of their fatness and their general image of their appearance.

The results of our examination concur with several other investigations in showing that more of the early-developed girls than of the late developers thought that they weighed too much. However, it should be noted that although menarcheal age was related to weight problems it was not connected with the evaluation of appearance nor with overall self-evaluation. These are critical findings. Although physical maturation brings with it more problems connected with body build, maturation did not seem to be linked to perceived general appearance or self-esteem. In the broader perspective, the results imply that we cannot assume a more or less direct path going from body image, through self-image, to self-evaluation in adolescent females. About the same results as these have been reported by Blyth, Simmons, and Zakin (1985), among others.

Another domain dealing with aspects of themselves with which the girls were dissatisfied, which differentiated among the menarcheal groups of girls, concerned temperamental issues. More of the early-developed girls than of the later-developed reported that they had difficulties in restraining their tempers. The examples that were given mainly concerned the parents as the targets for the girls' bad temper. These differences between early- and later-developed girls are of interest in view of the fact that the early-developed girls, as mentioned, blamed themselves for things they could not control. One interpretation that lies near at hand is that such self-accusations were the result of temperament outbursts against parents.

Other interpretations are also possible. The early-developed girls seemed to believe that they were more mature and more romantic, while also holding themselves responsible for things they could not control. These self-accusations might have their roots in the girls' opposite-sex contacts. They might indicate that the early developers felt more pressure to adjust to the social behavior of older boyfriends or of peers with more advanced behavior and for that reason did not feel themselves fully responsible for their behavior. On the other hand, it also might be the case that the loss of control aspect is an issue which is not directly linked with beliefs about oneself as romantic or mature but pertains to some other domain not yet examined. Attempts to pull together a fuller picture of these findings are done in subsequent chapters.

5

Interpersonal Relations

PHYSICAL GROWTH AND INTERPERSONAL RELATIONS

Three major changes with respect to social network conditions typify the period between late childhood and early adulthood. There is an orientation away from parents, commonly associated with the pursuit of more autonomy and freedom to make one's own decisions. There is also an orientation toward the peer group. From late childhood onward peer-group activities occupy more and more of girls' daily life (Bowerman & Kinch, 1959). New experiences are shared with peers, and new social patterns are learned in the peer group. The character of the peer group changes from close neighborhood contacts to relations with peers in the wider community, toward more clique formation and participation in crowds (Crocket, Losoff, & Petersen, 1984), and from unisexual to heterosexual groups (Csikszentmihalyi & Larson, 1984; Dunphy, 1963). Two features of gender differences in these respects seem to have general validity in the adolescent peer literature. There is a closer integration of interpersonal orientation and personal satisfaction among females than among males and a more intimate quality of the friendship (Crocket, Losoff, & Petersen, 1984; Douvan & Adelson, 1966; Hartup, 1970). Last, it is in this period of life that the first serious relations with members of the opposite sex are established.

Some developmental theorists have argued that peer reliance might complicate the relations to parents in this life phase (Coleman, 1961). Stronger adherence to values of the peer group and higher involvement in peer activities have been assumed to be accompanied by more strained relations to parents. There are empirical data suggesting that competing parent versus peer worlds are more likely to appear among girls than among boys (Montemayor, 1982).

The parent versus peer orientation has been criticized as being oversimplified. It is not a unidimensional phenomenon, but has different features (cf. Coleman, 1980). Kandel and Lesser (1972) observed that, whereas adolescent boys and girls preferred the company of their peers over that of their parents, overall they respected the opinions of their parents more than they did those of their peers. They were somewhat more likely to discuss personal problems with peers, but more often discussed career plans and school-related subjects with their parents. Peers and parents had different functions: Parent occupied the nurture and the control functions, whereas peers gradually replaced parents with respect to intimacy (Hunter & Youniss, 1982). Montemayor (1982) and Csikszentmihalyi and Larson (1984) found that the time spent with parents mainly concerned instrumental activities and watching television whereas the time spent with peers centered more around entertainment, talking, and playing games.

Generally, empirical data attest that teenagers do respect their parents and maintain a good relationship with them during adolescence (Bandura, 1972; Douvan & Adelson, 1966; Niles, 1979; Offer, 1969), as well as that adolescents share basic value systems with their parents, in particular with respect to issues connected with the adolescents' future (Berzonsky, 1981) and with moral issues (Young & Ferguson, 1979). Moreover, in important areas, such as the teenager's educational career, rather high concordance has been found between the values expressed by parents and those expressed by peers (Kandel & Lesser, 1972). It has also been shown that where value conflicts occur, the adolescents' values are often not contrary to the parents' but are more independent of them, being concentrated in particular domains such as dating, clothes, manners, and so on. The generation gap, often implied in the peer versus parent orientation, seems to have been exaggerated as well (Bandura, 1972). Kandel and Lesser (1972) found different types of orientations. There were adolescents who almost exclusively relied on their parents and adolescents who were extremely peer oriented. There were those who could be considered as neither peer- nor parent-dependent and those who were highly dependent on both simultaneously.

The interpersonal relations area has received comparatively strong attention as it relates to the timing and status of physical development among girls. The investigation of the interpersonal relations for same-age girls differing in pubertal growth has brought a new perspective to an issue that earlier was more or less approached as a purely social phenomenon. Prominence has been given to two of the three transition features among adolescent girls: family relations and the development of opposite-sex relations.

A critical factor in family relations is girls' strivings toward independence and autonomy from parents. The identification that accompanies early pubertal development, that of being a mature female, and the gradually deepening adult image may make the early-matured girl attempt to gain independence from her parents at a time when independence is not common for teenage girls generally. This process may contribute to conflicting parent–girl relations and strained emotional relationships. The hypothesis of more adjustment problems at home among the early-matured girls than among the late-developed has been advanced in the physical growth literature.

As to the issue of heterosexual relations, the earlier her sexual development, the sooner the girl has been presumed to develop contacts with boys, establish more stable relations, and engage in sexual behaviors. At a particular point in time in the adolescent life phase, there will be individual differences with respect to heterosexual interests and activities among girls, in part due to an earlier start on the part of the biologically early-developed girls.

No specific hypotheses with regard to the association between pubertal growth in girls and peer relations has been postulated, except as it concerns social prestige. To differ from the majority of peers physically in early adolescence has been assumed to be disadvantageous for the early-matured girl's social status in the peer group (Faust, 1960). As time elapses, physical maturity becomes a more salient and positively evaluated quality in peer interaction, thereby contributing to increased prestige for the early-matured girls relative to other girls.

A summary of the main findings with regard to peer relations, heterosexual relations, and family relations, as these pertain to individual differences in pubertal development among girls is given in this chapter.

PEER RELATIONS

Peer relations subsumes a number of different dimensions. It covers the girl's social status in her peer group, her public engagement, the emotional quality in her contacts with friends, the range of the circle of friends, the characteristics of those friends she usually meets and so forth. As mentioned, apart from the issue of *social prestige* in the peer group, few specific hypotheses have been advanced regarding how peer relations might be affected by girls' physical development.

To be early-matured at a time when most of one's same-sex peers are still prepubertal and when almost all boys their age have not yet started their pubertal development, is not a prestige-lending quality. This is also what was reported by Jones (1949). The early developers in the Oakland Growth study were characterized in middle adolescence as low in prestige and popularity and as reserved in interpersonal contacts. The hypothesis of low social status among the early-developed girls in early adolescence but increasing prestige, relative to other girls, over time, was the hypothesis which Faust (1960) tested cross sectionally. In an earlier review, she found the lowest social status concentrated among the early-developed girls in the sixth grade. During the seventh through ninth grades the situation was reversed. At these ages, the most physically matured girls were assigned the most prestigious qualities.

Earlier and later empirical investigations on social prestige have only partially supported Faust's data. Harper and Collins (1972) did not find any such age-related trend nor any other relationship between physical maturation and social status. Zakin, Blyth, and Simmons (1984) examined perceived popularity among same-sex peers for attractive and unattractive premenarcheal and postmenarcheal 12- to 13-year-old girls. They reported a strong effect of attractiveness, but no significant effect of pubertal status on the popularity measure. Jones and Mussen (1958), however, reported that the early-matured girls were rated above average in popularity the last year at high school. A tendency for the early-matured girls to gain in sociability over time was found by Everett (1943, cited in Eichorn, 1963). The early developers were rated equally cheerful and sociable as late-developed girls at the Grades 10 and 11, and more popular, whereas sociability traits were more concentrated to the late-developed girls at Grades 7 through 9.

For a sample of 15-year-old girls in the Oakland Growth Study, Jones (1958) reported a correlation of -.42 between the frequency of being mentioned in the high school newspaper and skeletal age. The lower public recognition among the early-developed girls seem at first glance to contradict the higher social prestige surrounding these girls in higher grades as reported by Faust. However, the status measure employed by Jones (1958) was probably more achievement-oriented than were the general social status measures used by Faust. Jones and Mussen's (1958) finding of higher need for recognition and need for achievement among the late-developed girls in the same sample supports this assumption that the negative correlation might be a function of higher achievement motivation among the late developers. If the rather high negative correlation between public recognition and pubertal timing reflected some other substantial characteristics that differentiate the early from the late maturers, one would have thought that it would have turned up in other studies as well. This has not been the case. Simmons, Blyth, and McKinney (1983), for example, found nonsignificant associations between pubertal timing and a measure of public engagement. Participation and leadership in school clubs and out-of-school clubs were not related to pubertal timing at any of the Grades 6, 7, 9, or 10. In this latter study, the authors did not find any consistent relations between perceived popularity and value of popularity, on the one hand, and pubertal development, on the other.

Few empirical studies have been made to investigate other aspects of peer relations besides the prestige issue. One exception is Crocket and Petersen's (1987) short-term longitudinal study. They compared the emotional quality in peer contacts between early- and late-developed girls. Neither the total peer relations scale, nor any specific questions on closeness and understanding between friends, were related to pubertal timing at Grade 6 through Grade

8. As a byproduct of the social prestige study, Faust (1960) obtained indications that the character of the peer group differed between the early- and the late-developed girls. At the seventh and the eighth grade she found the early-developed girls to engage more often with chronologically older peers than did the late-developed girls. Similarly, Money and Walker (1971) reported that most of a group of girls with precocious puberty typically engaged with chronologically older peers. Brooks-Gunn, Warren, Samelson, and Fox (1986) found that early-developed fifth-, sixth-, and seventh-grade girls had more friends than did later-matured girls.

One interesting study, conducted by Savin-Williams (1979, 1980), was designed as a naturalistic observational study of dominance hierarchies among male and female, 11- to 15-year-old teenagers during a 5-week camping trip. Behavior episodes containing dominance and submissiveness in different contexts were observed and recorded. It was found that pubertal development, for both sexes, was related to positions in the dominance hierarchy, such that early developers more often were leaders in the groups. Contrary to what might have been thought, chronological age was not significantly associated with hierarchical position in the peer group.

OPPOSITE-SEX RELATIONS

If, as psychodynamic theory suggests, puberty is accompanied by an increase in sexual drive, or psychosexual development generally coincides with hormonal changes tied to pubertal onset, then one would expect a marked increase in heterosexual contacts in this period among males and females and also a correspondence between menarcheal age and sexual behaviors. Further, one would expect that the prevalence of sexual intercourse among the youngest teenagers, seen over a longer period of time, would follow the pace of the secular trend toward earlier sexual maturation.

In sharp contrast to males, whose sexual activity culminated in mid-adolescence, the Kinsey report (Kinsey, Pomeroy, Martin, & Gebhard, 1953) found no such adolescent peak for females. Instead, there was a slow and gradual increase in the incidence of sex dreams, masturbation, and sexual intercourse over age, culminating in the mid-30s among females. From these data there is little support for the psychodynamic assumption of strongly intensified sexual urges with puberty among girls. One exceptional behavior was petting, which showed an incremental increase among females from age 15.

Kinsey reported little evidence of a connection between menarcheal age and sexual behaviors. Neither masturbation nor dreams with a sexual content occurred more frequently among the early developers, and slight differences, with early-matured girls reporting higher incidences, were found for petting and sexual intercourse. In the latter case, it was particularly the most early-matured girls who differed from the rest of the girls. With regard to precocious puberty, Money and Clopper (1974) concluded that there was little direct evidence that sexual behavior was more advanced in extremely early-developed girls, and Money and Walker (1971) speculated that

> there seems to be no evidence that a postpubertal level of circulating sex hormones propels the pubertally precocious young girls into feelings and desires of a romantic or sexual nature that one associates with the teenage adolescent. In fact, one may infer that the central nervous system program that subserves falling in love and mating behavior is separate from that which subserves pituitary-ovarian function, and that the two are not synchronized for at least the earliest years of premature puberty. (p. 59)

In agreement with the conclusion that there is little hormonal control of sexual behavior is Petersen and Boxer's (1982) observation that the marked increase in prevalence of sexual intercourse among young adolescent girls during the 1970s did not coincide with the much slower secular trend with regard to sexual maturation.

A study conducted in 1967, which examined whether or not early maturation implied a higher biological readiness to sexual behavior, was presented by Gagnon (1983). It involved 584 college and university females, ages 17 through 22, who responded to a variety of issues on sexual matters for their high school and college years. No consistent tendencies with regard to masturbation by menarcheal age was found. In fact, the earliest-developed girls reported lower incidence of masturbation in high school than did other girls. Neither was there found any relationship between the age at which the girls fell in love for the first time and menarcheal age. Small effects of menarcheal age on genital petting and dating were reported, such that a larger proportion of the earliest-developed girls reported such types of experiences earlier than did the other females. Finally, although a smaller percentage of the latest-developed girls reported having had sexual intercourse in high school, an anomalous finding was that a higher incidence of intercourse was found for these girls in comparison with other girls for the college years. Gagnon summarized the result by concluding that "from these data it would be difficult to support any of the hypotheses about the relation between menarche and sexual conduct except for the relatively weak assertion that early menses may contribute in some minor way to early start-up of heterosocial and heterosexual conduct" (pp. 182–183).

Dornbusch et al. (1981) reported data from the U.S. National Health Examination Survey of 12- to 17-year-old youths, totalling more than 6,000 subjects, on the issue of biological and age-normative influences on dating. Physical maturation was assessed by physicians according to the Tanner scheme. Separate analyses for chronological age and physical maturation for the females in the subject sample produced a correlation of .56 between the age of subjects and dating, and a correlation of .38 between physical maturation and dating. A stepwise regression analysis showed that virtually all of the variance that physical maturation shared with dating diminished once the age of the subject had entered the equation. Thus, chronological age made a rather good prognosis of dating, and biological maturation did not increase this prediction to any appreciable extent.

From these studies, the first dealing with sexual behavior among more than 5,000 females, the second with different aspects of sexual conduct among a sample of about 600 college and university students, and the third with dating patterns for more than 3,000 investigated female subjects, pubertal development seemed to be of little importance for relations to members of the opposite sex or for sexual behavior like masturbation. This conclusion must be somewhat amplified, however.

Effects of physical maturity on incidence of premarital coitus and petting was found in the Kinsey report, the effect being concentrated to the most early-matured girls. In support of the idea that early maturation is linked with an earlier entry into sexual behavior, Gagnon reported negative relationships between high school dating, genital petting, premarital coitus and college coital frequency, and menarcheal age. However, early maturation was not associated with higher incidence of masturbation in high school or college nor with incidence of sexual intercourse in college. In agreement with the Kinsey report, Gagnon's data showed overall that to the extent that there was an effect of menarcheal age on sexual behavior, it was weak and was limited to the most early-developed girls. Although the Gagnon study covers various areas of teenage sociosexual behavior, the investigation nevertheless suffers from two weaknesses. First, the study was retrospective, asking subjects to recall sexual experiences that happened years ago. Second, and more important, there were serious limitations with respect to the nature of the subject sample. They constituted a probability sample of college and university students who, presumably, had higher achievement motivation and were more career-oriented compared to lower education aspirations and higher family orientation that characterized the girls who did not continue into college and thus were never investigated.

The literature rather consistently demonstrates that less career-oriented females have more positive attitudes toward sexuality and are more involved in sexual activities. To give one example, Miller and Simon (1980) reported that the prevalence of intercourse among adolescents was twice as high for those who did not aspire to post-high school education in comparison with those who planned further schooling. Therefore, due to the fact that the investigation did not involve subjects for whom the effects of menarcheal age theoretically might be of more psychological and social significance, the nonconsistent results and the small effect of menarcheal age on sexual behavior in this study were perhaps predictable beforehand.

In Dornbusch et al.'s study, the dating variable was dichotomized as to whether or not the girl had ever had a date that precludes a more detailed analysis of how advanced the girls' dating habits were at different ages and whether or not advancement in such habits was associated with biological maturity, independent of age. Also, the prognosis was made for a rather wide age continuum, which makes the prognostic power of chronological age not surprising. Besides, to the extent that one aims at examining the evidence for greater heterosexual advancements of early physically matured females, other aspects of female psychosexual development, other than dating, are more likely to be governed by maturational factors (Meyer-Bahlburg et al., 1985).

Although the two large-scale national surveys and Gagnon's study demonstrate a limited role for menarcheal age in sexuality and heterosexual contacts, a different story is portrayed in the national survey reported by Zelnik, Kantner, and Ford (1981). They interviewed some 4,400 15- to 19-year-old females in 1971 and 2,200 females of the same age in 1976, concerning issues connected with sex and pregnancy. Among the prognostic factors for sexuality and fertility in these nationwide adolescent samples was a measure on menarcheal age. It was significantly related at the .01 level to prevalence of premarital intercourse among White subjects in the 1971 sample, with earlier pubertal timing being connected with greater prevalence of intercourse. The prevalence rates of sexual intercourse among White females were menarche ≤ 11 years: 31.8%; 12–13 years: 26.4%; 14–15 years: 20.2%; ≥ 16 years: 18.4%. More inconsistent findings were obtained for Black females. In the 1976 sample, menarcheal age was again significantly related to prevalence of intercourse among the White students ($p < .05$) but not among the Black females. Almost twice as many of the White girls who had menstruated before the age of 12 years had had sexual intercourse than were the case among girls who had had their menarche at 14–15 years of age. Menarcheal age was also found to be significantly related ($p < .01$) to the mean age of first premarital intercourse among both White and Black females in the 1971 as well as in the 1976 sample. White girls in the 1971 sample who menstruated before their 12th birthday had had their first intercourse about 1 year earlier than had girls who had menstruated at the age of 14 or later. In fact, menarcheal age was reported to constitute the variable, within both races, which had the greatest impact on age at first sexual intercourse. It was a more important prognostic factor than other social variables involved in the analyses: socioeconomic status, family stability, and religiosity.

In a study involving Black and White women, constituting a sample of about 3,000 females age around 30 years of age in low income neighborhoods, Udry (1979) also reported that an earlier age at menarche was connected with an earlier point in time for the first sexual intercourse. The relationship between menarcheal age and first intercourse was quite strong and was valid for both Black and White women. The study utilized a retrospective design, however, and is subject to many of the same problems that affect Gagnon's study.

Other studies have also reported physical maturation to be a powerful determinant for heterosexual interests and interactions. Stone and Barker (1937) reported that postmenarcheal girls more often preferred activities involving contacts with the opposite sex than did the premenarcheal girls. In this study the girls were an average 13 years of age, with a range from 12 to 15, and they were tested with the Pressey Interest-Attitude Test and the Sullivan Scale for Measuring Developmental Age. With the tentative results from this study as a guideline for future research, Stone and Barker (1939) incorporated new items to verify the earlier findings and to test new hypotheses. Nineteen of the 21 items subsumed under "heterosexual interests and activities," which indicated greater interests of this kind in the new instrument, were favored by a higher proportion of the postmenarcheal girls than of the premenarcheal girls, whereas of the 20 items indicating less interest in heterosexual relations, 18 were favored by a higher proportion of the premenarcheal girls. They also found the postmenarcheal girls to show higher interest for use of cosmetics and interest in clothes, hygiene, and manners.

Comparing the responses to the Mooney Problem Checklist, Garwood and Allen (1979) found the postmenarcheal 13-year-old girls to score higher for the problem area "Boy–Girl Relations" than did the same-age premenarcheal girls. For the same group of girls followed in Grades 6, 7, 9, and 10, Simmons, Blyth, and McKinney (1983) reported a rather consistent pattern across grades of more active boy–girl relations among the early-matured girls than among the later-matured. The pubertal timing measure was significant for perceiving oneself as popular among boys in the seventh and ninth grades, for showing concern about popularity among boys in seventh grade, for evaluating opposite-sex popularity higher than competence in Grades 7 and 9, and for dating in the 10th grade. In another study, Simmons, Blyth, Van Cleave, and Bush (1979) found that more of the earlier developers in the seventh grade had a boyfriend than did the later developed girls. In a follow-up of girls from Grade 6 to Grade 8, Crocket and Petersen (1987) reported that differences between the early-and the late-matured girls did not show up until the last grade investigated. At this age the early developers were more likely to have a boyfriend, to talk with boys on the phone, to date, and to "make-out." In seventh grade, one difference appeared: The early developers were more likely to talk with boys on the phone than were the late-developed girls. In a study of early, on-time, and late 11-, 12-, and 13-year-old girls (Brooks-Gunn, 1987), the early-developed girls were found to begin to date at an earlier time than did the later-developed. Finally, Jones and Mussen's (1958) investigation of extreme early- and extreme late-developed 17-year-old girls showed that the early maturers gave more romantic interpretations of the administered TAT cards.

One particularly interesting group of subjects in this context concerns girls with precocious puberty. Several studies have been presented in the literature on the impact on sexuality of precocious puberty. Either there are no reported differences between the extremely early-developed girls and other girls, or only modest effects emerge. An overview of these studies has been presented by Meyer-Bahlburg et al. (1985). These authors also reported findings from a match-control study of psychosexual development involving 16 pairs of females, with an age range from 13 to 21 years. They did not find any differences between the girls with precocious puberty and control subjects with respect to social aspects of psychosexual development (crush, dating, boyfriend, going steady, love), nor any significant differences as to the girls current sexual activities. However, girls with precocious puberty had experienced different forms of sexual activities at an earlier time than had their matched controls (kissing games, romantic kiss, necking, breast petting, intercourse, masturbation). For example, the mean difference between the two groups for intercourse was 1.75 years. The strongest differences were obtained for masturbation. Girls with precocious puberty reported having masturbated more than 5 years earlier, on the average, than had the matched-control subjects.

In summary, most of the studies on the relationships between pubertal development and heterosexual interests and activities suggest that there is such a relationship. Early-developed girls are involved in heterosexual behaviors at an earlier point in time compared with the late developers. There are exceptional studies, but even in some of these studies there are tendencies in the same direction circumscribed to the group of girls who enter puberty very early. Retrospective analyses and biased samples complicate the picture, but, across investigations, a rather consistent pattern emerges that pubertal development to some extent accelerates or delays the point in time for engaging in heterosexual activities. It should also be mentioned that when adolescents themselves think about what characterize as early-maturing and late-maturing girls, they spontaneously tend to connect physical maturation with opposite-sex contacts, with the early developer having an asset in these relationships (Faust, 1983).

FAMILY RELATIONS

A review of the empirical literature on the relationship between physical development and family adjustment suggests the following:

1. family conflicts tend to be more common among the early-developed

girls than among the late-matured. This finding of more frequent parent-girl disputes in families of the early-matured girls is rather consistent across studies.
2. more conflicting parental relations among the early developers compared with the late-matured girls are found in the early teenage years, as well as in mid-adolescence and in late adolescence.
3. the parent–daughter conflict is more a mother–daughter than a father–daughter conflict.

Early-matured girls have been reported to spend less time at home and to argue more frequently with their parents (Crocket & Petersen, 1987; Frisk, Tenhunen, Widholm, & Hortling, 1966; Garwood & Allen, 1979; Jones & Mussen, 1958) and to mention parents less often as the persons being important to them (Garbarino, Burston, Raber, Russel, & Crouter, 1978). The postmenarcheal 13-year-old girls in Garwood and Allen's (1979) study on problems and self-conceptions scored higher for the problem area "home and family" than did the premenstrual girls. For the projective material used by Smith and Powell (1956), postmenarcheal girls, 13 years of age, stated that they felt more deprived of attention by their parents than did the premenarcheal girls. Crocket and Petersen's (1987) short-term longitudinal study of the same group of girls followed from Grade 6 to Grade 8 is particularly informative on age trends in family adjustment. Significant effects of pubertal timing on family relations were found at Grade 6 and Grade 8. At the earliest age, willingness to talk to their mothers when having problems was more common among the late-matured than among the early-matured girls, and the subscale family relations showed a linear relationship to pubertal timing with the early developers reporting the least positive feelings. In the eighth grade, the frequency of arguing with parents was higher and the time spent with the family was less among the early-developed girls.

In the Milwaukee longitudinal study, Simmons, Blyth, and McKinney (1983) found that independence from parents was a more salient feature of the early-matured girls than of the late-developed. For specific behaviors they reported that early maturers were more apt to take a bus without being accompanied by an adult in the sixth grade, more likely to be left alone when parents were not at home in the seventh grade, and more likely to go to places without parents' permission in the ninth grade. A comparison in terms of pubertal status showed that postmenarcheal girls were babysitting more often in the sixth and in the seventh grades and also perceived that they could make their own decisions without parents permission in the sixth grade. They also showed greater concern about independence from parents than did the premenarcheal girls in the sixth grade. The observation that late developers are surrounded by more social restrictions and fewer privileges at home than are early-developed girls seems to be a feature of family relations that persists into late adolescence. In their TAT investigation of extreme early- and extreme late-developed 17-year-old girls, Jones and Mussen (1958) reported a tendency for the late maturers to perceive that they were forced by their parents to do things contrary to what they wished and that they were prevented from doing things that they wanted.

Just around the time of the menarche the mother is thought to be particularly important in the life of the developing girl. She is the natural confidante with whom to discuss the physical and personal problems associated with the first menstruation, and it is with her the girl can identify. The majority of girls tell their mother first of all people about their first menstruation (Koff, Rierdan, & Jacobson, 1981; Ruble & Brooks-Gunn, 1982; Whisnant & Zegans, 1975). Because the mother is the primary identification object and source of affiliation in daily life, she is likely to be the parent in the family who will come into focus in possible parent-girl conflicts (cf. Montemayor, 1982, who reported that 85% of 15-year-old girls' conflicts with parents occurred with their mother). In Crocket and Petersen's study (1987), the significant effect of pubertal timing on talking to parents about own problems in the sixth grade appeared for mothers but was less clear for fathers. Conflicting mother–girl relations among the early-developed girls was among the most striking findings in the study of a clinical sample of subjects reported by Frisk, Tenhunen, Widholm, and Hortling (1966).

An interesting study on family adjustment in the frame of reference of general and differential pubertal impact (see chapter 1), with assessments obtained both from the perspective of parents and from the adolescents themselves, was reported by Hill and Holmbeck (1987; Hill, Holmbeck, Marlow, Green, & Lynch, 1985). They expected that temporary family interaction conflicts would occur immediately after girls reached their menarche, independent of when this occurred. For a sample of 100 girls aged 13-years old, they also confirmed that the girls who just had had their first menstruation (menarche < 6 months earlier) reported less mother acceptance, more disagreement over rules, and less participation in family activities (mothers' reports), than did the premenarcheal girls and the girls who reached their menarche at an earlier point in time (menarche > 6 and < 12 months earlier). Overall, the data supported the authors' hypotheses of temporary "perturbations" connected with the menarche. However, the most early-developed girls (menarche > 12 months earlier) also reported low acceptance from their mothers and low parental influence. Their mothers reported low involvement in family activities on part of these girls and disagreement over rules and standards in the family.

Although the expectations of the authors were that girls generally would re-establish their family relations as time passed from the menarche, another path was suggested for the earliest-developed girls: "Our preferred interpretation is that early-maturing girls are at a risk for disordered family relations. . .Withdrawal and conflict may persist (or even increase) for "off-time," early-maturing girls, leading us to suspect, in a longitudinal study, curves that do not show the return to premenarcheal baselines we would anticipate for girls who are "on-time" (p. 215). Thus, a differential pubertal impact on family relations were assumed to operate for these early-developed girls. The results, overall, also confirmed other research showing that it was the mother–daughter relation that was affected by pubertal development in girls rather than the father–daughter relationship.

Certain collaborating evidence of parental conflicts related to early maturing among females, again measured from the perspective of both the parents and the adolescent females, was reported by Savin-Williams and Small (1986). Adolescents between 10 and 17, and their parents, were questioned with respect to different aspects of family interactions. The investigators reported an unclear result picture. However, among early developers, the parents of female adolescents reported more conflicts than did the parents of the males, while the opposite was the case among the late-maturing adolescents. The parents of the early-developed girls also reported more anxiety with respect to their offspring's social behavior than did the parents of the early-maturing boys. Furthermore, the parents, particularly the mother, of the early-developed females reported being more emotional upset than were the parents of on-time and later-maturing females. From the viewpoint of the adolescent, the early-developed females reported more parental conflicts than did the on-time or the late-maturing females. It should be observed that no control for actual chronological age was performed in this investigation.

Steinberg (1987) reported a similar study, encompassing a sample of first-born adolescents, 10 to 15 years of age, and their parents. The late-developing girls reported less frequent conflict with their mothers, experienced less authoritarian parental decisions, and experienced their mothers as less controlling than did other girls. The mothers of the late-developed girls reported less frequent conflicts with their daughters, and the fathers of the early-developed girls reported being less authoritarian than did the fathers of the other girls.

Overall, a wide variety of studies, utilizing different measurement techniques, have reported more conflicting parental relations among early- compared with late-developed girls.

THE PRESENT STUDY

The present investigation of the role of pubertal timing for the interpersonal relations of mid-adolescent girls entails a broad characterization, covering the girls' relations to their parents and to other representatives of the adult world, to the peer network and to members of the opposite sex. A closer examination also is made of the topic of parent versus peer orientation and how this orientation relates to physical development.

CHARACTERIZING THE PEER NETWORK

In contrast to much previous research on the differentiating features of early- and late-developed girls, great importance is placed on the peer network in the present research. Behind this focus on the peer culture is the hypothesis that girls in the mid-adolescent years tend to associate with peers who match their physical development. The synthesis of the expectations is that by virtue of their earlier pubertal development, the early-matured girls will be more likely to associate with peers who are outside the normal, same-age-same-class context of girls generally. They will more often associate with older peers, and with older peers who are already employed, with girls of their age who also are early developers, and more often in male contexts: in mixed-sex groups and in opposite-sex dyads. Five hypotheses are advanced regarding the interpersonal situation in mid-adolescence for the early-developed girl. In comparison with the later-developed girl the early-developer is more likely to:

1. associate with chronologically older peers;
2. associate with early-developed, same-age girls, rather than with late-developed girls of her own age;
3. be found in groups of males;
4. associate with peers who have quit school in favor of full-time employment; and
5. have established intimate relations with boys.

In addition to investigating these hypotheses, a comparison between early- and later-developed girls is made as to the affectional quality in relations to peers. A comparison also is made with regard to the issue of social prestige.

Data for peer network consisted of a self-report instrument administered to the subjects when they were 14.7 years of age. The girls were first asked to state the total number of close friends in their peer group. They were instructed to list their ordinary friends as distinguished from closer, opposite-sex contacts. Next, to ascertain the *age characteristics* of their friends, they were asked how many of the friends listed were at least 1 year younger, the same age, or at least 1 year older than themselves. Subsequently, in order to establish the *social characterization* of the friends that they had initially listed, the subjects were asked how many of the total number of different friends they had first declared were, like themselves, full-time students and how many friends were employed. Thus, after having stated the total number of different friends, they partitioned this number according to the chronological age of the peers and social characterization. Within the domain of age characteristics and within the domain of social characteristics, the categories are mutually exclusive and exhaustive. A further task involved stating how many of their school peers attended the same class as the subject herself.

A comparison among the four menarcheal groups of girls with respect to the total number of friends with whom they engaged at 14.7 years is presented in Table 5.1.

Almost all girls stated that they had at least one friend, and about 9 of 10 girls answered that they had two or more friends. As determined by an ANOVA analysis, there was a slight tendency ($p < .10$) for the earliest-developed group of girls to report a greater number of friends than did other girls. Generally, however, it can be concluded that few differences existed among the menarcheal groups of girls with regard to the *number of peers* with whom they engaged at this age.

Age Characterization

The number of peers who were younger, same-age, or older than themselves among the girls in the four menarcheal groups is reported in Table 5.2.

As might be expected, most of the friends of the girls were same-age peers. More than 9 out of 10 girls engaged with friends of their own age. Next common were chronologically older peers. Every second girl had older peers in their circle of friends. Girls in the sample associated with younger peers less frequently. Just one out of eight had friends who were younger than themselves, and only 4.5% of the girls had several younger peers with whom they engaged.

Younger peers in a girl's circle of friends were more common among the

TABLE 5.1
The Total Number of Friends of Girls in Four Menarcheal Groups at 14.7 Years

Age at menarche	Number of friends (percentages)			Mean number
	No one	One	Many	
−11 yrs	—	14.3	85.7	5.45
11–12 yrs	1.1	3.2	95.7	4.56
12–13 yrs	—	7.7	93.3	4.76
13– yrs	.09	12.4	86.7	4.25
Total	0.5	8.6	90.9	4.65

$\chi^2 (6, N = 417) = 9.42$, ns $F = 2.54, df = (3,416)$, ns

TABLE 5.2
Age Characterization of Friends for Girls in Four Menarcheal Groups
(14.7 years)

Age at menarche	Number of friends (percentages)			Mean number
	No one	One	Many	
		Younger friends		
−11 yrs	95.2	4.8	—	0.05
11–12 yrs	92.5	5.4	2.1	0.11
12–13 yrs	81.1	12.4	6.5	0.30
13– yrs	86.7	8.0	5.3	0.23
Total	86.6	8.9	4.5	0.21
	χ^2 (6, N = 417) = 10.59, ns			$F = 2,70$, $(df = 3,416)$, $p<.05$ post hoc: 1<3,4&2<3
		Same-age friends		
−11 yrs	7.1	28.6	64.3	2.29
11–12 yrs	4.3	15.1	80.6	2.89
12–13 yrs	3.6	16.6	79.8	2.95
13– yrs	5.3	17.7	77.0	3.03
Total	4.6	17.7	77.7	2.89
	χ^2 (6, N = 417) = 5.62, ns			$F = 1,89$, $(df = 3,416)$, ns
		Older friends		
−11 yrs	26.2	19.0	54.8	3.05
11–12 yrs	44.1	18.3	37.6	1.54
12–13 yrs	48.5	14.8	36.7	1.50
13– yrs	61.1	14.2	24.7	0.96
Total	48.7	15.8	35.5	1.52
	χ^2 (6, N = 417) = 17.50, $p < .01$			$F = 11.09$, $(df = 3,416)$ $p <.001$ post hoc: 1>2,3>4

later-developed girls than among the early-developed. Of the girls in the two latest developed groups 16.7% had at least one younger friend, in comparison with 6.7% of the girls in the two earliest developed groups.

No significant differences were obtained with respect to the number of *same-age friends* among the four menarcheal groups. A tendency for the most early-developed to have a more restricted same-age peer association, however, could be discerned.

In accordance with the hypothesis, there were marked differences among the menarcheal groups of girls with respect to engaging with chronologically *older peers*. The difference obtained among the four groups of girls with regard to the number of older friends was significant at the promille level. The earliest-matured girls reported twice as many chronologically older peers as did the later-developed girls. Although almost three out of four of the earliest-matured engaged with at least one older friend, this was the case for a minority (38.9%) of the latest developed.

With the partitioning method for characterizing the peer network follows certain benefits that do not apply to many other methods for data collection. One such advantageous property is the possibility of estimating the proportion of friends having a certain characteristic among the full circle of friends. For example, as can be seen in Table 5.2, most of the friends of the girls who had their first menstruation after 11 years were same-age peers. The peer network was quite another for the earliest-matured girls. Indeed, the majority of the more permanent friends among the earliest-developed girls were older peers (an average of 3.05 older vs. 2.29 same-age and 0.05 younger friends). This strongly supports our assumptions of the early-matured girl seeking out and being sought out by others who match her own developmental level.

Social Characterization

With the aim of investigating the hypothesis that more of the early- than of the late-matured girls would have friends outside school, the menarcheal groups of girls were compared for the number of school pupils and employed persons in their circle of friends. This comparison is shown in Table 5.3.

The most obvious impression of the data presented in Table 5.3 is the high proportion of friends who were, like the girls themselves, school pupils: either classmates or friends in other classes. Almost all girls had a friend who, like themselves, was a pupil at school, and nine out of ten girls had several such friends. By contrast, only 22.8% of all girls had friends who were employed.

There were no systematic, significant differences among the four menarcheal groups of girls with respect to having *friends at school*. A weak curvilinear tendency could be detected with the most early- and the latest-developed girls reporting fewer school friends.

Neither were there any significant differences with regard to the number of *friends in the same class* among the four groups of girls. It could be noted, however, that the earliest-matured girls reported the lowest number of such friends. Whereas 31% of the earliest-matured girls reported that they had no friend in their own class, this occurred for a lower proportion, 18.7% among the rest of the girls.

The most profound differences were obtained for *working friends*. On the average, the earliest-matured girls reported that they had four times as many employed friends as did the other girls. The difference was significant at a high level of significance. More than 40% of the earliest-matured girls reported that their circles of friends contained several working friends. This was the case for less than 10% of the other girls.

Perhaps the most striking differences among the menarcheal groups is

TABLE 5.3
Social Characterization of Friends for Girls in Four Menarcheal Groups

Age at menarche	Number of friends (percentages)			Mean number
	No one	One	Many	
		Friends in the same class		
−11 yrs	31.0	28.6	40.4	1.55
11–12 yrs	16.1	35.5	48.4	1.89
12–13 yrs	18.3	27.2	54.5	1.95
13– yrs	21.2	24.8	54.0	1.95
Total	19.9	28.5	51.6	1.90
	χ^2 (6, N = 417) = 7.37, ns			$F = 0.77$, $df = 3,416$, ns
		Friends at school		
−11 yrs	2.4	16.7	80.9	3.79
11–12 yrs	1.1	3.2	95.7	4.18
12–13 yrs	1.2	8.3	91.5	4.28
13– yrs	0.9	12.4	86.7	3.86
Total	1.2	9.1	89.7	4.09
	χ^2 (6, N = 417) = 9.08, ns			$F = 1.17$, $df = 3,416$, ns
		Working friends		
−11 yrs	54.8	4.8	40.4	1.36
11–12 yrs	78.5	10.8	10.7	0.38
12–13 yrs	80.5	10.7	8.8	0.34
13– yrs	79.6	15.9	4.5	0.30
Total	77.2	11.5	11.3	0.44
	χ^2 (6, N = 417) = 44.18, $p < .001$			$F = 14.34$, $df = 3,416$, $p <.001$ post hoc: 1>2,3,4

revealed when partitioning out of the proportion of employed friends from the total number of friends. The earliest-matured girls reported an average of 3.79 school friends and 1.36 working friends, meaning that one out of four friends for these girls were persons already in the labor market. The percentage of working friends among the total number of friends listed for the other girls was about 8%.

Same-Age–Same-Sex Peer Preferences

Does the early-maturing girl prefer her early-maturing classmates of the same chronological age? Does the late-maturing girl prefer her late-maturing school fellows? These questions concern the issue of friendship formation among same-age–same-class peers, and they were investigated using sociometric data.

All girls assigned a rank-order value of preference for all of her same-sex classmates at age 14.5 years. The girls were to imagine that they were to be transferred to another class, and they were asked to rank order all of their female classmates in the order in which they would like these girls to accompany them (Magnusson, Dunér, & Zetterblom, 1975). A data matrix was subsequently constructed for each class, with the raters represented by rows and the rank-order values of their rated peers represented by columns. Next, data were aggregated over subjects within each of the four menarcheal groups, both row-wise and column-wise. The result was a four by four rater-ratee matrix for each class, indicating how classmates in the four menarcheal groups were preferred by girls who differed in biological maturity. The resulting matrix, aggregated over classes, is presented in Table 5.4 A low score in the matrix means a high rank-order of preference.

Data in Table 5.4 show that the earliest- and the latest-developing girls tended to prefer other girls of the same age in their class in terms of the extent to which they matched their own level of biological development. The earliest-developed girls were assigned a lower rank (i.e., were the most preferred) by classmates who also were early-developing girls (5.3) than by the latest-maturing girls (8.5). Conversely, the late-developing girls were assigned a lower rank-order value by girls who themselves were late-developing (6.6) than by the most early-developed girls (7.4). Altogether, these data indicated that girls preferred same-age–same-sex peers who matched their own stage of biological development.

Friendship–Group Membership

As another measure of the peer network, friendship-group membership was investigated. A question in the Norm Inventory administered at 14.5 years asked if the girl belonged to a group of friends that "hung out" together. Of

TABLE 5.4
Mean Rank-Order Preference Assigned to Classmates Varying in Biological Maturation by Girls who Themselves Differ in Biological Maturation
(age 14.5 Years)

Girls who are "raters"	Girls who are "rated"			
	−11 yrs	11–12 yrs	12–13 yrs	13– yrs
−11 yrs	5.3	7.4	6.8	7.4
11–12 yrs	7.9	7.1	7.0	7.7
12–13 yrs	7.4	7.6	7.0	7.2
13– yrs	8.5	7.3	6.7	6.6
	$r = .18^{***}$	$r = .00$	$r = -.04$	$r = -.12^{**}$

**$p<.01$.   ***$p<.001$

the 427 females, 191 or 44.7% answered in the affirmative. No differences as to membership in friendship-groups was found among the menarcheal groups ($X^2 = 2.47$, $df = 3$, ns).

More to the point, a complementary question asked the girls who belonged to such a group about the characteristics of the group to which they belonged: Did it contain females only, males only, or members of both sexes. Although no overall significant differences among girls in the four menarcheal groups were obtained, there was an obvious tendency that early developers more often belonged to a group with males only (except for herself), and it was more common that later-developed girls were members of groups containing females only. A detailed account of the results is presented in Table 5.5

Comments. The data we have just presented show, in a very decisive way, the importance of taking into account the *characterization* of a girl's peer group when investigating the peer network of early- and later-matured girls. When the peers were amalgamated in one heterogeneous group—friends generally—few differences were obtained between the early- and late-matured girls. Few girls lacked a friend relation and most of the girls in the four menarcheal groups had several friends.

The more fine-graded subsequent analyses showed that there were marked differences among the four menarcheal groups of girls with respect to the kind of peers with whom they associated. The majority of the friends of the earliest-developed girls were chronologically older peers, whereas same-age peers were most common among the other girls. To associate with younger peers was rather uncommon among the girls in the sample, but when it occurred the younger peer contact was concentrated among the later-developed girls primarily. Also the social characterization of the peers differed between the early- and late-matured girls. Whereas most of the friends of both the early- and the later-developed girls were school pupils, a considerably higher proportion of the circle of peers among the earliest matured girls were peers who were employed. A further investigation was made of the role of match in

TABLE 5.5
Percentage of Girls in Four Menarcheal Groups who Belonged to Various Types of Friendship Groups at Age 14.5 Years

Age at menarche	Friendship-groups characterization		
	Girls only	Boys only	Both sexes
–11 yrs	—	9.5	90.5
11–12 yrs	8.5	6.4	85.1
12–13 yrs	10.0	1.3	88.8
13– yrs	14.3	—	85.7

$\chi^2(6, N = 190) = 9.66$, $p = .14$

biological maturation for preferences of same-age–same-sex peers. The results supported the hypothesis that girls preferred other girls in their own class who matched their own stage of biological maturity. Finally, an investigation of the characteristics of the gang to which girls belonged indicated that it was more common among the early-developed girls to participate in male-dominated groups, and less common to find these girls in female groups.

THE RELATIONAL QUALITY IN PEER CONTACTS

To measure the emotional quality in girls' relations to peers, early- and later-matured girls were compared for a "Peer Relations" scale, comprising 15 items. The scale covered feelings of shyness, hostility, understanding, liking, dominance, and disappointment in the contacts with peers. These items were taken from the Adjustment Screening Questionnaire administered at 14.10 years. The total scale had a mean inter-item correlation of .29 and an alpha reliability of .86.[1]

There were some indications that the relational quality in the peer contacts differed among the girls in the four menarcheal groups ($p < .10$). A curvilinear relationship between menarcheal timing and relations to peers (total scale) was obtained, such that the most early- and the most late-developed girls perceived themselves to have somewhat less satisfying peer relations than did the more average-developing girls. For specific items, a significant curvilinear relationship appeared for "Are your friends unjust towards you?", and "Do you think your friends are 'against' you?", and the same tendency, at the .10 level, was found for "Do you and your friends understand each other?".

Overall, the differences obtained, as they apply to mid-adolescence, were of low significance and appeared for just a few of the 15 items comprising the scale. Where they occurred, they indicated that the girls with their menarche more "on time" had somewhat more satisfying relations to their peers than the two "off-time" groups of girls.

---

[1] Included in the peer relations scale administered at 14.10 years were the items "Do you and your friends understand each other?", "How often are you angry and irritated with your friends?", "Do you feel shy and unsecure among peers?", "Do you trust your friends?", "Do you feel 'outside' when you are with your peers?", "Do you often get tired of your friends?", "Are your friends unjust towards you?", "Is it difficult for you to make friends?", "Is it important for you that your friends go along with what you decide?", "How often do you feel disappointed in your friends?", "Do you think your friends like you?", "How often do you fall out with your friends?", "Do you think your friends are 'against' you?", "Are you afraid your friends will get tired of you?", and "Has it happened that you have gone along with things just to have the opportunity to be with your friends?".

Information about peer relations was collected for the total sample via pupils' questionnaires and sociometric instruments from the start of the project. In the early school years, data for the range and quality of the peer-group contact, for perceived threat from peers, and for social status in the peer group, were of vital interest. With some indications that the girls who were more out of phase with their peers in physical development had slightly less satisfying emotional relations to peers in the mid-adolescent years, the question arises as to whether the same tendency already occurred at earlier ages or whether other peer relations patterns dominated the scene in these years. To answer this question, relevant data at the age of 9.9 years (Grade 3) and 12.9 years (Grade 6) were analyzed. At both ages, the specific questions were obtained from pupils's questionnaires.

Very few differences among the four menarcheal groups of girls were found at the age of 9.9 years. Two items differed at the 10% level of significance, and they both referred to perception of threat from peers ("Do your classmates make trouble in the breaks?" and "Are you afraid of any of the pupils during the breaks?"). The earliest-matured group of girls perceived less often than did other girls that their classmates made trouble during the breaks, and the latest-matured group of girls were more afraid of other pupils during the breaks at school. For other items, whether the girl liked her classmates, the intensity of contact with comrades, and other questions on peer threat, no significant differences were obtained between the early- and the later-matured girls.[2]

The same lack of strong differences among the menarcheal groups was found at age 12.9 years. One item differed significantly at the .10 level among the four groups of girls, and it asked about how frequently the girl met her peers during the leisure time ("How many evenings per week do you see your friends?"). The earlier the girl entered puberty, the more intensive was the peer contact during leisure time at this age.

In summary, as indicated by data on the emotional quality in girls' relations to their peers from age 10 up to mid-adolescence, few differences were obtained among the four menarcheal groups of girls. At the age of 14.10 years, there were some indications that the earliest- and the latest-matured groups of girls perceived themselves to have more problematic peer relations than did the

---

[2] The following items were included in the peer relations scale administered at 9.9 years: "Do you like your classmates?", "How many of your classmates do you really like?", "Do you take home comrades to play?", "Do your classmates make trouble on your way to and from school?", "Can you play with the comrades you like best?", "Are you afraid of any of the pupils during the breaks?", "Do your classmates make trouble in the breaks?". The items included in the Peer relations scale at 12.9 years were: "Do you have a really good friend in the class?", "Do you get along well with your classmates?", "How many of your classmates do you get along well with?", "How many evenings per week do you see your friends?", "Do other children inflict trouble on you during the breaks?".

more average-maturing girls. This situation was not observed at the earlier ages. At the age of 9.9 years there were some suggestions that early-matured girls perceived less threat from their classmates than did later-matured girls, and at age 12.9 years earlier-matured girls reported somewhat more intensive contacts with peers than did late matured.

SOCIAL STATUS

As described in the introduction to this chapter, social prestige has been the peer relations issue that has received the greatest interest in the physical maturity literature. For this reason it was deemed important to examine the social status of the girls in the four menarcheal groups from the early school years up into mid-adolescence.

The same measure of the girl's social status in her class was collected at three ages; 9.9 years, 12.9 years, and 14.5 years. It was the sociometric measure described on page 148. The obtained preference values were transformed to standard scores, and the mean of these z-values was calculated for each girl. A constant of 3 was added to avoid negative scores (see Magnusson, Dunér, & Zetterblom, 1975, for a more detailed description). The changes over time for the four menarcheal groups of girls for this sociometric measure is graphically depicted in Fig. 5.1.

At all three ages examined, there were differentiating tendencies in social

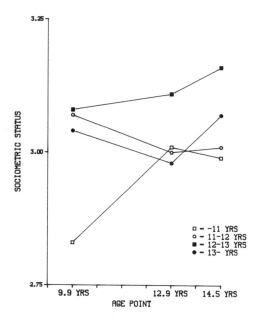

FIG. 5.1. Social status of girls in four menarcheal groups at 9.9, 12.9, and 14.5 years of age.

status among the menarcheal groups, but the nature of these differences varied over ages. At the age of 9.9 years, significant differences at the 5% level were obtained among the four groups ($F = 3.24$, $df = 3, 352$, $p < .05$). Post hoc tests showed the earliest matured girls to be rated lower in status than each of the other three groups of girls. At age 12.9 years, the differences among the groups were weaker ($F = 2.29$, $df = 3, 391$, $p < .10$), and at this age it was the girls who had their first menstruation between 12 and 13 years who were more favorably rated by their peers than the other groups of girls. About the same pattern was found at age 14.10 years ($F = 2.88$, $df = 3, 445$, $p < .05$). Post hoc analyses showed the girls with their menarche between 12 and 13 years to be rated significantly higher than were the girls in the first two menarcheal groups.

These results support earlier findings of low social prestige among early-matured girls in early adolescence. The results also support the increase in popularity with the passage of time for the early-developed group of girls relative to other girls. As in Faust's investigation, the earliest-matured girls were found here to be the group of girls lowest in popularity at age 9.9. Their status increased considerably from 9.9 to 12.9 years. At all ages, but especially at the two later test occasions, the average-matured girls were found to receive the highest ratings of social status. This finding differs from the results presented by Faust, who reported that the earliest-matured girls received the highest prestige scores at these latter years. Whatever explains the difference in obtained results between studies, it seems warranted to conclude that early physical maturation is detrimental for a girl's social prestige in early adolescence but that their status increases with time.

OPPOSITE-SEX RELATIONS

Six items in the Adjustment Screening Questionnaire administered at 14.10 years comprised an Opposite-Sex Relations scale. The mean interitem correlation was .27 and the alpha reliability was estimated at .69. Differences among the girls in the four menarcheal groups, with respect to contacts with boys, are presented in Table 5.6.

There were very marked and highly significant differences among the four groups of girls with respect to heterosexual interactions. Early-matured girls were generally more advanced in their contacts with boys than were the later matured. There was a monotonic negative relation between menarcheal age and opposite-sex relations such that the girls who were earliest-developed were significantly more advanced than other girls, the girls who had their first menstruation somewhat later were significantly more advanced than the later-developed, and so on. Only one item in the scale, "Have you ever really been in love?", did not differ significantly among the menarcheal groups.

TABLE 5.6
Differences Among Girls in Four Menarcheal Groups with Respect to Opposite-Sex Relations at Age 14.10 Years

	F	post hoc
Have you ever really been in love?	1.07	
Are you popular among boys?	3.91**	1,2,3>4&2>3
Have you now or have you had a steady relation to a boy?	10.68***	1>2>3>4
Have you had sexual intercourse?	13.86***	1>2>3,4
Do you feel sexually more experienced than your age-mates?	15.12***	1>2>3,4
Do you feel sexually less experienced than your age-mates?	5.00**	1,2,3<4&2<3
Scale: Opposite Sex Relations	17.84***	1,2>3>4

**$p<.01$. ***$p<.001$

TABLE 5.7
Percentage of Girls in Four Menarcheal Groups Who Were Going Steady at 14.10 Years

	Do you now or have you had a steady relation with a boy?		
Age at menarche	No	Have had	Have now
−11 yrs	17.0	44.7	38.3
11–12 yrs	28.6	39.0	32.4
12–13 yrs	37.4	41.8	20.8
13– yrs	48.0	40.5	11.6
Total	36.0	41.1	22.8

$\chi^2(6, N = 455) = 27.59, p < = .001$

Table 5.7 presents a detailed account of the result from examining the item in the opposite-sex relations scale that dealt with steady relations to boys. About half of the latest-matured girls had never had such a steady relation to a boy, and slightly more than 1 out of 10 had such a relation at the precise time the inventory was administered. By contrast, more than 8 out of 10 of the earliest-matured girls had gone steady with a boy, and close to 4 out of 10 had such an established boy relation when the question was asked at 14.10 years. Table 5.7 shows that steady relations with boys were not concentrated only to the most early-matured girls. There was a step-by-step increase of such relations from the latest- to the early-developed girls.

More detailed information from the question about sexual intercourse is presented in Table 5.8.

TABLE 5.8
The Percentage of Girls in Four Menarcheal Groups Who Had Had Sexual Intercourse at 14.10 Years

	Have you had sexual intercourse?		
Age at menarche	No	Once	Several times
−11 yrs	55.1	10.2	34.7
11–12 yrs	64.5	15.9	19.6
12–13 yrs	84.8	3.8	11.4
13– yrs	89.2	5.8	5.0
Total	78.0	7.8	14.1

$\chi^2(6, N = 460) = 47.44, p < .001$

Altogether 22% of all girls had had sexual intercourse once or several times at 14.10 years. There were very strong differences among the four menarcheal groups of girls with respect to prevalence of sexual intercourse. About 10% of the latest-matured girls had had sexual intercourse at 14.10 years. The same figure for the earliest-matured girls was around 45%. Of the earliest developers every third girl had had sexual intercourse several times in comparison with every fifth girl who had had their menarche between 11 and 12 years, every 10th girl of girls who had had their first menstruation between 12 and 13 years, and every 20th girl who had had their menarche at 13 years of age or later.

Attitudes Toward Sexuality

The observation of a relationship between pubertal timing and sexual intercourse raises a number of questions. At the time the question was administered, all but 5.2% of the girls were menstruating, and many of the girls had approached sexual reproductive capacity at this point in time. Against the background of great differences in prevalence rates in sexual intercourse between the early and later developers, one interesting question is if, and to what extent, the early and the late-developed girls differed in their attitudes toward sexuality. It can be argued, for example, that the early-matured girls at this time in mid-adolescence were more prepared and less afraid of such contacts due to the fact that they had had more time to integrate their physical development and feminine role and for this reason had had more time to be acquainted with the idea of sex.

To examine differences between the early and the late developers with regard to attitudes toward sexuality, the menarcheal groups of girls were compared with respect to five attitude questions, given in the Adjustment Screening Questionnaire at 14.10 years.:

1. Do you worry that something might be wrong with your body, that it does not work as it should?
2. Do you think of sex as something that is frightening or unpleasant?
3. Are you afraid that you might be drawn into sexual matters without wanting it?
4. Do you think you ought to know more about sexual issues?
5. How often do you think about things connected with sex?

Worry: Of the girls in the sample, 13.9% stated that they rather often or very often worried that something might be wrong with their bodies. No relationship to menarcheal age was found ($F = 0.89$, $df = 3,457$, ns).

Sex frightening: Few girls, 1.9% of all girls in the sample, thought sex was unpleasant or frightening. A weak tendency for the late-matured girls to think that way more than did the early-matured girls could be observed, but there were no significant differences among the four menarcheal groups ($F = 1.54$, $df = 3,454$, ns).

Afraid to be drawn into sex: 5.7% of the girls stated that they rather often or very often were afraid of being drawn into sexual matters unwillingly. No differences among the four menarcheal groups of girls were obtained ($F = 1.24$, $df = 3,453$, ns).

Information about sex: 11.9% of the girls wanted more information about sexual issues rather much or very much. This information was somewhat more wanted by the early maturers than by the late ones. Differences among the four menarcheal groups reached statistical significance at the 10% level ($F = 2.47$, $df = 3,457$, $p < .10$).

Thinking about sex: That the issue of sex cannot be said to be of central concern to girls at this age was evident after an inspection of the data on this last question. Every second girl stated that she never or just occasionally thought about sexual issues, 10.6% stated that they thought rather often about such things, and only 1.1% of the girls reported that they very often thought about such issues. We have no comparable data for boys, but if such existed they would very probably show a quite higher interest pattern than that reported by girls (cf. Schofield, 1965).

Thoughts about sex were somewhat related to pubertal timing, with the early-matured girls thinking more about sex than the late-matured ($F = 2.11$, $df = 3,457$, $p < .10$).

Taking all these attitude questions together, the idea that the later-matured girls were more afraid than the early-developed of engaging in sexual behavior had no support in our data, neither for worrying over bodily malfunctions, connecting sex with fear and unpleasant thoughts, nor regarding fears of being drawn into sexual behavior. If the late-matured girls felt themselves less prepared for engaging in intimate heterosexual contacts, one might perhaps expect that they would actively seek this information. In fact, the reverse tendency with regard to pubertal maturation was observed, which intuitively makes sense in view of the fact that more of the early- than of the late-matured girls were engaged in steady relations with boys at this age. It was of more central concern for these girls to have knowledge of sexual issues in their everyday contacts with boyfriends. That the early-matured girls thought more often about things connected with sex supports this interpretation that sexual issues were more salient for them.

The attitude data showed that the late-matured girls were not more afraid of engaging in sex. There are other data which indicate that the late-matured girls did not avoid heterosexual contacts in the mid-adolescent years, but indeed actively sought this contact. An open-ended question was given as the final question in the Adjustment Screening Test at 14.10 years. The girls in the research group were asked "What do you consider as your greatest problem right now?", and they were free to describe whatever they thought was important for them personally.

Two out of three girls wrote "I don't know really," "Have no major problems right now," "I only have comparatively small problems," or the like. Of the girls who wrote down a problem description, the most common domain mentioned concerned relations to boys. But the character of the boy-relation problems were quite different for the early- and the late-matured girls. Of the 16 girls in the earliest-matured group of girls who admitted that they had problems, and described them, 3 mentioned problems connected with boys, and all 3 of them mentioned that something might come up that might lead to breaking off the relationship with their boyfriend. Of the 46 latest-developed girls who answered, 19 girls, or more than 40%, mentioned problems connected with boys. Of the 19 girls, all but 1 wrote that they had difficulties in establishing contacts with boys. Common comments were "My problem is that I have no boyfriend." "My best friend has a boyfriend, so I have to stay at home when they are together. I also would like to meet a boy," "I am not allowed to be out late in evenings, so I don't know how to get in contact with the boy I like," or "The boy I like does not even look at me." So, whereas the early-matured girls seemed to have rather established relations with boys and were afraid that these might be broken at this time in mid-adolescence, the late-developed girls had not yet formed such contacts and experienced this as distressing.[3]

In conclusion, strong differences between early- and late-developed girls were obtained with respect to opposite-sex contacts and sexual intercourse. The early developers were more advanced in heterosexual contacts at this time in adolescence than were girls who were later matured. The early developers also tended to think more about sexual issues and wanted more information about such things than did the later-developed girls. The differences obtained between early- and late-developed girls did not seem to occur because the late developers were more afraid of or expressed more worry in connection with sex. On the contrary, it seemed as if the late-developed girls actively sought contacts with boys at this time. Some of the reasons suggested from the girls' free descriptions were that the late developers could not attract boys' attention or had fewer possibilities to establish such contacts because their parents had strict rules about times to be at home in evenings for these girls.

INDIVIDUAL FRIENDSHIP PROFILES

We place great emphasis on the characteristics of the peer network among girls, and on the level of sophistication in the social behavior of girls' reference groups of peers, when mapping the social adjustment of early- and late-developed girls. Much of the differences in social development in adolescence between early and late maturers are proposed to be attributable to the differen-

---

[3] The nature of the most prominent problem expressed by the earliest- and the latest-developed girls at 14.10 years.

Earliest-developed (n= 16)		Latest-developed (n = 46)	
Boys (break of an established relation)	3	Boys (difficulty establishing contact)	18
Appearance	2	Appearance	4
Not allowed freedom at home	2	School achievement	4
Worries for parents	2	Peer problems	4
Nervousness, shyness	2	Illness	3
Other problems[a]	5	Not allowed freedom at home	2
		Nervousness, shyness	2
		Other problems[b]	9

[a] Existential death, feel different, summer plans, peer problems, grades
[b] Move to another town, not participating in gymnastics, mother starts working, summer plans, feelings of being immature, worries for parents, tired of school, boys (break of an established relation)

tial association with peers who are outside the normal same-sex–same-school sphere.

Being principally interested in the peers who are outside the same-age–same-sex circle—older peers, younger peers, working peers, and opposite-sex relations—the task before us in this section is to make a more detailed analysis of the patterning of friendships among the girls. Is it the same girl who associates with older peers who also has working peers in her circle of friends? Does the same girl who engages with younger peers also associate with chronologically older friends? To what extent do nonconventional types of peers (older peers, working peers, opposite-sex relation) cluster together in the same individual? Stated in this manner, the question at hand concerns the peer profiles for individual subjects.

One way to look at peer associations is in terms of peer configurations. Table 5.9 presents all the combinations of peer types that occurred among the girls in the research group, along with their observed frequencies.

Table 5.9 was composed using the peer types as dichotomized measures. A dichotomized score was calculated for the variables older peers, younger peers,

TABLE 5.9
Peer Profiles for Girls in the Research Group ($N = 407$)

Peer profile					
Y	O	W	B[a]	Frequency	Percentage of Ss
0	0	0	0	146	35.9
0	0	0	1	19	4.7
0	0	1	0	4	1.0
0	0	1	1	0	0.0
0	1	0	0	69	17.0
0	1	0	1	32	7.9
0	1	1	0	48	11.8
0	1	1	1	33	8.1
1	0	0	0	26	6.4
1	0	0	1	5	1.2
1	0	1	0	1	0.2
1	0	1	1	0	0.0
1	1	0	0	15	3.7
1	1	0	1	4	1.0
1	1	1	0	4	1.0
1	1	1	1	1	0.2
56	206	91	94	407	
14%	51%	22%	23%		

[a] Y = Younger peers
O = Older peers
W = Working peers
B = Boyfriend

and working peers with "0" representing no contacts with respective peer type and "1" representing one or more such peers in one's circle of peers. As to opposite sex relations, girls who reported that they had a steady boyfriend contact at 14.10 years were assigned the value "1" and the other girls the scale value "0." Of all the 466 girls with menarcheal information, 407 had complete data for these four peer-type variables. It will be remembered that because the peer categories were developed independent of each other, we cannot determine the extent to which the different peer types in one peer profile referred to the same individual. This is one limitation of the analysis.

There are totally 16 peer profiles possible for the four peer types. Eight of them gathered just a few cases. The remaining eight peer profiles arranged in decreasing order of how common they were among the girls in the research group were:

- No contacts with anyone of the four peer types (35.9%)
- Older peers (17%)
- Older peers and working peers (11.8%)
- Older peers, working peers, and heterosexual relation (8.1%)
- Older peers and heterosexual relations (7.9%)
- Younger peers (6.4%)
- Heterosexual relation (4.7%)
- Younger and older peers (3.7%)

The following remarks can be made with regard to these peer configurations.

The most common peer configuration among girls in mid-adolescence was to have no close relations with neither younger, older, or working peers nor with heterosexual relations. Although the dominant peer configuration category, this configuration was expressed by a *minority* of the girls. Altogether 35.9% of all girls lacked relations with peers outside the expected peer frame. That a minority of girls had conventional peer contacts exclusively should be recognized in the light of previous researchers' assumptions that same-age–same-school female peers constitute the statistically normative peer configuration in adolescence (Kandel & Lesser, 1972).

About every second girl reported that she had one or more older peers in her circle of friends. Of the girls who reported associating with older friends, the most common peer configuration was to have older friends without engaging with the other peer types. However, it was more common to find that if a girl associated with older peers, she also had other nonconventional peer types.

The same pattern applies to associating with working peers and having heterosexual contacts. Almost all girls who said that they had friends who were employed also associated with older peers or had both older peers and a boyfriend contact. Three out of four girls who reported having a steady relation to a boy also reported associating with older peers or with both older peers and working peers.

By contrast, two out of three girls who declared that they associated with younger peers had this type of peer contact only. Very few of the girls who associated with younger peers had a steady relation to a boy.

Individual Peer Profiles and Pubertal Development

Moving from a description of peer configurations pertaining to the total sample of girls to peer profiles for the early- and late-developed girls, the question may be raised: Is there a connection between belongingness to a particular peer profile and pubertal timing, such that certain peer profiles are more characteristic of the early- whereas others are more representative of the later-developed girls?

In order to make an examination of this issue, the following procedure was used. A decision was made to circumscribe the analysis only to peer configurations that occurred relatively frequently among the subjects. This meant in the present case that the further detailed investigation was circumscribed to the eight profiles in Table 5.9 that involved more than five individuals. The remaining peer profiles, involving five or fewer persons, were gathered in a "residue" category ($n = 19$). Next, the eight peer configurations, together with the residue category, were recorded as nine values in a common "profile" variable. A two-way contingency table was subsequently formed with the four menarcheal groups as columns and the nine profile categories represented by rows.

The subsequent issue concerned the methodology for data analysis. The ordinary chi-square statistic is obviously not suitable in this analysis because it only computes the overall relationship between the two categorical variables. For our analysis we needed a more elaborate testing of the particular cells in the contingency table. For this reason we used a program developed by Bergman and El-Khouri (1986) for exact tests of single-cell frequencies utilizing the hypergeometric distribution. The obtained significance tests for respective cells were adjusted for the fact that several dependent tests of significance were performed. The results of the analysis are presented in Table 5.10.

Three of the peer profiles in Table 5.10 are of special interest. Two of these can be termed *minimal exposure* peer configurations. A common feature of these profiles is that they contain no associations with nonconventional peers: peers whose presence in the circle of friends might have a negative influence on a girl's social adjustment. These are the profiles with the four-digit codes 0000 and 1000 (i.e., girls who had no close contacts with any of

TABLE 5.10
Exact Tests of Differences in Peer-Type Profiles Among Girls
in Four Menarcheal Groups

					Age at menarche						
					−11 yrs		11–12 yrs		12–13 yrs		13– yrs
Y	O	W	B[a]	n	Obs	Exp	Obs	Exp	Obs	Exp	Obs	Exp
0	0	0	0	146	10	14.3	30	32.2	52	59.5	54	39.8
0	0	0	1	19	0	1.9	5	4.2	11	7.7	3	5.2
0	1	0	0	69	6	6.8	20	15.3	30	28.1	13	18.8
0	1	0	1	32	5	3.1	10	7.1	13	13.1	4	8.7
0	1	1	0	48	7	4.7	6	10.6	18	19.6	17	13.1
0	1	1	1	33	10	3.2*	12	7.3	8	13.5	3	9.0*
1	0	0	0	26	0	2.6	3	5.7	15	10.6	8	7.1
1	1	1	1	15	0	1.5	0	3.2	11	6.1	4	4.1
E	L	S	E	19	2	1.9	4	4.2	8	7.5	5	5.2

[a] Y = Younger peers
O = Older peers
W = Working peers
B = Boyfriends

*$p<.05$ Corrected for multiple comparisons

the four peer types included, and girls who associated with younger peers exclusively). The third profile of special interest can be labeled a *maximal exposure* peer configuration for the influence of nonconventional peers. It will be the 0111 profile, referring to the girls who had all the nonconventional peer types in their circles of friends.

The main findings of the comparison of peer-type profiles among the four menarcheal groups of girls supported the assumption that the early-maturing girls would be overrepresented in the maximal exposure peer profile, while the late-developing girls would be overrepresented in the minimal exposure profiles. As depicted in Table 5.10, the latest developed girls were overrepresented on the minimal exposure peer configuration [0000] of having no close contacts with any of the included four peer types. In contrast, more of the earliest developed and less of the latest developed girls reported the maximal exposure peer configuration [0111] than could be expected by chance. These results generally support the assumption of higher exposure for early developers and less exposure for late developers to the influence of nonconventional peers.

RELATIONS TO ADULTS

In this section, a comparison among the four menarcheal groups of girls will be made with respect to their relations to adults: parents and teachers. Given the different roles the mother and the father are assumed to play in the influence of pubertal development on family relations, separate accounts are made for girls' relations to their fathers and to their mothers.

Family Relations

Mother Relations. Six items in the Adjustment Screening Questionnaire administered at 14.10 years made up a scale labeled "mother relations." The specific items included, and a comparison among the menarcheal groups, are presented in Table 5.11. The scale had a mean interitem correlation of .47 and an alpha reliability of .83.

Disregarding for a moment the issue of menarcheal timing and mother relations, a broader examination of how the total group of girls generally answered the questions comprising the scale revealed that the majority of girls experienced themselves to have rather good relations to their mothers. Almost every second girl stated that she almost never felt disappointed with her mother, more than 3 out of 4 said that they and their mothers understood each other very well or rather well, more than 8 out of 10 girls stated that they never or just occasionally wished their mothers to be different, and 6 out of 10 girls were seldom or just occasionally irritated with their mothers. This is not to say that there did not exist mother–girl conflicts among girls in the sample. It is rather that these conflicts were expressed by a considerably lower proportion of girls than might have been expected from a generation gap or a *sturm und drang* view of mid-adolescence. For example, only 1 out of 8 girls quarrelled very often, or rather often, with their mothers, 6% very often expressed disappointment with their mother, 7% stated that they and their mothers understood each other poorly or very poorly, 4% often wished their

TABLE 5.11
Differences Among Girls in Four Menarcheal Groups with respect to Mother Relations at Age 14.10 Years

	F	post hoc
How much would you as adult like to be similar to your mother?	0.74	
Are you often disappointed in your mother?	2.99*	1,2>3,4
Do you and your mother understand each other?	5.25**	1>2,3,4&3>4
Do you wish your mother to be different from the way she is?	1.81	
Do you and your mother often quarrel?	5.43**	1>2>3,4
Are you often irritated with your mother?	1.05	
Scale: Mother Relations	3.66*	1>2,3,4&2>4

*$p<.05$.   **$p<.01$.

mothers to be different, and 12% were rather often or very often irritated with their mothers.

It is against this background of generally rather satisfying mother–girl relations that one has to interpret the results from the comparison between the early- and the later-matured girls. For the most part, the obtained differences refer to differences localized on the positive side of the mother relations continuum.

As is shown in Table 5.11, there were significant mean differences among the four menarcheal groups of girls for the total mother relations scale. Subsequent post hoc tests showed significant differences between the earliest-matured group and each one of the other three groups of girls, as well as between girls with their first menstruation between 11 and 12 years and the latest-matured girls. The nature of these differences were worse mother relations among the early-matured girls.

It is of interest to take a closer look at those items for which the menarcheal groups differed from each other. This analysis showed that for those relational aspects which referred to the mother as an identification object—being similar to mother and wishing mother to be different—few differences appeared between the early and the late developers. Instead, it was for items portraying mother–girl understanding—often disappointed with mother, mutual understanding, and quarreling—that differences between the menarcheal groups emerged. This suggests that early-matured girls did not actively reject their mothers but disagreed with their opinions (cf. Rutter, Graham, Chadwick, & Yule, 1976).

Father Relations. Six items, identical in wording to the items in the mother relations scale, except that "mother" was substituted with "father" made up the "father relations" scale. The mean interitem correlation for this scale was .53, and there was an alpha reliability of .87.

The overall impression from the girls' answers to the items in the father relations scale was that most girls perceived themselves to have good relations with their fathers. Two out of three girls seldom or just occasionally wished their fathers to be different, and about as many were seldom or just occasionally irritated with them. Two out of three girls also stated that there was very good or rather good mutual understanding, and three out of four seldom or just occasionally quarreled with their fathers.

There were no significant mean differences among the four menarcheal groups of girl for the total scale. One item differed at the .10 level of significance among the groups of girls, and this item referred to quarreling with the father. Early-matured girls stated that they quarreled more often with their fathers than did the late-matured girls. The suggestion that emerged from the item analysis of the mother relations scale, that early-matured girls differed by having other opinions on things than their parents, received further support here.

Parental Relations. To obtain a view of the girls' relations to their parents as a couple, nine questions in the Adjustment Screening Questionnaire at 14.10 years addressed the girls' relations to their parents together, rather than specifically designating one parent or the other. The included items made up the parent relations' scale that is presented in Table 5.12 along with a comparison among the four menarcheal groups of girls. The mean interitem correlation was .32, and the alpha reliability was estimated at .82.

When girls view their relationships to their parents, it is most often their mothers who first come to their minds. This conclusion can be arrived at from an inspection of the correlations of the mother relations and father relations scales to the parent relations scale. The correlation between mother relations and parent relations was .73, whereas the similar coefficient for the relationship between father relations and parent relations was lower, .48. Both coefficients were significant at a high level ($p < .001$).

The impression from the former analyses, that most girls experience having adequate relations to their mothers and fathers, received further support for the items included in the parent relations scale. More than 7 out of 10 girls

TABLE 5.12
Differences Among Girls in Four Menarcheal Groups with Respect to Parental Relations at Age 14.10 Years

	F	post hoc
Do your parents listen to you and care about your opinion?	4.09**	1>2,3>4
Do you feel you stand in your parents' way or cause them trouble?	3.99**	1>2,3,4 & 2>3
Do you often feel like doing the opposite of what your parents suggest?	1.91	
Do you think your parents are angry or irritated with you?	5.15**	1>2,3>4
Do you think your parents are disappointed in you?	4.78**	1>2,3>4
How often are your parents the ones you care for most?	0.91	
Do you often do the opposite of what your parents want?	6.06***	1>2,3,4
Do you feel criticized by your parents?	2.98*	1>2>3,4
How often is what your parents think what you care most about?	0.41	
Scale: Parental relations	5.39**	1>2,3,4

*$p<.05$.   **$p<.01$.   ***$p<.001$

stated that their parents most often, or rather often, listened to them and cared about their opinions. The absolute majority said that they almost never felt themselves criticized by their parents, close to three out of four girls said that they almost never or just occasionally felt themselves standing in their parents' way, and about the same proportion of girls almost never or just occasionally felt that their parents were disappointed in them.

A one-way analysis of variance revealed significant differences at the .01 level among the four menarcheal groups of girls for the total scale. Post hoc analysis showed the effects to be concentrated to the most early-matured girls; they perceived themselves to have poorer parental relations than did the other girls.

The suggestion that the higher conflict level with parents among the early-matured girls were a matter of differences of opinions rather than encompassing more fundamental issues, such as rejection of the parents as identification objects, persisted after an inspection of the specific items that differed among the four groups of girls in the parent relations scale. The earliest-matured girls admitted more often than the other girls that their parents were irritated with them (17.4% vs. 10.3% stated very often or rather often); that they thought their parents were disappointed with them (13.1% vs. 4.8% stated very often or rather often); that their parents seldom listened to them (20.4% vs. 7.9% stated just occasionally or almost never); that they felt they were standing in their parents' way (6.6% vs. 3.3% stated rather often or very often); and that they more often did the opposite of what their parents wanted (25% vs. 6.3% stated rather often or very often). However, they thought as often as other girls that the parents were the ones they cared for most, that the parents' opinions should be cared about, and that they actually did not feel like doing things in opposition to their parents.

Observe here that the earliest-matured girls stated to the same extent as did other girls that they did not feel like doing the opposite of what their parents suggested. Still, more than four times as many girls among the earliest-matured than among the other girls admitted that they did just that!

Teacher Relations

The girls' relations to their teachers at mid-adolescence were measured by seven items in the Adjustment Screening Test. These items are presented in Table 5.13, along with a comparison between the early- and the late-developed girls. For the total teacher relations scale a mean interitem correlation of .39 was obtained, and the scale had an alpha reliability of .81.

Almost three out of four girls thought most of their teachers were fair to them, 37.7% thought most of their teachers liked them, 53.7% that some teachers liked them, and only 8.7% of the girls thought that hardly any of

TABLE 5.13
Differences Among Girls in Four Menarcheal Groups With Respect to
Teacher Relations at Age 14.10 Years

	F	post hoc
Do you like your teachers?	1.90	
Do you feel obstinate toward your teachers?	6.00***	1>2,3,4
Are your teachers fair to you?	3.02*	1>2,3,4
Do your teachers like you?	2.15	
Do your teachers understand you?	3.46*	1>2,3,4
Are you used to contradicting your teachers?	5.22**	1>2,3>4
Do your teachers treat you correctly?	1.33	
Scale: Teacher Relations	5.64***	1>2,3,4

*$p<.05$. **$p<.01$. ***$p<.001$

the teachers liked them. From the point of view of the girls, 50.9% of the total group of girls liked most of their teachers and 42.5% said that they liked some of their teachers, whereas only a minority of 6.7% of the girls stated that they did not like hardly any, or none, of their teachers. These examples are given to show that girls generally experienced rather trouble-free contacts with their teachers.

The earliest-developed girls were the group of girls who experienced the poorest teacher relations. They differed significantly from the rest of the girls for the total teacher relations scale (four out of seven items in the scale differed significantly among the menarcheal groups). The two items that pictured obstinate tendencies toward the teacher were the ones that most significantly differentiated among the four menarcheal groups. Among the earliest matured, 17.1% felt themselves obstinate against most of their teachers in comparison with 2.2% of the other girls, and whereas 18.4% of the earliest-matured contradicted their teachers rather often or very often, this occurred for 7% among the rest of the girls. To some extent the earliest-matured girls also perceived their teachers to be less fair to them, to show less understanding and liking. However, their conception of overall correct treatment by their teachers was the same as the other girls', and they liked about as many of their teachers as did their later-matured peers.

The overall picture that emerges is one of a rebellious attitude toward their teachers among the earliest-developed girls, without this being particularly connected with feelings of being more unjustly treated in general or being less liked by the teacher. The results that were presented for parental relations show resemblances with these. The item in Table 5.12 that differed most significantly between the early- and the late-matured girls was "Do you often do the opposite of what your parents want?". The main difference between the early maturers and the late maturers with respect to parent and teacher relations thus seems to be connected with what the girls did toward parents and teachers, and less with what parents and teachers did toward them. Overall, the differences obtained between early- and late-matured girls with respect to adult authority is one characterized by a unidirectional rebellion by the early-matured girls, manifested in discipline problems and quarreling within the own family and refractoriness toward the teachers.

PEER VERSUS PARENT ORIENTATION

In the former sections of this chapter we dealt with describing the relations to parents separately from the relations to peers. Now we bring together these two aspects, addressing the following issue: In what way is a parent versus peer orientation in mid-adolescence related to a girl's maturational stage?

Our first task is to take a broader view of the matter, analyzing the peer versus parent orientation for the total group of girls. Is a girl's relations to her parents connected with her association with peers? One might discern at least two general views of this subject. On the one hand, one might argue that the emotional quality in a girl's relations to her parents is likely to "spread over" to the contacts with peers. This argument would suggest a positive correlation between peer and parent relations. On the other hand, one might argue that girls in this phase of adolescence are placed in two antagonistic cultures. There is the peer culture, characterized by social attitudes and norms that are liberal and not necessarily coincident with the adult society's. There are also the parents, representatives of the more restricted moral code of society. If a girl adopts the more permissive evaluations and norms of the peer culture she is likely to face problems in her everyday life contacts with her parents. Hence, this would suggest a negative correlation between parental relations and the extensiveness of girls' peer contacts.

Observe the type of relationships the two points of view deal with. The first hypothesis concerns mainly the emotional quality of the contact. It says something about the character of the relation. A warm relation to parents is likely to concur with a warm relation to peers. The second argument is more concerned with the characteristics of the peers. It assumes that if a girl associates with peers who hold values that are in opposition to those of the adult world, this would very probably cause troubles in the girl's contacts with her parents. Thus, the distinction between what characterizes the relations and what characterizes the peers is a necessary one to make. Actually, both of the hypotheses we have derived—one predicting a positive correlation between parent and peer relations and the other predicting a negative one—might both simultaneously be true. Testing them presupposes different types of data, due to the fact that they refer to different aspects of the peer contact.

To get some insight into this matter, we first analyzed the coefficients of correlation between, on the one hand, the scales mother relations, father relations, and parent relations, and, on the other, the peer relations scale. These scales basically measure the emotional quality of the contact. All three of the family relations scales were positively correlated at the promille level of significance with the scale measuring peer relations. There was a correlation coefficient of .19 ($p < .001$) between mother relations and peer relations, a correlation of .21 ($p < .001$) between father relations and peer relations, and a correlation coefficient of .25 ($p < .001$) between parent relations and the peer relations scale. The direction of the relationships suggest that the kind of emotionality involved in contacts with parents tended to reappear in the associations with peers.

Against the background of what seems to be a similar emotional quality in relations to parents as in relations to peers, we were interested in whether or not the relationships were the same, independent of the type of peers involved. One way this might be estimated is to analyze the relationship between the girls' emotional relation to their parents and the range of friends represented as different types of peers. Thus, correlation coefficients were computed between the parent relations scales (mother relations and father relations) and each one of the peer-type variables (range of friends generally, younger, same-age, and older friends, as well as school friends, number of friends in the same class, and number of employed friends).

There were overall low correlations between the different types of friends and the emotional contact with the girls' mothers and fathers. Father relations was generally positively correlated with the number of friends of various character, and it reached significance ($p < .05$) for the total number of friends and the number of friends the girls had at school. There was virtually no relationship between the number of friends the girls had totally and her affectional relationship to her mother. However, to have more younger friends and fewer working friends were significantly connected with better mother relations ($p < .05$). Thus, it appears as if having more conventional and less potentially harmful contacts with peers was beneficial for the mother–girl contact.

It is probably against this background of having more employed peers and relatively fewer younger peers in one's circle of friends among the early-developed girls, in comparison with the later-developed girls, that the greater conflicts with their mothers among the early-developed girls should be seen. A more detailed analysis of the role of peers for parental relations is presented in following chapters.

In addition to treating parents and peers as two independently derived measurement entities, we wanted to apply a more confrontative strategy, contrasting parents with peers, in order to illuminate the parent versus peer orientation for early- and late-matured girls. With that purpose two questions asked in the pupil's questionnaire at 14.5 years were analyzed. The first one dealt with how the girls would react if parents and peers wanted them to do different things. The answers to this question concerning the parent versus peer orientation for the girls in the four menarcheal groups is presented in Table 5.14.

From the results presented in Table 5.14 it can be concluded that most girls favored neither parents nor peers. They were indecisive, either reflecting that whether to follow parents or peers dependent on the concurrent situational circumstance or that they did not perceive any opposition between parents and peers. A somewhat higher proportion of the girls in the sample (24.1%) sided with parents over peers (16.8%). Few differences appeared among the four menarcheal groups of girls, and the relationship between menarcheal age and the parent versus peer question was not significant. However, a tendency could be observed for a somewhat higher proportion of the early-matured girls than of the more late-matured girls favoring peers.

The second question in the pupils' questionnaire asked whether parents or peers understood them best. The results, with the girls subdivided into the four menarcheal groups, are presented in Table 5.15.

The most conspicuous result in Table 5.15 is that girls perceived that peers understood them better than parents did. In fact, more than twice as many girls reported that they were better understood by their friends than by their parents.

That peers understand them best was especially characteristic of the earliest-matured group of girls in the present sample. However, about as large proportion of the early-matured girls as of the other girls stated that parents understood them best. The overall relationship between the menarcheal age grouping and the girls' answers to this question was not significant.

From the data we have presented here there was some evidence, although not particularly strong, that the early-matured girls, to a higher extent than

TABLE 5.14
A Comparison of the Peer Versus Parent Orientation Among Girls in Four Menarcheal Groups at 14.5 Years

	If your parents want you to do one thing while your peers want you to do another—how does it usually end?		
Age at menarche	As my parents want	I am not sure	As my peers want
−11 yrs	23.9	47.8	28.3
11–12 yrs	25.0	56.3	18.8
12–13 yrs	26.3	59.4	14.3
13– years	19.8	65.8	14.4
Total	24.1	59.1	16.8

$\chi^2(6, N = 428) = 8.03$, ns

TABLE 5.15
A Comparison Among Girls in Four Menarcheal Groups With Respect to
Parent Versus Peer Understanding at Age 14.5 Years

	Who seems to know you best—parents or peers?		
Age at menarche	Parents better	Equally well	Peers better
−11 yrs	21.3	25.5	53.2
11–12 yrs	13.4	42.3	44.3
12–13 yrs	18.6	41.8	39.5
13– yrs	21.4	37.5	41.1
Total	18.5	39.0	42.5

$\chi^2(6, N = 433) = 6.71$, ns

the late-matured girls, preferred their peers over their parents in situations when parents and peers desire opposite responses. There was also a higher proportion of the early-matured girls who considered that their peers understood them best. For both these issues of parent–peer orientation the effect was concentrated mainly in the most early-matured group of girls.

SUMMARY AND DISCUSSION

We opened this chapter by describing three aspects of developmental changes in social network conditions connected with the transition from childhood to adult life. The striving of independence from home restrictions, an orientation toward peers and the peer culture, and the establishment of heterosexual relations are major interpersonal shifts that commonly are thought to characterize adolescent girls in their normal development toward adult status.

The early-matured girls in the research group were found to have more problematic parental relations than the later-matured girls. This would, if judged by the belief of an opposition between the adult world and the peer world, lead us to suspect that early-developed girls, compared to the later-developed, would have:

1. more close and intimate peer contacts,
2. a higher peer orientation,
3. a greater circle of friends, and
4. more peer intensive contacts.

However, only weak support for these assumptions was found. In fact, at this time in mid-adolescence, the data indicated that the most early- and most late-developed girls had somewhat less affectionate peer relations relative to the more average-developing girls.

A slight tendency for the early-matured girls to favor peers more often than did the later-matured was obtained for the two questions contrasting parents with peers. However, for neither of these questions were there any significant overall relationships between parent–peer orientation and pubertal timing. Moreover, the early-matured favored parents about as often as did the late-matured girls for these questions.

There was a tendency for the early-matured girls to have a wider circle of friends. The tendency was concentrated among the most early-developed girls, but the relation between the average number of friends and menarcheal age was significant only at the 10% level.

At the age of 13, the early-matured girls saw their friends somewhat more often in the evenings than did the later-matured girls ($p < .10$). On a similar question at age 14.5 years, asking how many evenings per week the girls stayed at home, no significant differences among the menarcheal groups of girls were found ($F = 1.01$, $df = 3,452$, ns). Nonsignificant differences were found for the same question asked at 14.10 years ($F = 1.16$, $df = 3,399$, ns).

The lack of substantial differences in peer interaction in the aforementioned respects was in line with our general expectations. The theoretical framework for our position on the connection between pubertal development and peer relations suggested that we had to look elsewhere. It was not to be found primarily in the quality of the peer contact, in the range of peers, or in the intensity of the peer interaction: it had to do with the characteristics of the peers. It was assumed that individual differences in physical maturity would be connected with individual differences in associating with peers outside the normal peer frame of girls generally. This theoretical position received substantial empirical support. An overview of the results are given in Table 5.16.

As shown in Table 5.16 almost all girls had friends who, like themselves, were school pupils. Most of the girls also had a friend in their own class. Most girls had one or several friends who were at the same age as themselves. These results, of friendship formation based on age and environmental proximity, are also what has been found in other studies in the peer literature. As was assumed from the present approach, no differences as to peer contacts between early- and later-developed females were to be obtained in these respects.

Rather, differences in peer association appeared, as hypothesized, for associating with out-of-school peers and with peers in different age strata. Early-developed girls were found to engage more frequently with chronologically older peers and also had more employed peers in their circle of friends than had the later-developed girls. In line with the hypothesis of peer association based on level of pubertal development, later-developed girls had more chronologically younger peers in their circles of friends than had earlier-matured girls.

SUMMARY AND DISCUSSION

TABLE 5.16
Peer Characterization of Early- and Late-Developed Girls: A Summary.
The Percentage of Girls in Four Menarcheal Groups Who Had Different Types
of Peers in Mid-adolescence

	Age at menarche (years)				
	−11	11–12	12–13	13–	p
	Social characterization				
Friends at school	97.6	98.9	98.8	99.1	ns
(in the same class)	69.0	83.9	81.7	78.8	ns
Working peers	45.2	21.5	19.5	20.4	.001
	Age characterization				
Younger peers	4.8	7.5	18.9	13.3	.05
Same age peers	92.9	95.7	96.4	94.7	ns
Older peers	73.8	55.9	51.5	38.9	.001
	Heterosexual relations				
Boyfriend relation	38.3	32.4	20.8	11.6	.001
	Friendship group membership				
Male friendship groups	9.5	6.4	1.3	—	ns
Female friendship groups	—	8.5	10.0	14.3	ns

An earlier sexual maturation, connected with an earlier identification and awareness of self as a matured female, were assumed to have the consequence of more extensive contacts with boys in mid-adolescence among the early-developed girls. The early developers also were found to have more established relations with boys in midadolescence than did the late-matured girls. Sexual experiences were also more common among the early-matured than among the late-matured girls at this age. It should also be mentioned that the friendship groups to which the later-developed girls belonged often contained exclusively girls, whereas the friendship groups of the early-developed girls more often contained males.

The great differences in heterosexual contacts and experiences among the four menarcheal groups of girls led us to examine whether these differences in sexual behaviors had counterparts in differences in attitudes towards sexual issues. Not unexpectedly in the light of the imminence of sexual experiences among the early developers, some tendencies were found for these girls to have thoughts about sex more often than did other girls, and for them to desire more information about sexual issues. However, no differences were obtained among the four menarcheal groups of girls with respect to being afraid of engaging in sexual behaviors, whether worrying about bodily malfunctions, connecting sex with fear and unpleasant thoughts, or being afraid of being drawn into sex unwillingly. Furthermore, self-descriptions of problems among the early- and late-matured girls revealed that the late-matured girls actively sought stable contacts with boys but could not attract the boys' attention or were prevented by parents in obtaining this contact (for example by not being allowed by stay out late).

Before leaving the discussion on peer relations we comment on one aspect of the girls' peer interaction often discussed in previous literature, namely, the social prestige among peers. In agreement with Faust (1960), we found the early-matured girls to be rated lowest in social status in early adolescence at the age of 10. Earlier findings of an increase in status over time for this group of girls was also confirmed. However, in opposition to Faust, the average-developed girls were found to be highest in social status at all ages investigated. By contrast, Faust reported the earliest-matured girls to be the group of girls highest in prestige at higher ages in the adolescent period.

The differences in results from the present study and Faust's investigation do not lend themselves to any easy interpretation. Somewhat different specific techniques were used, but the two studies have in common a same-sex–same-class sociometric investigation. Both Faust's and the present investigation had a limited scope: it applied to the *status in one's class*. When investigating social status and its changing patterns over time, we must remind ourselves of the sort of measures on which we base our conclusions. In our opinion social status measures obtained with the class as the frame of reference are systematically biased to the disadvantage of the early-matured girls. Because early-developed girls have a circle of friends different from that of later-developed girls—often localized outside the school—any class-based measure cannot assess the support, interest, or prestige devoted to these early-matured girls by outside-school peers. From the point of view of the classroom setting, they might appear to have low prestige and to be isolated. From the point of view of the peer group in a more general sense, quite another picture of the social status of these girls might be revealed. This argument is supported by several observations.

First, of the total number of peers reported by the earliest-matured girls, the majority were chronologically older, in contrast with the same-age peers most often reported by the later-developed girls. Second, the earliest-matured girls were found to associate with peers who had quit school and were already employed to a considerably higher extent than did the later-matured. Third, although the earliest-developed girls reported that they had, on the average, more friends than did the later-matured, they simultaneously claimed a fewer number of classmates as friends than did the later-matured girls. These data support the argument that prestige measures obtained within the class capture just a fragment of the social prestige of the early-matured girl.

A second set of observations adds another dimension to the issue. The early-matured girls thought of themselves as being more popular among boys than did the late developers. Other data support these differences in perceived popularity among boys. The early-matured had more experiences of steady dating, they had more sexual experiences, and whereas the late-matured girls' major problem concerned difficulties in establishing the boy–girl relations, the early-matured girls who already had such heterosexual relations declared that their problem was fear of separation from their boyfriends.

If these results illustrate the actual situation with respect to heterosexual relations, which we believe they do (additional support from independent sources are given in chapter 6), then the most reasonable outcome of a sociometric investigation of status, using *boys* as the raters, would be higher social status for the early-matured girls than for the late-matured. However, this did not turn out to be the case for the present sample of girls. At the same time as the same-sex ratings of popularity were administered at 14.5 years, opposite-sex ratings, with the same instructions, were also collected. The boys were asked to rank all girls in the order they wanted them to move to a new class which was to be formed. The results of the sociometrics are depicted in Fig. 5.2

As can be seen in Fig. 5.2, the boys in the class rated their female classmates very much the same as the girls themselves rated their same-sex classmates. The most early- and the most late-developed girls received lower rank order values than did the more average-matured girls ($F = 2.94$, $df = 3, 460$, $p < .05$). The correlation between boys' ratings and the girls' ratings was quite high, .50 ($n = 446$, $p < .001$).

These data show the limited potential for sociometric investigations in

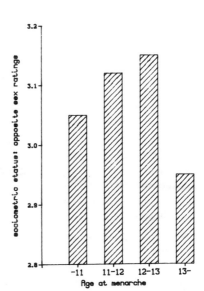

FIG. 5.2. Social status of girls in four menarcheal groups at 14.5 years as rated by same-class boys.

capturing the prestige dimension if the analysis is conducted with the school class as the frame of reference. If these data had been presented in isolation, the fact that they were classroom-based would probably have been neglected, and the interpretation that average-matured girls were more favored socially by boys than were the off-time-matured groups of girls would be the generalized main finding.

Boys and girls in the same class are constantly interacting with each other and thus develop similar frames of references based on intensive personal contact. They know each other's strengths and weaknesses well, and they utilize this information that is not easily accessible for outside, more casual observers. Their opinions of each other socially might coincide with the opinions also expressed by boys and girls outside the class, but not necessarily. Classmates will base their opinions of each other on various aspects of personal qualifications and probably relatively less on physical attributes whereas the latter quality would be more salient for the outside-classroom peers. The finding of high agreement in ratings of popularity between boys and girls in the class, coupled with a substantially different social interaction pattern for the early-matured girls when taking into account the outside-class peers, suggest that the classroom situation is a world of its own with features that might, but need not, coincide with the outside world.

What has been said here concerning the limited applicability of class-based sociometric techniques to capture the social status of early-matured girls also applies to the use of such measures for the study of social status among teenagers generally. Every school- or class-based sociometric measure is based on the assumption that the school or the classroom is a miniature of the wider peer world. In many cases this is perhaps not far from the truth. That the best friends of adolescent girls often are girls in the same class speaks in favor of this assumption. But for those adolescents whose friends are primarily localized outside the classroom, problems with regard to estimating their social status in the peer group begin to arise. Because we cannot a priori determine the effect of outside-classroom peer influences, researchers utilizing this technique must be careful when generalizing their results to the general effect of adolescent peer networks. It is very likely that commonly accepted conclusions regarding the low social status of early-matured girls, have been misleading either because measures of prestige have been obtained from the opinions of classmates only or because observations have been conducted in the school setting only. For example, in the Oakland Growth Study observations were made in the playground of the school and while awaiting physical examinations. Jones and Mussen (1958), commenting on the lower status scores for the early developers, remarked that "it is not unlikely that if these girls of more mature status had been observed in social groups of their own choosing (presumably outside of school) the behavior picture might have been more favorable" (p. 497).

Parenthetically, one common approach to the study of the influence of peers on the behavior of teenagers has involved a matching of best friends. The subject is asked to name his or her best friend and a subject-peer dyad is formed to the extent that the named friend reciprocates by also naming the subject (Kandel, 1978; Kandel & Lesser, 1972). Socially maladapted teenagers are often rejected in their class, and are less often chosen in turn by subjects whom they mention as best friends. This group of adolescents, probably one of the most important when screening for social problems in adolescent samples, might be assumed to recruit their friends outside school in order to satisfy their interpersonal needs. Thus, the finding that they are underrepresented in matched samples (Kandel, Kessler, & Margulies, 1978) when the best friend is sought in the same school, is not surprising.

Our final comments concern the differences between early- and later-developed females with regard to relations to adults. The data that were presented in this chapter demonstrated that most of the mid-adolescent Swedish girls in the present sample had on the whole rather harmonious relations with their parents. Most of the girls were seldom or never disappointed with the parents, thought that their parents listened to them and cared about their opinions, seldom felt themselves in their parents' way at home or that their parents were disappointed at them. Most also thought that they and their parents understood each other well. When evaluating the findings concerning the impact of early versus later pubertal maturation on relations to parents, the obtained differences must be viewed against the background of these satisfying parental relations characterizing most of the girls in the sample.

Empirically our approach involved a comparison of the four menarcheal groups of girls for three parent relations scales: mother relations, father relations, and parent relations. The menarcheal groups were also compared for a teacher relations scale. To summarize the differences obtained with regard to relations to adults in midadolescence between the early- and the late-matured girls, four features stand out:

1. Most girls in the sample reported that they had rather satisfying relations to their parents (as well as their teachers).
2. Significant differences among the four menarcheal groups of girls were obtained for mother relations and teacher relations, with the early-matured girls indicating less satisfying interactions.
3. The impact of early versus late maturation seemed to be connected to everyday life controversies, but not with parents as identification objects or with the more basic affectionate parent–daughter (teacher–girl) climate.
4. The parent–daughter and the teacher–girl conflicts among the early-developed girls seemed to be unidirectionally initiated with the early-

matured girls seeing themselves as the source of the controversies rather than the other way around.

If one were to synthesize these findings, it appears that the early-matured girls were particularly sensitive to adult authority. Their daily life contained features that set these girls in opposition to the adult standards of living. However, it should be observed that there were no differences among the four menarcheal groups of girls with respect to whether the girls felt like doing the opposite of what their parents suggested. At the same time, the earliest-matured girls admitted to a considerably higher extent than later-developed girls that they often in fact did do the opposite of what their parents wanted. The finding in chapter 4 that the early developers, more often than the late, reported difficulties in restraining their temper, particularly with parents as the targets for the girls' irritation, are in agreement with the present results of the early developers' differing from other girls in reporting more often actually doing the opposite of what their parents wanted even though they appeared to have no intention of doing so. Their opposition seemed to be more behavioral than attitudinal.

This finding that the early developers tended to instigate the conflicts with their parents was even more pronounced for the early-developed girls' teacher relations. No significant differences among the four menarcheal groups of girls was obtained for whether they thought their teachers treated them correctly nor for whether they liked their teachers. The item that differentiated most between the four groups of girls was a question of whether the girls felt themselves obstinate toward their teachers. The second most differentiating item was a question of whether the girls contradicted their teachers. For both these items the earliest-matured girls answered that this was the case more often than did the later matured girls.

6
Social and Emotional Adjustment

FEMALE PROBLEM BEHAVIOR

It is one of the curious facts of life that in the growth toward adult status and responsibility many people go through a period of social adjustment problems during adolescence. With the work by Jessor and Jessor (1977) being a notable exception, there is a scarcity of basic facts about the course and the variations of social problems among females during this period. We suffer from the lack of solid empirical investigations designed to study the processes that influence the direction of social behavior during the adolescent years for females and of data that permit comparisons between the sexes. Hitherto, the overwhelming majority of cross-sectional and longitudinal studies in the literature, from their beginnings, either have focused their investigation on boys and have excluded girls in reporting the results or have presented their data for the sexes pooled rather than separately (Miller, 1979). It comes as no surprise that the two most recent reviews of the etiology of criminal behavior, Loeber (1982) and Loeber and Dishion (1983), focus their attention exclusively on early prognostic factors for boys.

Our chief concern with respect to social adjustment in the present chapter is with those social behaviors that signify breaches of informal rules or with those behaviors that depart from the socially expected. Jessor and Jessor's (1977) conception of "problem behavior" captures our focus: "The concept of problem behavior, rather than connoting some valuative stance, refers to behavior that is socially defined as a problem, a source of concern, or as undesirable by the norms of conventional society and the institutions of adult authority, and its occurrence usually elicits some kind of social control response" (p. 33).

What we call problem behaviors, socially maladaptive behavior, misconduct, and so forth, is a constellation of social behaviors that characteristically appear in early adolescence, around 12–13 years of age. They increase in frequency through the adolescent years and will encompass the vast majority of teenagers until they level off in late adolescence. In their longitudinal study of three linked high school cohorts (7th through 10th grade, 8th through 11th grade, and 9th through 12th grade), together with a college population followed from the year they were 19 and annually for 4 years, Jessor and Jessor (1977) had the opportunity to plot the direction of social behavior over time. For variables from the personality domain, they reported declines across time in the value of academic achievement, intolerance of deviance, alienation and religiosity, and increased evaluations of independence and social criticism for both males and females from Grade 7. During high school an increase in perceived parents' and friends' approval of problem behavior was documented, as was an increase in friends providing models for such behavior and a decrease in parental controls.

The findings pointed clearly in the direction of less conventionality with increasing age, although some of the variables appeared to stabilize in college. The change over time in specific behaviors pointed in the same direction. Marijuana use increased all the way from Grade 7 until the end of college. By the age of 22, 75% of all males and 62% of all females had used marijuana at lease once. Sexual intercourse experiences showed a similar trajectory, with 21% of the males and 26% of all females reporting such experiences at the age of 17, whereas 82% of the males and 85% of the females had had sexual experiences by the age of 22. Drinking and problem drinking also increased during the years in high school. A stabilization then occurred for the measure of problem drinking in college. Almost all subjects had used alcohol by the age of 22. Finally, general deviant behavior, involving lying, cheating, stealing, aggressiveness, and vandalism, increased for both sexes over the high school years but tapered off in college. Altogether, these data suggested increasing social problems among teenagers, at least up to the last years in high school.

Few sex differences in prevalence rates were obtained in the Jessor and Jessor (1977) study for marijuana use, sexual intercourse, drinking, and problem drinking. However, differences between the sexes were found for frequent engagement in such activities. For behavior that reflected general deviance (stealing, lying, property destruction, disruptive behavior, and aggression) significant sex differences existed, with males reporting higher engagement for all the high school years. Thus, the more frequent the engagement in problem activity, and the more serious the acts, the greater were the sex differences.

Nesselroade and Baltes (1974) reported similar age-related findings from examining four cohorts of 12- to 17-year-old subjects, each one followed over a 2-year period. There was an increase over time in independence and decreases in super-ego strength and achievement. Thus, Jessor and Jessor's data as well as the study by Nesselroade and Baltes indicated that the increase in problem behavior tends to go hand in hand with a decline in behavior expressing a conventional orientation—achievement and religiosity.

Problem behavior in the adolescent period among males involves both what Loeber and Schmaling (1985) denoted as covert antisocial activities, which by their nature are concealed from adults (stealing, lying, truancy, running away, drug and alcohol use, etc.) and overt, confrontative actions aimed at gaining control over others (bullying, fighting, demanding, arguing, disobedience, etc.). Among females, covert problem behavior tends to dominate. Overt antisocial activities surely will characterize some females, but is not the antisocial pattern that typifies the misconduct of normal adolescent girls.

According to present-day legal statistics in Sweden, the most common crimes for females were, in descending order, petty theft; carelessness in traffic; offenses against the Road Traffic Ordinance Act; theft; unlawful driving; offense against other statutes concerning road traffic; fraud; arbitrary conduct; petty smuggling of goods; and offenses against the Road Traffic Tax Act. Similar types of offenses were also common among males. However, among the 10 most common crimes for males were also included assault and offenses inflicting damage. That few females commit violent crimes was also evident in the present longitudinal population. Of all males who had committed one or several offenses, 25% had committed offenses that were classified as violent offenses against other persons, and 17% had committed property damage. By contrast, of all females who had committed a crime only 10% had committed violent offenses against other persons, and 3% were recorded for property damage (Stattin, Magnusson, & Reichel, in press). The findings that sustained antisocial activity among females, at least as they can be found in clinical samples, tends to involve sexual promiscuity, drunkenness, and property crimes (Robins, 1966), provide further support for the more covert feature of social problems among females.

Another aspect of the official data on delinquency for the present sample of subjects is worth mentioning. When we collected detailed information about the offenses and the surrounding circumstances, as recorded in the preliminary investigations by the local police, we were surprised how often it happened that girls were involved without ever being charged or without their involvement's being brought to the attention of the social welfare authorities. They "happened" to be passing by during a theft, to be passengers in stolen cars, visitors to the owner of the apartment where the police found narcotics, and so on. They were overall passively involved, and few found their way into official delinquency registers.

The covert nature of problem behaviors among females has commonly been connected with the feminine role and the feminine identity, as well as with gender-specific activity patterns. Starting from a very early age, boys are more physically active than girls and are more socially disorganized and inclined toward irritability, temper tantrums, hyperactivity, and competitiveness (MacFarlane, Allen, & Honzik, 1954). Not only are these types of behavior more frequent among males, it is also typical for the same behavior to take different manifest forms between the sexes. A case in point is aggressive behavior. Males, in general, show an acting-out form of aggressiveness, usually of a physical and a direct nature, whereas females tend to exhibit indirect aggressiveness that takes a more verbal form (Pulkkinen, 1983b). The differences in mode of expression must be taken into account in the light of the often-reported higher aggressiveness level of males compared to females.

Problem Behavior and the Normative Age-Graded Perspective

A general theoretical consensus is that adolescent problem behavior is part of age-prescribed and age-normative social behaviors occurring at different points in time during adolescence. This view reflects a "normal developmental" approach to problem behavior among teenagers (cf. Jessor & Jessor, 1977).

Certain logical implications follow from such an outlook. Forming a judgment of whether a certain social behavior in an individual is to be regarded as normal or as a risk for future social maladaptation requires setting the individual's behavior against the background of the age-normative problem behavior rate. For instance, regular established drinking habits in a 12-year-old girl or boy is socially deviant relative to the drinking pattern among same-age peers, but the same drinking habits for a 16-year-old subject are more what should be expected, because drinking among adolescents at this age is a statistically normative behavior.

The growing number of norm violations with increasing age from early adolescence has been interpreted, from the normal "growing-up" viewpoint, in part as a feature of the developmental progression through which the individual establishes a personal and independent norm system. With more permissiveness from parents and intensified striving toward independence and autonomy, adolescents' conceptions of right and wrong are put to the test. The absence of parental norms to guide behavior in these years result in what Covington (1982) called an "anomie of age."

Rather than expressing rejection of common standards, norm breaking in adolescence might signify adherence to goals that may have a developmentally adaptive function for the teenager. As part of research in the present longitudinal program, Henricson (1973) studied, among other things, the correspondence between teenagers' evaluations of various forms of norm breaking and their actual behavioral intentions in situations leading to breaches of such norms. When asked about their evaluations of ignoring parents' prohibitions, only 16% of the 14-year-old girls and boys thought that it was acceptable to ignore these prohibitions. Nevertheless, as many as 55% of the boys and girls would, in fact, defy their parents and go to a party where their friends were invited. Similarly, only 2 out of 10 teenagers (23% of the males and 21% of the females) reported an accepting attitude toward getting drunk, whereas almost twice as many (41% of the males and 39% of the females) stated that they definitely would get drunk in an actual situation where they were offered alcohol by close peers.

This example mirrors the "situational ethic" (Gold & Petronio, 1980) of everyday behavior among teenagers. Norm breaking occurs, not from a deviance motive primarily, but in order to fulfill central, personal motives in adolescence. The example also illustrates another important feature: namely, that teenagers, overall, tend to regard engagement in these types of behavior as *not* acceptable. However, behavior of this kind might nevertheless occur in order to adopt to the social behavior patterns of the peer group.

The adaptive functioning of teenage behaviors that are normally designated as problem behaviors or misconduct has been investigated by Silbereisen and Noack (1986). Using an ecological approach, they showed that in order to maintain relations with peers, to make new peer contacts, and to establish relations with the opposite sex, teenagers were forced into certain types of leisure-time settings in which behaviors such as smoking and using alcohol were reinforced.

Moreover, some of the social behavior that adults view as deviant might, in fact, reflect an adaptation to the behavior patterns of the adults themselves, because these behaviors connote a more "mature" status. Bloch and Niederhoffer (1958) proposed that deviant behavior might be a way of reaching sociological maturity.

Jessor and Jessor (1977) showed that problem behavior was closely tied to a pattern of changes toward maturity. Empirically, they verified that when teenagers engaged in social transition behaviors they simultaneously tended to adopt various forms of problem-prone behaviors. For example, teenagers who started to drink tended, at the same time, to engage in deviant behaviors like vandalism and lying, value academic achievement less, and engage with friends who had a similar tolerant attitude toward deviance, and the like. It was not that adolescents adopted a series of specific problem behaviors. Rather, there were simultaneous changes in personality, in the perceived environment, and in behavior:

> This fact provides the basis for talking about transition in social-psychological "status" rather than mere adoption of a particular new behavior. A "constellation" of developmental changes appears to characterize the process of becoming a drinker, or a marijuana user, or a nonvirgin, and the taking on this pattern

of characteristics seems to be involved in making a transition. Transition behaviors, rather than being isolated events, probably represent a syndrome of activities oriented toward accession to a developmentally later status. (pp. 205–206)

Overall, the different views of problem behavior in adolescence, seen from an normative age-related perspective, are views that do not necessarily coincide with a delinquency perspective. Our expectations with regard to norm breaking among teenagers from the latter point of view is, most commonly, that something in the socialization must have failed. The roots of norm breaking are sought in negative upbringing conditions or in concurrent association with "bad" peers. The norm breaker might have had antisocial parents, have had serious separation experiences, or have come from bad socioeconomic conditions: furthermore, perhaps, he or she is being raised in an unloving, punitive home, lives in a "bad" neighborhood, has an aggressive acting-out character, is hyperactive, has failed in school, and so on. We tend to overlook the possibility that teenagers in general condemn antisocial activities but might engage in them if other more important issues are at stake.

The forthcoming analyses focus on features of the social life situation for adolescent females that may, but need not be, connected with a criminal career or other future social adjustment problems. The long-range effects of problem behavior among adolescent females is an empirical issue that is addressed later on.

Problem Behavior and Interpersonal Relations

Any attempt to understand adaptation among teenage girls, whether social or emotional, cannot bypass the strong interpersonal orientation affecting the feminine self-concept and reality orientation.

Much discussion has centered around the issue of what constitutes the core features of identity development among adolescents. Identity is commonly defined as a person's sense of life's meaning and the individual's commitment to long-range goals and aspirations. A substantial part of the research on the identity concept is built on Erikson's model. His framework has been particularly attentive to the different ways individuals cope with the domains of occupation and ideology, both of which are assumed to be important identity-related issues for teenagers.

The question has been raised as to whether or not identity development, as it can be expressed in these two domains, actually mirrors what is central in the development among females. Are the life domains of occupation and ideology as important for the identity development among girls as they seem to be for boys? Marcia (1980) expressed these doubts in the following way:

the predominant concerns of most adolescent girls are not with occupation and ideology. Rather, they are concerned with the establishment and maintenance of interpersonal relationships. . . . I would like to make the suggestion here for approaches to research on identity with women. The first is, that if an identity status approach should be taken, the areas around which crisis and commitment are to be determined should be those around which women are expected, initially, to form an identity: the establishment and maintenance of relationships. I do not mean by this that the "male" concerns of occupation and ideology are not available as identity-constellating issues for women. They are. But as society is currently structured, women who go on this route, particularly if they do so to the exclusion of the interpersonal one, will pay a price in the lack of extensive social support. (p. 179)

In her discussion of the issue of identity, Gilligan (1982), in a similar vein, linked female identity to intimacy and connectedness, to responsibilities for others, and to caring and attachment. In contrast to the close connection between self and autonomy, individual achievement, individuation and self-realization among males, self, for females, was portrayed as an inseparable part of the connection with others.

Similar views have been reported at several instances in the adolescent literature (cf. Deutsch, 1944; Rosenberg & Simmons, 1975). Dusek and Flaherty (1981) found that adolescent girls' self-concept was more associated with congeniality/sociability relative to boys, whereas boys' self-concept relatively more reflected achievement-leadership. Douvan and Adelson (1966) concluded from a large-scale interview study of adolescents that

> friendship, dating, popularity and the understanding and management of interpersonal crisis hold the key to adolescent growth and integration for the girl. The internalization of feminine goals also has important implications for the girl's development. Here too the goal is to form a lasting tie to another, and is not an individual achievement in the sense that the boy's vocational goal is. (p. 350)

A consequence of this is a possible difference between those girls whose identity development is built on interpersonal relations and those whose identity is based on typically male-related domains of occupational career and ideology.

Finding that those aspects associated with popularity among early to mid-adolescent girls were related to physical appearance and personality, whereas popularity was more related to athletic achievement among boys, Crocket, Losoff, and Petersen (1984) concluded the following:

> From our findings it would seem that the ascriptive|achievement distinction still holds among young adolescent boys and girls. A girl is still more likely to feel

that her standing in the peer group is largely fixed by her looks and her personality, attributes over which she has relatively little control, whereas a boy is more likely to feel that he can earn peer recognition by developing his athletic (and perhaps academic) competence. (p. 169)

On the other hand, sex-related interests, with boys more achievement oriented and girls more interpersonally oriented, are not features unaffected by changes in the larger society. Competence and achievement are becoming increasingly salient for adolescent girls (cf. Bush, Simmons, Hutchinson, & Blyth, 1977–1978).

As is well-documented in the developmental literature, gaining autonomy in adolescence tends to go hand in hand with a shift from parents, as the primary source of comfort and interaction partners, to peers, as those with whom the adolescent shares his or her daily experiences (Montemayor, 1982; Coleman, 1961; Andersson, 1969). The reliance on peers, especially best-friend relations, is more pronounced in girls than in boys. In the last chapter it was found that of the mid-adolescent females in the research sample a substantially higher proportion reported that their peers rather than their parents understood them best. A comparison with the boys in the longitudinal population makes salient how important the peer contact is for girls. For the girls in our research groups at age 14.5 years, 42.5% reported that peers understood them best, 30% said that both their peers and their parents understood them best, whereas only 18.5% of the girls stated that parents understood them best. The situation was different for the males of the same age. A considerably lower proportion of the boys than of the girls, 26%, reported that their peers understood them best, 43.5% said that both parents and peers understood them best, whereas 30.% thought their parents understood them best. The boy–girl comparison was significant at a high level ($p < .001$). Other studies in the social network literature support these findings. Coleman (1961) and Andersson (1969), among others, have presented convincing empirical evidence of greater peer reliance among girls than among boys in the mid-adolescent years. Part of these sex differences are probably a consequence of the fact that girls reach puberty somewhat earlier than boys.

The strong interpersonal component in the female social role and the female identity, and the subsequent sensitivity to interpersonal sanctions, seems to speak against social deviation in adolescent girls (Coleman, 1961). For example, Miller (1979) made the point that

> Since popular images of the male role stress the acquisation of goods and dominance over others, the young boy who commits offences against property or the person is behaving in a way that is consistent with the adult role to which he aspires. Where the adolescent girl is concerned, however, any involvement in delinquency, far from expressing valued elements of the female gender role,

is actually inconsistent with that role. The female role stresses popularity and heterosexual success in adolescence. (p. 124)

This seems reasonable. Still, the strong interpersonal component in the psychology of the female probably functions in two directions, both *inhibiting* problem behavior but also, in pursuit of satisfying of an interpersonal motive, *facilitating* norm breaking. To the extent that interpersonal relations are of central concern to the teenage girl, and a shift in interpersonal orientation toward peers is at the core of adolescent development among females, we would suspect that the characteristics of peers and their social behaviors would greatly contribute to the understanding of problem behavior among females. (In this context, it should be noted that when female problem behavior occurs, such as alcohol drinking in adolescence, it tends to be quite strongly related to such habits in the girl's peer group; Downs, 1985; Pulkkinen, 1983a.)

Extensive research work has documented the role of peers in adolescent problem behavior. For example, Jessor and Jessor (1977) based much of their framework for studying problem behavior on the parent versus peer orientation, stating that "when a youth is located in a peer rather than a parental context, it defines greater proneness to problem behavior" (p. 24). However, they used the term *peer orientation* as a gross denominator and did not investigate which features of the peer group coincided with a higher risk for engaging in problem behavior. Actually, measuring "peers" in a highly general sense, thus yielding few possibilities to discern more differential peer influence patterns, is a characteristics feature of the methodology that has guided much of the research on peer influences. That a peer orientation on the part of adolescents implies a higher risk for engaging in deviant activities, would, if true, mean that adolescent girls, who are more peer oriented than boys, would be expected to be more likely to engage in deviancy. But they are not. Denominators like "the peer culture" or "peer orientation" are too ambiguous to describe the interpersonal environment of adolescent girls and boys, and they omit the specific characteristics of the peers as well as the features of peer group activities. Such designations seem to suggest that peers have similar implications for behavior, lumping together in an overall peer category the variety of peer contacts of adolescents: distal peers and best friends, younger, same-age, and older peers. Furthermore, we do not know the intensity of the peer contact, whether the teenagers are members of friendship groups or not, and if so, the characteristics of the friendship group.

The framework of our approach claims a central role for the understanding of individual differences in social behavior in mid-adolescence to be the *characteristics* of peers and the adolescent's adjustment to the social manners, supports, norms, sanction policies, and expectations typical of the particular peer context the individual experiences. Although much previous peer research has proceeded from the premise that the important peer influence involves peers who are associated with the subjects through physical, social, and chronological age proximity, the present view suggests that contacts with peers outside of this "expected circle" of friends will have particular significance for individual differences in norm-breaking activity among adolescent girls. Key transmitters of more sophisticated leisure-time activities outside of the expected same-sex–same-age–same-class circle of friends are chronologically older peers, employed peers, and members of the opposite sex: peers that we have labeled *nonconventional peer groups*. The more nonconventional peers in the circle of friends among mid-adolescent girls generally, the higher the assumed risk for engaging in norm-breaking activities.

If the age-related interpretation of problem behavior among females during adolescence and the role played by peers for progression in social behavior in adolescence are valid assumptions, then, given a match between girls' level of biological maturation and the characteristics of their peer network, the early-matured girls, who predominantly associate with nonconventional peers, should be the group of girls who are particularly prone to engaging in advanced social behavior in mid-adolescence, including the breaching of formal and informal rules. The involvement in this type of behavior is assumed to be less for the later-developed girls. Among other things, their more frequent association with chronological younger peers, with their more childish social life patterns and conventional attitudes, will protect these girls from engaging in more socially deviant activities. The proposition that early-developed girls are more prone and later-developed girls less prone to engage in problem behavior in mid-adolescence is investigated below.

Previous Research on the Issue of Pubertal Development and Problem Behavior

Previous research in pubertal development has yielded conflicting evidence supporting a presumed connection between physical maturation and problem behavior in adolescence. For example, in a study of 11- to 12-year-old early adolescent females, Lenerz, Kucher, East, Lerner and Lerner (1987) obtained no significant relationship between physical status and conduct behavior. If pubertal development is thought to be connected with common transition problems with regard to social adjustment, in this study the early age of the subjects and the limited subject sample ($N = 45$), might be obvious reasons for the nonsignificant results. However, Duncan, Ritter, Dornbusch, Gross, and Carlsmith (1985) also found no connection between pubertal timing of 12- to 17-year-old females and deviant behavior (law contact, running away, smoking, school absenteeism, and discipline problems) in their study on a national sample of more than 5,000 adolescents.

On the other hand, confirming evidence was reported in the Milwaukee longitudinal study (Simmons, Blyth, & McKinney, 1983). The same group of girls was tested in various aspects in Grades 6, 7, 9 and 10 (see pp. 25–26 for problems of interpretation of data from the upper grades). Problem behavior, as it is manifested at school, differed among the postmenarcheal and the premenarcheal girls in Grades 6 and 7, with the postmenarcheal girls evidencing more problems: getting into trouble, being sent to the principal's office for doing something wrong, being placed on school probation, and skipping school. In a follow-up analysis of factors contributing to self-esteem in teenagers, Simmons, Carlton-Ford, and Blyth (1987) reported a direct relationship between pubertal development and school problem behavior, subsequently linked to low self-esteem among the girls in Grade 6 and 7. They also noted that early independence from home was associated with problem behavior, subsequently affecting self-esteem negatively.

Results in the same direction were reported by Frisk (1968) and Frisk, Tenhunen, Widholm, and Hortling (1966). In their outpatient clinic sample, the early-matured girls were characterized by sexual problems and early sexual contacts, by school fatigue, by defiance, by identity problems, and by insecurity. It was reported that 73% of the early-developed females seen in the clinic manifested some aspect of failed adjustment (parental conflicts, school maladjustment, and asocial behavior), compared to 33% of the very late-developed girls. The peers of the early-developed girls were often older than the girls themselves. The early-developers' adjustment problems were most intensive around 14–15 years.

To the authors' knowledge, only two studies have been presented in the literature relating pubertal development to delinquency. Neither of them found any significant connections. Wadsworth (1979) reported data from the National Survey of Health and Development, involving 11- to 15-year-old female subjects. Only a tendency for the early developers to have somewhat more registered offenses was noted. Gold and Petronio (1980) reported on a study investigating the relation between measures of skeletal age and self-reported delinquency among 11- to 18-year-old boys and girls. For neither sex, and at no age, was there any relationship between pubertal development and delinquent behavior.

As to studies on girls with precocious puberty, higher conduct disturbances among these girls than among comparable groups of normal-developed girls have been reported in a few studies (cf. Comite et al., 1987; Ehrhardt et al., 1984). Case studies in the literature show similar findings.

THE PRESENT STUDY

Social Adjustment and Pubertal Development

Data for a comparison of problem behavior between early- and later-matured girls in mid-adolescence were collected from two questionnaires—a Norm Inventory administered at 14.5 years and the Adjustment Screening Test given at 14.10 years. The former inventory contained descriptions of concrete instances of norm breaking. Situations picturing different types of norm-violative activity—truancy, shoplifting, alcohol drinking, and the like—were described. Thereafter, the girls were asked a number of questions, among other things their actual breaches of norms of the type the situation described up to the point in time when the instrument was administered. An example of a situation description picturing cheating at school is the following: "They are going to have an exam, and Gunilla has not had time to prepare herself. It is important for Gunilla to succeed. She has brought a scrap of paper with notes. She hesitates over whether to take up the note and cheat."

The girls' answers were analyzed for the question concerning how many times they had actually violated such rules. Answers were given on 5 point Likert scales with the alternatives: (1) never; (2) once; (3) 2–3 times; (4) 4–10 times; and (5) more than 10 times (see Magnusson, Dunér, & Zetterblom, 1975, pp. 100–103).

Data from the Adjustment Screening Test consisted of eight questions asking whether the girls had ever run away from home, provoked other people, stolen, and so on. The alternative replies were the same as those in the Norm Inventory.

Altogether, data on norm-breaking activity covered various forms of norm violations *at home:* (1) Ignoring parents' prohibitions, (2) staying out late without permission, (3) running away from home, *at school:* (4) cheating on an exam, (5) playing truant, and *during leisure time:* (6) smoking hashish, (7) using harder narcotics, (8) getting drunk, (9) provoking and afflicting other people, (10) committing vandalism, (11) stealing other peoples' money, (12) running away from payment, (13) shoplifting, (14) stealing goods, (15) forging signatures, and (16) loitering in town every evening.

In order to get a grasp of how common norm breaking among girls was at this age, the prevalence of violating the 16 types of norm breaches was first analyzed for the total group of girls. The results are presented in Table 6.1.

As can be seen in Table 6.1, it was rather common among the females at this age to have violated different types of formal and informal rules, but more uncommon that this behavior occurred frequently. The majority of girls had, at least once, cheated on an exam, stayed out late without their parents' permission, ignored parents' prohibitions, loitered in town during the evening, or stolen other people's money. However, fewer than 10% of the girls had

TABLE 6.1
The Prevalence of Different Types of Norm Breaking Among
14-Year-Old Girls

Norm-breaking behaviors	At least once	Frequently (4+)
Home		
Ignoring parents' prohibitions	58.0	5.5
Staying out late without permission	63.1	9.2
Running away from home*	8.3	0.2
School		
Cheating on an exam	65.6	6.9
Playing truant	47.6	11.7
Leisure time		
Smoking hashish	3.0	0.2
Using harder narcotics*	3.5	1.3
Getting drunk	40.7	13.3
Provoking and afflicting other people*	18.6	1.5
Committing vandalism*	22.3	2.6
Stealing other people's money*	50.8	8.7
Running away from payment	13.3	1.1
Stealing goods*	31.7	7.8
Forging signatures*	12.8	1.6
Loitering in town in evenings	53.5	8.5
Shoplifting	37.9	5.1

*Item from age 14.10 yrs. Other items from 14.5 yrs.

engaged in these respective norm-violative activities frequently. The two particular forms of norm breaking in which girls engaged most frequently were truancy and drunkenness.

An appreciation of the range of norm-breaking activities of the girls can be obtained from Table 6.2. Each type of norm-breaking activity was dichotomized, with the girls who had not been engaged scoring 0 and the girls who had been involved in such activities at least once scoring 1. For the girls who had complete data for each of the 16 types of norm-breaking activities, items were summed to a total "norm-breaking range" score varying from 0 to 16.

Almost all girls acknowledged that they had violated one or several of the 16 norm-breaking activities. Only 5.7% stated that they had never violated any of the norms specified. On the other hand, it was also rather uncommon for girls to have been involved in many of these types of norm-violating activities. Most common among the girls was a modest number of rule breaches. The mean number of types of norm-breaking activities was 5.16 for the total sample of girls. However, 17.1% of all girls had breached a majority of the norm types that were specified.

TABLE 6.2
The Range of Occurence of 16 Types of Norm-breaking Activities

Number of times	N	Ever violated %	cum%
0	20	5.7	5.7
1	31	8.8	14.5
2	37	10.5	25.1
3	30	8.5	33.6
4	44	12.5	46.2
5	37	10.5	56.7
6	37	10.5	67.2
7	33	9.4	76.6
8	22	6.3	82.9
9	22	6.3	89.2
10	11	3.1	92.3
11	13	3.7	96.0
12	10	2.8	98.9
13	1	0.3	99.1
14	2	0.6	99.7
15	1	0.3	100.0
16	—	—	—

In summary, norm violations among the sample of the girls at 14 years of age was not something rare. More than 9 out of 10 girls had violated at least one of the norms that were specified in Table 6.1, and the average girl had violated around 5.

Norm-Violative Activity

With these population-describing analyses as a background, let us now turn to the issue of the relation between pubertal timing and norm breaking. Table 6.3 presents the results of testing mean differences among the four menarcheal groups of girls with respect to the different types of norm violations. Table 6.4 shows the percentage of girls in each of the four menarcheal groups who reported frequent norm breaking (four times or more) for each norm-breaking behavior.

The results comparing the incidence of norm breaking among the four menarcheal groups in Table 6.3 show a clear-cut association between pubertal timing and norm breaking and indicate that variations in norm violations in the mid-adolescent years were highly related to a girl's physical maturation. The tests of differences of the scope of norm breaking were significant in 11 out of the 16 instances, often at a high level of significance. The direction of the differences between the girls was that more frequent norm violations were found among the early-matured girls than among the later-matured.

As can be seen in Table 6.4, there was also a considerably higher percentage

TABLE 6.3
A Comparison of Rate of Norm Breaking at 14.5 and 14.10 Years Among Girls in Four Menarcheal Groups

	F	p	post hoc
Home			
Ignoring parents' prohibitions	4.84	.01	1>2,3,4&2>4
Staying out late without permission	9.96	.001	1>2>3>4
Running away from home	7.00	.001	1>2,3,4
School			
Cheating on an exam	0.48	ns	
Playing truant	12.40	.001	1>2>3,4
Leisure time			
Smoking hashish	8.43	.001	1>3,4
Using harder narcotics*	3.25	.05	1>2,3,4
Getting drunk	11.43	.001	1>2>3>4
Provoking and afflicting other people*	0.63	ns	
Committing vandalism*	1.77	ns	
Stealing people's money*	0.91	ns	
Running away from payment*	1.75	ns	
Stealing goods*	6.54	.001	1>2,3,4
Forging signatures*	4.67	.001	1>2,3,4
Loitering in town every evening	2.69	.05	1>2,3,4
Shoplifting	3.61	.05	1>2,3,4&2>4

*Item from age 14.10 yrs. Other items from 14.5 yrs.

of the earliest-matured girls, compared to the later-matured, who reported frequent violations of norms at home, at school, and during their leisure time. Significant differences for frequent norm breaking among the four menarcheal groups emerged for 10 of the 16 types of norm-violative activity.

As can be observed in Tables 6.3 and 6.4, there was a direct relation between norm breaking and age of menarche. But it is also obvious that the relation for many of the included norm breaking types was not linear. The earliest-developing group of girls differed, often markedly, in both the scope and the frequency of norm breaches from the other, later-developing girls.

Finally, an investigation of how many types of norm-breaking behaviors that girls in the menarcheal groups on the average had been involved in frequently, showed strong differences among the four groups [$F(3,342) = 9.47, p < .001$].

To summarize, whether we look at the incidence of norm breaking per se,

TABLE 6.4
Percentage of Girls in Four Menarcheal Groups Reporting Frequent
Norm Breaking at 14.5 and 14.10 Years

	Age of menarche					
	−11 yrs	11–12 yrs	12–13 yrs	−13 yrs	F	p
Home						
Ignoring parents' prohibitions	16.7	7.1	2.8	3.6	15.09	.01
Staying out late without permission	27.1	12.2	5.6	4.5	25.27	.001
Running away from home	0.0	1.0	0.0	0.0	0.32	ns
School						
Cheating on an exam	17.0	5.1	5.1	7.3	8.92	ns
Playing truant	39.6	14.3	5.6	7.1	45.20	.001
Leisure time						
Smoking hashish	12.0	4.1	1.1	0.9	8.10	.001
Using harder narcotics*	4.1	0.9	1.1	0.9	3.22	ns
Getting drunk	35.4	20.0	7.9	6.3	33.87	.001
Provoking and afflicting other people*	2.1	2.9	1.1	0.8	1.95	ns
Committing vandalism*	8.3	1.9	1.7	2.5	7.10	ns
Stealing people's money*	14.9	7.5	8.7	7.4	2.71	ns
Running away from payment*	6.4	1.9	0.0	0.0	16.20	.001
Stealing goods*	20.4	9.3	5.9	4.1	14.43	.01
Forging signatures*	8.5	0.0	1.1	0.9	16.93	.001
Loitering in town every evening	18.8	9.3	8.5	3.6	10.03	.05
Shoplifting	14.6	5.2	4.5	1.8	11.68	.01

*Item from age 14.10 yrs. Other items from 14.5 yrs.

or at the range of norm-breaking activities, the results are conclusive: Early-matured girls, particularly the most early-developed, had been more involved in such activities than had later-developed girls in the mid-adolescent years. The tests of significance among the four menarcheal groups of girls often reached a high level of significance.

Alcohol Use

Due to the fact that alcohol drinking among teenagers particularly illuminates issues of social transition in adolescence, a closer analysis is made here of the relationship presented in Tables 6.1 and 6.2 between menarcheal age and drunkenness.

Alcohol drinking is associated with several, often contrary, cultural messages, with social maturation as well as with social maladaptation. In its normal course, adolescent drinking, if not prescribed by society, is a social custom connected with approaching a more mature social life pattern and is generally a component of the "natural" growing-up process. On the other hand, that alcohol use among adolescents is not harmless, but is connected with earlier, concurrent, and later social maladaptation, has much empirical support in the literature. Results have been presented indicating, for example, that the adult alcoholic starts drinking at an earlier age than do others and drinks more frequently during adolescence than do teenagers in general (Cahalan, Cicin, & Crossley, 1969; Helgasson & Asmundsson, 1975). Furthermore, the heavy drinker and the antisocial teenager tend to be the same person (Donovan & Jessor, 1978; Wechsler & Thum, 1973; Zucker & Devoe, 1975).

Figure 6.1 presents a more detailed picture of the relationship between drinking behavior and menarcheal age than was given in previous tables. It shows the prevalence of drunkenness among girls in the four menarcheal groups at age 14.5 years and the percentage figures for frequent drunkenness (four times or more).

The results presented in Fig. 6.1 show a clear-cut association between the age of biological maturation and having been drunk at least once at 14.5 years. The prevalence of drunkenness was considerably higher among early-matured girls than among later-matured. More than twice as many girls among the earliest-matured had had such experiences than among the girls in the latest-developed group of girls. The test of differences in drunkenness among the four menarcheal groups was highly significant $(F = 6.80, df = 3,431, p < .001)$.

The differences among the menarcheal groups of girls were even more

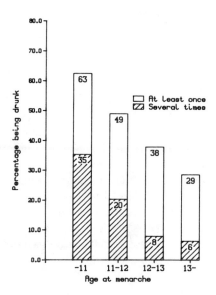

FIG. 6.1. Percentage of subjects in the four menarcheal groups who had been drunk and been drunk frequently (four times or more) at 14.5 years.

marked for frequent drunkenness. The percentage of girls with frequent drunkenness experiences was more than five times higher among the earliest-matured girls than among the latest-matured. Again the relationship was statistically significant at a high level ($F = 12.13$, $df = 3,431$, $p < .001$).

Delinquency

To analyze more severe forms of norm breaking, official data on criminal offenses up to age 18 were compared among the menarcheal groups of girls. Criminality data consisted of registered offending including:

1. committed offenses leading to any form of proceedings taken by the social authorities (the Child Welfare Committee) before the age of criminal responsibility (i.e., age 15),
2. offenses leading to public prosecution and conviction (after the age of criminal responsibility), and
3. offenses leading to arrests, detention, and temporary custody by the local police.

Information on criminal activities committed before the subjects' 15th birthday was supplied by the local police and by social authorities. All police authorities and all social authorities in the police districts in which the girls had lived up to 15 years of age were contacted, and data were collected from manual files. Information on offenses committed by the subjects after their 15th birthday was supplied by the National Police Board. These data cover all offenses leading to public prosecution. Data on arrests and temporary custody were supplied by the local police in all those police districts in which the subjects had lived up to age 18. A description of the procedure for data collection and information about criminal activity in different ages has been presented by Stattin, Magnusson, and Reichel (in press). Data on registered offenses were collected for all subjects.

Only 3 girls of the 466 with menarcheal data were registered for some offense up to age 15 years. Two of them belonged to the two early-developed groups of girls who had had their menarche before 12 years. Seven girls had official delinquency records up to age 18. Four of them belonged to the girls who had had their menarche before age 12. Of the four girls with several recorded offenses, three belonged to these early-matured groups of girls. The results tend to support, at least with respect to recurrent offenses, the data for self-reported norm breaking. However, due to the small number of girls who were registered for offenses, the results on registered delinquency must be interpreted with caution.

Complementary Analyses

In the light of considerable differences in norm-violative behavior among the menarcheal groups of girls at the age of 14.10 years, the question of whether these tendencies were already present in the form of problem behavior and conduct disturbances in earlier years becomes important. Data are reported for teacher-rated conduct disturbances and parents' reports of problems with their daughters from earlier grades.

Teacher Reports. Data on conduct disturbances at school were collected through teacher reports at the age of 9.9 and 12.9 years. The teachers, most of whom had served as the principal teachers of the pupils for 3 years (Grade 1 through Grade 3, and Grade 3 through Grade 6) and, consequently, knew the subjects well, rated the pupils' school behavior on 7-point scales with verbal descriptions of the end points. The girls in the class served as the reference group. The same teacher ratings were given at 9.9 as at 12.9 years (see Magnusson, Dunér, & Zetterblom, 1975, pp. 76–79, for more detailed descriptions of the data). The classroom behaviors and their respective definitions were:

- Aggressiveness
    Low: They work in harmony with the teacher and have positive contacts with classmates. Their relations to others easily becomes warm and affectionate.
    High: They are aggressive against teachers and classmates. They may, for example, be impertinent and impudent, actively obstructive or incite to rebellion. They like disturbing and quarreling with classmates.
    Most children are between these two extremes.

- Motor restlessness
    Low: They have no difficulty at all in satisfying even great demands for silence and quiet.
    High: They find it very difficult to sit still during lessons. They fidget uneasily in their seats or wish to move about in the classroom, even during lessons. They may also be talkative and noisy.
    Most children are between these two extremes.

- Concentration difficulties
    Low: They have a marked ability to concentrate on a task and persevere with it. They never allow themselves to be distracted and do not give up as long as a task suits their level of intelligence.
    High: They cannot concentrate on their work, but are occupied with irrelevant things, or sit daydreaming. For a few moments they

may work, but are soon lost in other thoughts again. They usually give up quickly, even when the work is suited to their level of intelligence.

Most children are between these extremes.

- Timidity

   Low: Characteristic of these students is that they are always open and frank.

   High: The behavior of such children is characterized by bashfulness and shyness. They seem to have poor self-esteem. They are inhibited and afraid to express themselves.

   Most children are neither particularly inhibited or markedly constantly open.

- Disharmony

   Low: They seem to be very harmonious and well-balanced, and are seldom involved in serious conflicts with their surroundings or themselves. They seem to be emotionally "at home" in school.

   High: They seem to be very disharmonous and unhappy. They are often in restrained or open conflict with their surroundings or themselves.

   Most children are somewhere between these extremes.

No significant differences among the four menarcheal groups appeared for any of the five classroom behaviors at the age of 9.9 years in Grade 3. For two behaviors, motor restlessness and concentration difficulties, significant differences were obtained at the age of 12.9 years in Grade 6 (motor restlessness: $F = 6.71$, $df = 3,407$, $p < .001$; concentration difficulties: $F = 3.22$, $df = 3,407$, $p < .05$). For motor restlessness there was a step-by step increase in disturbance the earlier-matured the girls were (mean values: -11 years: 4.07; 11–12 years: 3.85; 12–13 years: 3.30; 13–years: 3.10). A post hoc analysis showed the two earliest-developed groups of girls to be rated higher than girls in the latest menarcheal group. The earliest-developed girls were also rated by their teachers as higher in motor restlessness than were girls having their menarche between 12 and 13 years. A post hoc analysis for concentration difficulties revealed that the earliest-developed group of girls were rated as having significantly more concentration difficulties than did girls reaching menarche between 12 and 13 years. (The latest group of girls had a higher mean level than did girls reaching menarche between 12 and 13 years.)

It should further be stated that certain tendencies for higher motor disturbances among the early-developed girls were already present at the age of 9.9 years. At this age there was a significant tendency at the 10% level for the early developers to show more motor restlessness than did the other girls ($F = 2.55$, $df = 3,352$, $p < .10$). Overall, the results suggested some tendencies that conduct problems were more common among early-developed girls relative to other girls by late childhood and that the differences between the early- and the later-developed girls increased over time.

Parent Reports. At the ages of 9.9 and 12.9 years the parents were asked to mark, for several specified alternatives, which adjustment problems characterized their relationship with their daughters (trouble with pocket money, TV watching, homework, evening curfew time, peer relations, making the daughter obey).

At both these ages the parents of the early-matured girls reported more problems with their daughters than did the parents of the other girls. At the age of 10, the parents of the early maturers had more problems related to their daughters' pocket money ($p <. 10$), TV watching ($p < .05$), and evening curfew time ($p < .05$), whereas they had more problems with respect to making their daughters obey at age 13 ($p < .10$). Again these results suggest that a higher rate of problem behavior among the early-developed girls could already be found in late childhood. However, no increasing problem tendencies could be traced over time. Rather, the parents of the early-developed girls reported more problems at the age of 10 than at 13, compared with the parents of the later-developed girls.

School Adjustment and Pubertal Development

School adjustment among teenage girls has been the topic of much discussion in the developmental literature, and it has been connected with the social life of the girls at these years, with self-concept issues, and with the gender role. As to the latter aspects, Petersen and Wittig (1979) concluded the following:

> Motivation for high academic achievement is, in many instances, viewed as unfeminine and inappropriate for women. . . . Prior to adolescence, differential socialization pressures on girls focus mainly on personality variables such as dependency and activity. With adolescence, however, pressures related to sex role and future roles such as child-bearing become more important while educational success begins to recede in importance. Even today, with our societal consciousness raised by the women's movement, girls get a strong message about what level of achievement is appropriate and in what areas. (pp. 50, 56–57)

Despite many attempts at changing the attitudes toward various occupations, females tend to avoid technical occupations and prefer working in the social service domain or in occupations involving care and contacts. As a consequence, females often appear in occupational domains with low status and salaries. These gender-role-based occupational aspirations appear very early. When our group of girls were 13 years of age, 49% of them wished to work within the occupational domain contacts and welfare, whereas 4.3% aspired to have future work within the domains technical-natural science or industry-trade. By contrast, 35.6% of the boys wanted their future occupation to be within the two latter domains, and only 1.7% chose contacts and welfare (Dunér, 1972). In the ninth grade, an investigation of occupational aspirations was made utilizing the semantic differential technique with different occupations as the "targets." A closer analysis of the "like to be–not like to be" scale showed boys to rank order occupations mainly according to status, whereas girls ranked them according to occupational domain.

From the viewpoint of a more feminine orientation and a stronger establishment of the gender role among the early-developed girls, one might argue that the early developers would not be as achievement oriented, motivated for higher academic studies, or status oriented, as the later-developed girls. This hypothesis, based on self-conception and gender-role identity, suggests that differences in school achievement and school motivation might occur among the four groups of menarcheal girls in mid-adolescence, with the early-matured girls showing lower school interest, aspirations, and achievement.

From the perspective of other viewpoints lower school adjustment among the early-matured girls can also be expected. Jessor and Jessor (1977) postulated that an orientation away from conventional standards, such as religiosity and achievement, tended to be closely connected with an increase in problem behavior among teenagers. With the earlier-matured girls showing evidence of more marked norm violations in the mid-adolescent years, a lower school motivation on part of these girls would seem to be expected.

Third, the common findings of decreasing school motivation with increasing age, connected with the higher affiliation with chronologically older peers and employed peers among the earlier-matured girls, as well as more advanced contacts with boys, suggest that the early developers are surrounded by a social network that is more negative to school than are the girls who are later-developed. Thus, they can be expected to be less motivated for schoolwork and to show lower achievement than the later-matured girls.

The data presented earlier constitute another reason for expecting less school interest among the early-developed girls. In Table 6.3 and 6.4 we presented a comparison of incidence of norm violations among the four menarcheal groups that showed that the early-matured had committed more frequent norm breaking than had the later-matured girls. The two items that dealt with norm breaking at school, cheating and playing truant, showed the same tendency as the other items. In chapter 5, the relations to their teachers were found to be less satisfying for the earliest-matured group of girls compared with the other girls in the research group.

In an earlier study in the longitudinal project, Andersson, Dunér, & Magnusson (1981) found support for greater school maladjustment among the early-maturing girls. They compared a group of early-matured girls with a normative control group of the same chronological age on various aspects of school adjustment at the ages 13 and 15 years (e.g., feelings of dissatisfaction at school, educational aspirations, and burden of work in school as perceived by the girls themselves, as well as teacher ratings of concentration difficulties, school motivation, and aspiration). At both ages, early-matured girls showed more negative attitudes toward school and school work, and they were rated more unfavorably by their teachers. The differences in school adjustment between the two groups were stronger at the latter age.

Other studies have also found early-matured girls to be less motivated toward school work. Davies (1977) reported less favorable attitudes toward school among early-menstruating girls than among later-matured. The average age at the time of the testing in this study was 14.1 years, and the girls were matched for chronological age, verbal IQ, and socioeconomic status. In the outpatient clinic sample investigated by Frisk (1968), the main diagnosis for the early-matured girls was school-related problems, mainly school fatigue. More than one out of four girls among the extremely advanced girls in Frisk, Tenhunen, Widholm, and Hortling's (1966) investigation showed evidences of school failure in comparison with 6% of the extremely late-developed girls. From responses to the TAT material of extreme-early- and extreme-late-developed 17-year-old girls, used by Jones and Mussen (1958) the late developers were found to score higher on need for achievement.

Results in a similar direction were reported in the longitudinal study by Simmons, Blyth, and McKinney (1983; Simmons, Blyth, Van Cleave, & Bush, 1979). They found the main differences between the early- and late-matured girls to appear at the Grades 6 and 7, but not later (Grades 9 and 10). Early developers in this study had lower grade-point averages and lower scores on achievement tests in reading and mathematics, were less likely to perceive themselves as good students, and had lower aspirations for further education. In addition to performance variables, the early-developed girls showed more problem behavior at school and felt more often inclined to skip school.

An exception to this trend of findings of lower school adjustment among the early developers was reported by Crocket and Petersen (1987) in their longitudinal study covering Grades 6 through 8. At no grade did they find any relation between physical maturation and course grades. Nor did Duke et al. (1982; see also Duncan, Ritter, Dornbusch, Gross, & Carlsmith, 1985) obtain any systematic differences in school achievement between early-, average-, and late-matured girls among the more than 5,000 girls in the National Health Examination Survey. The same lack of correspondence between menarcheal age and academic achievement was reported by Viteles

(1929). Actually, the opposite pattern with early-matured girls displaying better school adjustment and achievement has been reported for British subject samples (Douglas, 1964; Douglas & Ross, 1964; Poppleton, 1968). For a selected sample of extreme-early and extreme-late developers, Abernethy (1925) reported accelerated school progress and higher achievement among the early-developed girls and more school adjustment problems concentrated among the extreme-late developers. A possible bias in some of these studies is that a selected, bright group of subjects has been investigated when achievement was compared between early and late developers in schools beyond the required education. An even more serious problem, which is rather general across studies, is the lack of control of intelligence when comparing early and late developers on achievement in school.

With the purpose of making a broad screening of the girls' school adjustment in the present research, it was decided to examine the following four aspects:

1. achievement as it could be measured by marks and results on standardized achievement tests,
2. school attendance as measured by records of school absence,
3. subjective experience of school satisfaction, and
4. parents' reports of their daughters' school adjustment.

School Achievement

The evaluation of achievement in Swedish schools is made by the teacher by means of ratings on a 5-point scale. Standardized achievement tests in core subjects are given to the pupils all over the country every third year. The mean score for a class on these tests is a guide for the teacher in his or her evaluation of the standard of the class relative to the national average.

Marks. Marks were available for comparison between early- and later-matured girls at four ages: 10, 13, 15, and 16 years. A grade-point average for the theoretical subjects that existed at respective ages was calculated. Different courses (variants of vocational preparatory and academic preparatory courses) were given in the upper grades (ages 15 and 16). Therefore, the relationship between pubertal development and marks is subject to certain problems at these ages. Nevertheless, with these limitations in mind, a tendency consistently emerged over ages, such that girls with their menarche between age 12 and 13 received higher marks than did both the two earliest- and the latest-developed groups of girls. Of the latter groups the early-developed had the lowest grade-point averages. The differences among the four menarcheal groups were significant at the 10% level of confidence at the ages of 10 and 15, and just above this confidence level at age 13. After controlling for intelligence (intelligence tests were given at age 10, 13, and 15, and these measures were introduced as covariates before estimating the relationships between menarcheal age and grade point average), these tendencies disappeared.

Standardized Achievement. Standardized achievement tests in Swedish and mathematics were given at the ages of 10 and 13. Achievements tests in Swedish were also given to the pupils at age 15. At this age two different forms of standardized achievement tests in mathematics, one vocational preparatory and one academic preparatory, were distributed. Testings of differences among the four menarcheal groups of girls with respect to results on these standardized achievement tests are presented in Table 6.5.

Girls who had their menarche between 12 and 13 years had higher scores than both of the two earliest- and the latest-developed group of girls on the standardized achievement tests in Swedish at the ages of 10 and 13, but the differences were not statistically significant ($p < .10$ at both ages). At the age of 15, pronounced differences ($p < .001$) among the four menarcheal groups were obtained. Post hoc tests showed that the girls who had had their menarche between 11 and 13 years scored higher than the group of girls who reached menarche at age 13 and later. After controlling for individual differences in

TABLE 6.5
A Comparison of Results from Standardized Achievement Tests at the Age of 10, 13, and 15 Years Among Girls in Four Menarcheal Groups

	Before control for intelligence			After control for intelligence		
	F	p	post hoc	F	p	post hoc
Age 10						
Standardized achievement: Swedish	2.02	ns		2.71	.05	
Math	0.95	ns		0.73	ns	
Age 13						
Standardized achievement: Swedish	2.32	ns		0.85	ns	
Math	3.17	.05		2.20	ns	
Age 15						
Standardized achievement: Swedish	6.08	.001	2,3>4	3.27	.05	
Math[a]	0.46	ns		0.66	ns	
Math[b]	6.39	.001	3,4>1	7.23	.001	3,4>1

[a] vocational preparatory
[b] academic preparatory

intelligence, significant differences among the four menarcheal groups were obtained at the ages of 10 and 15 ($p < .05$, respectively).

With respect to results on achievement tests in mathematics, no significant differences were obtained at the age 10. At the age of 13, significant differences were found, with the girls reaching menarche between 12 and 13 scoring higher than the other girls ($p < .05$). At age 15, no significant differences among the menarcheal groups were obtained for the girls who had chosen the vocational preparatory course in mathematics. Among the girls who chose the academic preparatory course, highly significant differences among the four menarcheal groups existed ($p < .001$). Post hoc testing showed that the early-matured girls had lower test scores than did girls in the menarcheal groups three and four. After controlling for intelligence, these significant differences still remained high ($p < .001$).

To summarize, a tendency could be found that the girls who reached their menarche between 12 and 13 years were somewhat better off with regard to school achievement than were either the earliest- or the latest-developed girls. There were already some indications in this direction in early school years. However, the differences were most evident at the age of 15.

School Absenteeism

Recorded hours of absence were available from the respective schools in the town at age 10, 13, and 15 years. No significant differences appeared among the four menarcheal groups of girls with respect to hours absent from school at the ages of 10 and 13. At age 15, however, significant differences were obtained ($p < .05$), with the earliest-matured girls being more often absent from school (an average of 69.1 hours) than were the rest of the girls (48.2 hours). So, as might have been expected from the self-reports of truancy reported earlier, objective data for hours absent from school showed the same results, with the early-developed girls being absent more frequently.

School Motivation

At all grades at which information about the girls' school achievement was collected, the girls' own experiences of their school situation were investigated along with the more objective data. This was performed by means of pupils' questionnaires and vocational questionnaires. At age 10 and 13, questions on the girls' liking of school, experiences of burden of work, and so on, were included in the pupils' questionnaires. Factor analyses extracted a "general school satisfaction" factor at respective ages. The eight items making up this factor at the age of 10 had a mean interitem correlation of .16, and the school adjustment scale that was subsequently formed from these items had an alpha reliability of .61. A similar scale was formed from the 18 items making up the school adjustment factor at age 13. The mean interitem correlation in this case was .27 and the alpha reliability was calculated to .89. In the upper grades information about the girls liking of school was collected through a vocational questionnaire at Grade 8 (14.10 years) and a pupils' questionnaire at Grade 9 (15.10 years). With few items at these respective grades to make up a specific school adjustment factor, data from both these grades were combined to form an analogous school adjustment measure as in the lower grades. The 11 items from Grade 8 and 9 had a mean interitem correlation of .22.

Tests of differences among the four menarcheal groups of girls with respect to their experienced school motivation at the ages of 10, 13, and 15–16, are presented in Table 6.6.

No significant differences among the four menarcheal groups with respect to different indicators of school satisfaction appeared at age 10. Some differences were found at age 13, the most conspicuous being that the early-matured girls felt more stomach trouble before going to school than did the later-developed girls ($p < .01$). The earliest-matured also thought their homework took too long time ($p < .05$) and they were less anxious about getting bad results than were the other girls ($p < .05$). However, for the total school adjustment factor, no statistically significant differences related to pubertal maturation were obtained.

Strong differences in school motivation among the four menarcheal groups of girls appeared in the upper grades. The most early-developed group of girls experienced their school situation considerably more negatively than did the other girls. Both the school adjustment items at age 15 and at age 16 showed the earliest-matured girls to have a harder time in school than did the other girls. They were less satisfied with their schoolwork, did not do their best, were more tired after school, did not work as hard before examinations, and did not think it was particularly important to be a good student. For a school adjustment factor formed by items from both grades, the difference among the four menarcheal groups of girls was significant at the .001 level, with the most early-developed girls expressing more negative school satisfaction than did the two latest-developed groups of girls.

Parent Reports

The lower school satisfaction in the upper grades, among the early-developed girls, in addition to their higher absenteeism and their tendency for lower school achievement, raises one's curiosity as to whether or not these school-related problems were reflected in the experiences of their daughters' school situation among the parents of the early-matured girls. Data on this issue were collected via parent questionnaires when the girls were around 10, 13, and 16 years of age. The parents were asked whether or not their daughters seemed to have a good time in school, and in the earlier ages they were also

TABLE 6.6
Comparison of School Adjustment at Age 10, 13, and 15–16 Years Among Girls in Four Menarcheal Groups

	F	p	post hoc
Age 10			
Have a good time at school?	1.26	ns	
Like school?	0.88	ns	
Headache before going to school?	1.11	ns	
Stomach trouble before going to school?	1.88	ns	
Feel sick before going to school?	0.23	ns	
Difficult to get schoolwork done as fast as the school teacher wants you to?	1.01	ns	
Burden of work at school	0.40	ns	
Tired after school?	0.65	ns	
Scale: School adjustment age 10	0.77	ns	
Age 13			
Have a good time at school?	1.93	ns	
Think about other things during lessons?	2.14	ns	
Like school?	0.41	ns	
School felt as a compulsion?	0.74	ns	
Find lessons interesting?	0.42	ns	
Fun to go back to lessons after breaks?	0.28	ns	
Too many hours spent at school?	0.79	ns	
Stomach trouble before going to school?	5.16	.07	1>2,3
Difficult to keep up with the working place?	0.98	ns	
Difficult to find the right answer when asked?	1.38	ns	
Feel sick before going to school?	1.79	ns	
Have to work hard at school?	0.49	ns	
Too many tests and written examinations?	0.30	ns	
Too much homework?	1.20	ns	
Homework takes too long a time?	2.72	.05	
Anxious about getting bad test results?	2.64	.05	
Tired after school?	0.52	ns	
Burden of work at school	1.36	ns	
Scale: School adjustment age 13	0.25	ns	
Age 15–16			
Have a good time at school?	2.74	.05	1>3
Satisfied with schoolwork?	3.17	.05	1>3,4
Burden of work at school*	1.85	ns	
Tired after school?*	3.06	.05	
Schoolwork meaningful for future?	1.84	ns	
Do your best at school?	4.17	.01	1>4
Work hard before examinations?*	5.02	.01	1>2,3,4
Important to get good marks?*	2.95	.05	1>4
Make an effort to achieve?*	1.32	ns	
Anxious about getting bad test results?*	0.33	ns	
Important to feel competent at school?	2.76	.05	1>4
Scale: School adjustment age 15–16	6.01	.001	1>3,4

*Item at age 16

asked whether or not they had observed any worries for going to school among their daughters.[1] Tests of differences in these respects showed that the parents of the girls in the four menarcheal groups did not differ with regard to their experiences of their daughters' school satisfaction at any age. The parents of the early-developed girls had about the same conception of their daughters' school situation as did the parents of the later-matured girls, and the parents of the early-developed girls had not observed that their daughters had headaches, stomach trouble, or worries before going to school any more than did the parents of the late developed girls.

That the early-developed girls, as indicated by several measures, showed less school interest, but that their parents held the same opinion of their school interest as the parents of the later-developed girls, might be due to lower expectations for academic success among the parents of early-maturing girls. One could, for example, argue that signs of school dissatisfaction among the early-developed girls and their greater interest in outdoor activities, caused their parents, gradually over time, to lower their aspirations for their daughters. Asked about their daughters' school interests, the parents of the early-developed girls thus perceived their daughters to get on equally well given their qualifications as the parents of the later-developed girls perceived their daughters' school interest from their qualifications.

In order to control for the possibility of lower parental aspirations for their daughters among the parents of the early developers, we used available data on parental school interest and aspirations for their daughters at the three age levels. Relevant questions to the parents were compared, for example, how often the parents helped their daughters with their homework, whether they visited school at parent–teacher meetings or at school-closing days, whether they wished their daughters to choose a theoretical or practical education, whether a theoretical or practical education suited their daughter, whether they had discussed with their daughters' future education with her, whether they and their daughters agreed on her future studies, and whether they wanted their daughters to have a long education. The results showed that the parents of the early-developed girls had about the same educational aspirations for their daughters and showed about the same interest in their school work as did the parents of the later-developed girls. No significant differences were obtained at any age or for any measure of parental interests and aspirations among the four menarcheal groups of girls.

From these results, it does not seem likely that the equality in the parents'

---

[1] The following items were included:
Age 10: "Does the pupil have a good time at school?", "Does the pupil get headaches before school?", "Does the pupil get stomach trouble before school?".
Age 13: "Does the pupil have a good time at school?", "Does the pupil get headaches before school?", "Does the pupil get stomach trouble before school?", "Is the pupil tired after school?", "Is the pupil worried about something at school?".
Age 16: "Does the pupil have a good time at school?"

opinions of their daughters school adjustment was a result of the parents of the early-developing girls having lower aspirations for them or showing less interest in their school work. There is, of course, also the possibility that the parents of the early-developed girls hesitated and were embarrassed to mention that their daughters did not seem to get along well at school. However, this case of underreporting of problems does not seem to be a likely explanation. For example, parents' reports of problems with their daughters at the age of 9.9 and 12.9 years reviewed in this chapter, showed that more of the parents of the early developers than of the late-maturing girls reported having problems with their daughters in late childhood and early adolescence.

An alternative explanation, that the early maturers tended to hold their school situation to themselves, is more likely. It should be observed that the *actual achievement* in school of the early developers in the present research group was not so much different from that of other girls at the ages of 10 and 13. So, as judged by objective data available to the parents of the early-developed girls for their achievement in school from earlier ages, their daughters' school achievement was similar to that of other girls. Furthermore, although the early-developed girls scored lower on standardized achievement tests at the age of 15 than did the later-matured girls, they did not receive lower marks. However, the early-developing girls differed in the upper grades primarily from other girls in their *attitudes*, by having lower school motivation and showing less interest in further education. It is of interest in this connection to note that Stone and Barker (1939) found that a higher proportion of the postmenarcheal 12- to 15-year-old girls than of the premenarcheal stated that they engaged in imaginative and secretive activities, liked to have secrets adults should not know about, had secret ambitions and desires, imagined things not told to others, and pretended to be older than their actual age.

Teacher Reports

No information based on teachers' reports of the girls' school behavior was available in the upper grades in the compulsory school. However, this information was collected for the sample at the average age of 9.9 and 12.9 years. When subjects were these ages, teachers were asked to rate the pupils' school motivation on a 7-point scale with verbal descriptions of the endpoints:

Low: They give an impression of feeling averse to learning and to the subject, and they seem to experience a general feeling of discomfort in the school, a feeling of being "fed up with" school. They are uninterested, and it is very difficult to get them to take part in ordinary schoolwork.

High: Their school motivation is strong and they feel at home in the school environment.

Most students feel neither "fed up with" school nor strongly motivated.

At the age of 9.9 in Grade 3, no significant differences in teacher-rated school motivation among the four menarcheal groups of girls was at hand ($F = 1.78$, $df = 3{,}352$, ns). Differences at the 10% level of significance was obtained at the age of 12.9 in Grade 6 ($F = 2.40$, $df = 4{,}406$, $p = .07$). The earliest-developed girls were rated as more lacking in school motivation than were the other girls. That some differences appeared at the later age, coupled with the earlier findings that similar tendencies over age were found for standardized achievement, self-reported school motivation, and school absenteeism, indicates a growing dissatisfaction with school with increasing age among the early-developing girls.

Emotional adjustment and pubertal development

The strains caused by living and adhering to multiple social networks with different norms, values, supports, and expectations, is assumed to be stressful for the early-matured girl. To use the term Stonequist (1937) coined, the early maturer will be a *marginal person* with one foot in the same-age peer culture and one foot in the peer culture of older adolescents. Not only does she have to confront the developmental problems that are characteristic of girls in her own chronological age strata, she also has to confront the types of problems more common among chronologically older peers. In addition, she has to adjust to expectations placed upon her, with same-age girls as the norm, as well as to expectations similar to those placed upon chronologically older females. Her engagement with members of the opposite sex, often males a couple of years older than herself, also involves considerable adjustment to more advanced leisure time activities and attitudes. This marginal position or double burden of adjusting to conventional and more nonconventional peer groups, who often can be assumed to reflect conflicting expectations thereby creating a cross-pressure condition, is a major reason for expecting more emotional problems among the early-matured girls than among the later-matured.

This general hypothesis has limited support in earlier research. Reviewing the literature on the relationship between pubertal maturation and personality and affective problems, conflicting findings and generally marginal influences of the maturational factor is what most often has been reported.

Stone and Barker (1934) correlated age at menarche for girls 16 to 24 years with their responses to neuroticism, self-sufficiency, introversion, and dominance, but they did not obtain any significant relationships. For another sample of pre- and postmenarcheal girls 12 to 15 years of age (Stone & Barker, 1937), nonsignificant correlations were again found. Crocket and Petersen

(1987) related measures of impulse control, emotional tone, and psychopathology to the pubertal status of girls in the sixth, seventh, and the eighth grade, but did not find any significant differences between the pre- and postmenarcheal girls. No differences between pre- and postmenarcheal 13-year-old females with regard to anxiety was found by Rierdan and Koff (1980b) utilizing the Draw-a-Person test. Neither did Stone and Barker (1939) find any differences between pre- and postmenarcheal 12- to 15-year-old girls for "general worries, fears, and anxieties" and "fears and worries about health, death, and morbid subjects."

Garwood and Allen (1979) found the postmenarcheal 13-year-old girls somewhat better off with regard to the scales General Maladjustment and Psychosis in the Tennessee Self-Concept Scale than were the premenarcheal girls of the same age. For a small sample of 26 post- and premenarcheal 12-year-old girls, Davidson and Gottlieb (1955) reported results indicating greater spontaneity in the postmenarcheal girls. Their study, utilizing Rorschach, comprised an investigation of the relationship between biological development and emotional maturity. Livson and Bronson (1961) reported a study on ego control for another small sample of twenty 11- to 13-year-old females. They found overcontrol of impulses to characterize girls with late menarche (and girls with early growth spurt). In contrast to the predominance of social maladaptation, identity problems and insecurity among the extremely early developers in Frisk, Tenhunen, Widholm, and Hortling's (1966) clinical sample (see also Frisk, 1968), the extreme-late-developers were particularly characterized by somatic and neurotic symptoms.

MacFarlane, Allen, and Honzik (1954) reported data from the Guidance Study. They found more behavioral problems related to personality and sleeping habits among the early-developed girls than among the late-developed. Some problems (emotional dependence, shyness, fears, somberness, and irritability) tended to peak 1 year before the girls reached menarche, indicating an increased tension before the first menstruation. For a subsample of 33 females in the Guidance Study, Peskin and Livson (1972) reported less impulse control around the time of pubertal onset among the early-developed females. Brooks-Gunn and Warren (1985b) reported a curvilinear relationship of pubertal timing to psychopathology, with the earliest-developed girls exhibiting the highest scores. In a study of 103 girls ages 10 to 14, Brooks-Gunn and Warren (1987) reported a significant interaction effect between negative life events and pubertal development on depressive reactions, such that among girls who had experienced negative life events, depressive reactions were more common among the late-developed, whereas it was more common among the early developers for the girls without negative life events. An interesting study by Susman et al. (1985) on a sample of 9- to 14-year-old girls, reported that high-for-age level of follicle stimulating hormone was connected with more emotional problems (higher psychopathology scores) and more sadness (emotional tone).

A study by Berzonsky (1981), concerning the correspondence between individual variations in pubertal development and the resolution of identity crises, is of interest for its bearing on transition problems generally and for its investigation of whether or not early-developing girls pass through more stressful adolescent transitions than do late-developing girls. Berzonsky tested the hypothesis that girls in late adolescence who were early-developed would have gone through more personal crises than adolescent females who were late maturers would have experienced. There were 114 subjects, 50 male and 64 female college freshmen, who were interviewed on their identity status using a method developed by Marcia. Four areas of investigation were covered: political, sexual, religious, and occupational. The subjects were also asked to state their age at sexual maturation. This measure of pubertal development was weak. It was left to the individuals to make an overall judgment. In agreement with the hypothesis, the early-developed girls reported on the average more evidence of identity crises than did the late maturers. It was concluded that at this time in late adolescence the early-developed girls were most characterized by having undergone or by being about to encounter personal crises, thus designating a more conflict-filled transition for these girls.

In summary, although early pubertal maturation in a few studies has been found to be connected with emotional problems, inconsistent findings seem to characterize the literature. Effects that have emerged are for the most part unreplicated, and, consequently, must be regarded as tentative. A particular problem is the limited subject samples, with their associated high risk for chance findings.

Self-Reported Emotional Problems

The occurrence of affectional problems and symptoms in the four menarcheal groups was investigated by means of a comparison of relevant item groups in the Adjustment Screening Test given at 14.10 years. Item groups were organized under the headings of "psychosomatic reactions," "anxiousness," "depression," and "hyperactive behavior." The comparison of the girls' responses for these types of emotional problems is presented in Table 6.7.

Two types of emotional problems were significantly connected with the pubertal timing in girls: psychosomatic reactions and depressive reactions. The early-developed girls scored higher than the later-developed for both these types of emotional reactions. The results showed a step-by-step increase of these types of problems the more matured the girls were. Among the items reflecting psychosomatic reactions, headache, nervous stomach, and sleeping difficulties were more common among the early-matured than among the later-matured girls. In terms of depressive reactions, two items that differed significantly among the four groups were those concerning suicidal thoughts and feelings of sadness (In addition, the item "feel it wonderful to live

TABLE 6.7
Differences in Emotional Adjustment at 14.10 Years Among Girls
in Four Menarcheal Groups

	F	p	post hoc
Psychosomatic reactions			
Headache	6.06	.001	1>2>3,4
Bad appetite	1.62	ns	
Nervous stomach	3.70	.05	1,2,4>3 & 2>4
Difficulties in falling asleep	5.00	.001	1,2>3,4
Disturbed sleep	6.94	.001	1,2>3,4
Anxiousness			
Afraid without knowing why	1.58	ns	
Nail-biting	0.60	ns	
Nightmares	1.46	ns	
Stammering	1.67	ns	
Thumb-sucking	0.05	ns	
Depressive reactions			
Feel sad and blue	4.18	.01	1>2>3,4
Have felt that I didn't want to live any more	7.97	.001	1>2>3,4
Feel it wonderful to live (rev.)	2.52	ns	
Feel dull and uncomfortable	0.67	ns	
Hyperactive behavior			
Restless	0.84	ns	
Concentration difficulties	1.62	ns	

[reversed]" differed at the .10 level of significance among the menarcheal groups.) The item "have felt that I didn't want to live any more" in particular was the one most differentiating among the menarcheal groups. In this connection one is reminded of the findings by Susman et al. (1985) that high-for-age level of follicle stimulating hormone was related to sadness (and, generally, to more emotional problems) and of those found by Money and Walker (1971), that depressed mood states tended to characterize girls with precocious puberty.

That the items under the heading of Anxiousness in the present research did not differentiate between the early- and later-matured girls is in agreement with previous studies on anxiety (Rierdan & Koff, 1980b; Stone & Barker, 1939). According to the girls' self-reports, the early-developed girls did not report higher restlessness and concentration difficulties than did the later-developed girls. No independent data for hyperactive behavior was available at the age of 15 years. However, it will be remembered that teachers rated the early-developed girls higher in restlessness and concentration difficulties at age 13 than they rated the later-developed girls.

Psychiatric Problems

In order to make a further investigation of problems related to mental health between early- and later-developed girls, a comparison among the menarcheal groups of girls was made with regard to their contacts with psychiatric clinics and counseling services in the town during the upper classes of the comprehensive school.

For the present research group, in the mid-adolescent years, there existed in the town three professional networks that girls might consult when having serious personal (and/or social and academic problems): (a) the Counseling Service of the Schools, (b) the Child and Youth Psychiatric Services, and (c) the local Social Welfare Authorities (the Child and Youth Welfare Committee). The girls went or were referred to the school psychologist or the outpatient clinics for various reasons; predominantly for adjustment problems at home, adjustment problems at school, school fatigue, identity problems, nervousness, worry, concentration difficulties, feelings of insecurity, depressive reactions, advanced and delayed puberty, somatic disturbances (such as headache, fatigue, indisposition and stomach trouble), gynecological problems, or unwanted pregnancies.

The Counseling Service of the Schools, as its name suggests, treats mainly school-related problems. The girls themselves might seek help for such problems. Equally often, teachers refer the girls to the clinic for frequent truancy or school fatigue. Of the girls in the research group who consulted the counseling service, or were referred, more than three out of four were registered for school-motivation problems. One out of eight was registered for reasons of unwanted pregnancy, and a few percent sought advice about nervous problems, peer problems, and home problems.

Nervous problems (26%), asocial behaviors (26%), applications for abortions (29%), and home problems (13%) were the main reasons given for seeking help at the Child and Youth Psychiatric Services in the town. Among the asocial problems, alcohol abuse and drug use dominated. More than half of all asocial problems concerned acute intoxication after use of tablets or alcohol. Loitering in town, running away from home, and sexual promiscuity accounted for one third of the referrals, and shoplifting for one fifth.

The local Child and Youth Welfare Committee has a service function to the parents, offering therapeutic help to the girl or to the total family. Antisocial behavior on the part of the girl accounted for 65% of the referrals to the local Child and Youth Welfare Committee, whereas 16% were due to home problems, 16% were formal applications for placing the girl temporarily or more permanently in the custody of another family, and 2% concerned school problems.

Information about contact, diagnosis, and treatment were collected from the three networks for the girls' 3 latest years in the comprehensive school: Grade 7 through 9. This time period corresponds to the interval from age 13.3 to 16.3 years for the average girl. The collection of register information was done by administrative staff at the respective networks. As an aid they had a readymade questionnaire asking:

- the point(s) in time for contact with the girl;
- the reason why she sought help or was referred;
- present and earlier problems in different areas;
- relevant information about the girl's social life situation, including home life, etc.;
- a description of her present symptoms;
- the seriousness of problems;
- her future prognosis; and
- the treatment given (see Crafoord, 1986, for more detailed information).

The coding of the data first involved a classification of the girl as to contact/ no contact with each of the three professional networks. Next was coded the particular problem(s) the girl sought help (or was referred), as well as the problem areas that arose during the treatment. To characterize the girl in broader terms, all available information, the anamnesis and present information, was summed to a total judgment on whether she showed predominantly extrinsic problems, predominantly intrinsic problems, or a combination of extrinsic and intrinsic problems. Subsumed under extrinsic problems were school-related problems, antisocial behaviors, and most of the peer-related problems. Subsumed under intrinsic problems were all such problems that, through their emotional expression or experience conditions, were connected with strong subjective suffering on the part of the person. In this category were included problems of nervousness, self-esteem, and such peer-related problems that signified that the girls could not or would not reach other peers.

The seriousness of a girl's problem panorama was coded according to the duration of the problems, the number of problem areas affected, and a subjective judgment of whether her problems were relatively mild or more severe. An overall judgment was formed with the categories (a) *formality* (formal application, for example, an application to do one's school duties in a vocational school, (b) *mild problems* (only solitary problems with short duration and a less severe nature, (c) *average problemacy* (one or two judgments of having enduring problems; many mild problems; or one "serious problem"), and (d) *severe problems* (enduring and serious problems in several problem areas).

Crafoord (1986) reported an interrater agreement of 90% for the classification of the total material. There was 97% interrater agreement for the problem categories, and 91% agreement on the seriousness of the girls' problems.

Problem Prevalence. Fifty-five, 11.8% of the girls in our research group were registered at one or several of the three professional networks. These 55 girls, distributed on the four menarcheal groups, are graphically depicted under the heading of "original population" in Fig. 6.2.

Figure 6.2 shows a curvilinear relationship between pubertal development and mental health. The results indicated more problems among both the early-matured and the latest-matured group of girls compared to the girls with their menarche on the average expected time for girls (12–13 years). Particularly the two earliest-matured groups of girls were overrepresented among the girls with contacts with the professional networks.

The 55 registered girls included cases for whom the contact with the professional networks was a formal application for change of school, for transference to a vocational training education, for change of custody, or for abortion. If these girls with formal applications were excluded, isolating the girls with documented personal and social disturbances, the reduced population consisted of 40 girls or 8.6% of all the girls in the research group. The distribution of these girls on the four menarcheal groups is presented under the heading "reduced population" in Fig. 6.2.

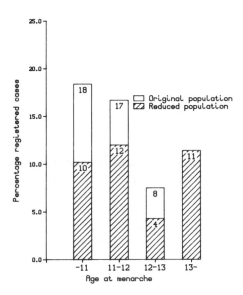

FIG. 6.2. Percentage of subjects in the four menarcheal groups who were registered at psychiatric clinics or counseling services in Grades 7 through 9.

Even with the formal application cases excluded, the early-matured and the latest-matured groups of girls contained higher proportions of girls who had undergone psychiatric or psychological treatment of some type than did the average-maturing group of girls. More than twice as many girls in the first, the second, and the fourth menarcheal group were registered for psychiatric problems as compared to the girls with their menarche between 12 and 13 years of age.

Seriousness of Problems. Excluding the girls with formal applications, more than half of all registered girls (52.5%) were classified into the average problematic category. More than one third (35%) showed occasional problems with short duration and of a milder kind. One out of eight girls (12.5%) were judged to have enduring and serious problems in different problem areas.

There were some differences among the four menarcheal groups with respect to the seriousness of their disturbances, but due to the small number of girls, a definite tendency in the data could not be established. All of the problems among the most early-matured girls were classified as being of average problemacy or as heavy problems; none of them had only mild problems. Among the girls with their menarche between 11 and 12 years, the situation was the reverse. Of the 13 girls, 9 were judged to have only mild problems. Among the two later-matured groups of girls, disturbances of average problemacy dominated. Twenty-five percent of the girls with their menarche between 12 and 13 years of age and 21.4% of those with their menarche at age 13 or later had mild problems. Observe that girls in the third menarcheal group were not only less frequently represented among the registered girls, but that, in addition, none of them had a problem history classified as severe.

Extrinsic and Intrinsic Forms of Problems. The number and the percentage of the girls in the four menarcheal groups who had primarily extrinsic problems, primarily intrinsic problems, or both extrinsic and intrinsic problems, is shown in Table 6.8. Due to the limited number of subjects registered and the low frequency of very-early-maturing girls, the first two menarcheal groups of girls were pooled into one group of early-developing girls (menarche before age 12).

The most common form of psychiatric disturbance was to have a combination of extrinsic and intrinsic kinds of problems. Forty percent of all network-registered girls were so classified. Slightly less common, involving 35% of the girls, was a problem history of exclusively extrinsic problems, and least common among the girls, involving 25% of the registered girls, was the exclusively intrinsic form of psychiatric disturbances.

The occurrence of intrinsic problems exclusively (nervous problems, contact difficulties with peers, problems with self-esteem) was slightly higher among later-developed girls than among the earlier-matured. For exclusively

TABLE 6.8
The Incidence of Extrinsic and Intrinsic Forms of Psychiatric Disturbances
Among Early and Late Developing Girls

Age at menarche	Intrinsic problems			Extrinsic problems			Extrinsic and intrinsic problems		
	N	% of psychiatric cases	incidence	N	% of psychiatric cases	incidence	N	% of psychiatric cases	incidence
−12 yrs	3	16.7	1.9	10	55.6	6.4	5	27.8	3.2
12–13 yrs	4	50.0	2.2	—	—	—	4	50.0	2.2
13– yrs	3	21.4	3.1	4	28.6	3.3	7	50.0	5.7
Total	10	25.0	2.1	14	35.0	3.0	16	40.0	3.4

extrinsic problems, the rate was highest for the earliest-matured group of girls, almost double the rate for girls who reached menarche at the age of 13 or later. It can also be observed that none of the more average-developing girls (menarche 12–13 years) were classified as having predominantly extrinsic forms of problems. Contrary to our belief that both extrinsic and intrinsic problem histories would characterize the early-developed girls, the occurrence of extrinsic *and* intrinsic problem types was highest for the latest-developed group of girls. A closer analysis on the individual level showed that the later-matured girls with both extrinsic and intrinsic problems were often girls with self-esteem problems manifested in difficulties in peer contact and in low school motivation, whereas the early-matured girls with both types of problems were more prone to acting-out and to showing nervousness or depressive symptoms.

To make a more refined analysis of the girls who were classified as displaying some form of extrinsic problems (either exclusively extrinsic or in combination with intrinsic problems), we decided to look more closely into how many had had a history of antisocial behaviors. Out of 30 girls, 8 had records involving antisocial behavior. Of the 8 girls, 5 belonged to the two earliest-matured groups of girls. The number of subjects is too limited to permit any definite conclusions. However, the results suggest that antisocial disturbances are more frequently occurring among the early-matured girls than among the later-matured.

Unwanted Pregnancy. The earlier in time steady dating occurs and the earlier in time sexual activity begins, the greater the likelihood of pregnancy (de Anda, 1983; Zelnik, Kantner, & Ford, 1981). Early developers attain reproductive maturity at an earlier time than do later-developed girls, and because the early developers in our investigation were found to have more advanced contacts with boys, and the prevalence of sexual intercourse was markedly higher among the early-matured than among the later matured girls, it was expected that a higher proportion of these girls than of other girls would experience an unplanned pregnancy in the teenage years and thus would be overrepresented among girls who applied for abortions.

How reasonable this expectation of an effect of menarcheal age on abortions may seem, it has little support in the literature. Earlier research on the role of pubertal timing for pregnancy in adolescent girls has produced equivocal results, supporting early dating as a factor influencing the occurrence of pregnancy in girls, but lending no substantial role to when the first menstruation occurs (Gottschalk, Titchener, Piker & Stewart, 1964; Presser, 1978).

In Sweden, practically all abortions are legal. In 1974 a law on free abortions was passed declaring that the female herself has the right to decide whether she wants to give birth to the child or not. Abortions are granted upon request before the 19th week of pregnancy. At the time the Abortion Bill was passed, the number of abortions in Sweden was 20 per 1,000 fertile females. It has not changed much up to the present time. It is estimated that one third of all females during their life time will have had an abortion (Trost, 1982).

Before 1974, legal abortions were only performed if special circumstances (medical, sociomedical, or humanitarian reasons) were present. One such humanitarian reason was becoming pregnant at a young age. For an abortion to be performed, either permission from the National Social Welfare Board was required, or the girls had to apply for a so-called "two-doctors" certificate. In both cases, a medical and psychiatric examination was obligatory.

Of the 466 girls in our research group, 28 subjects, or 6%, applied for permission to have an abortion performed up to and including Grade 9 (16.3 years), that is, up to 1971. Nine of the girls, or 1.9% of all girls, applied several times.

As can be seen in Fig. 6.3, pubertal timing had a definite relationship with applications for abortions. The occurrence of abortions increased monotonically the earlier physically matured the girls were. More than every seventh girl, 14.3% of the earliest-matured had once or at several occasions applied for an abortion. This figure was twice as high as for the girls who had their first menstruation between 11 and 12 years (7.4%), and about six times as high as the incidence of abortions among the latest-matured (2.4%).

Of the nine girls who applied at more than one occasions, six of them were early-matured girls, who had had their menarche before age 12, and none belonged to the latest-matured group of girls. Altogether, these data show a strong association between pubertal development and unwanted pregnancy.

SUMMARY AND CONCLUSIONS

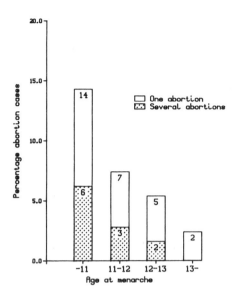

FIG. 6.3. Percentage of subjects in the four menarcheal groups who had undergone one or several abortions up through Grade 9 (16.3 years).

SUMMARY AND CONCLUSIONS

This chapter examined certain hypotheses regarding girls' social and emotional adjustment in mid-adolescence, from the perspective of biological development. From earlier-documented findings that the early-developed girls in mid-adolescence, to a higher extent than later-developed girls associated with friends whose behavior, for the girls' age, was socially advanced, it was hypothesized that early developers would be more prone to engage in norm-breaking behavior during the mid-adolescent years than would other girls. Second, living in multiple peer networks, with different norms, values, expectations, interests, and social behaviors, was expected to cause distress and emotional problems on the part of the early-developed girls relative to the late-developed.

These two main hypotheses were investigated utilizing a variety of data: self-reported norm violations, registered juvenile delinquency, self-reported emotional problems, teacher ratings of conduct disturbances, official data on contacts with professional psychiatric and social institutions, marks, standardized achievement tests, recorded absence from school, self-reported and teacher-rated school motivation, and parents' reports of their daughters' school adjustment.

The findings, supporting the assumption that early-developed girls, to a greater extent than later-developed, were found to break conventional rules in mid-adolescence, were:

1. Data on self-reported norm breaking indicated considerably higher norm violations among the early-matured than among the later-matured girls. This was true for different types of breaches of norms at home, at school, and during leisure time. The number of girls having been engaged in these activities was higher among the early-matured than among the later-matured girls. The frequency of norm breaking was higher, and the range of norm breaking across different types of norm-violating situations was also higher among the early-matured. The differences among the four menarcheal groups were more pronounced when making the analysis for frequent norm breaking rather than for norm breaking per se.

2. There were some indications that the early-developed girls were overrepresented among the girls with official delinquency records up to age 18.

3. The percentage of the early-matured girls who were classified as having exclusively extrinsic forms of problems (school maladjustment, antisocial behaviors, and some forms of peer-related problems) in their contacts with social and psychiatric institutions was higher among the early- than among the later-matured girls.

4. A considerably higher frequency of unwanted pregnancies in adolescence was found among the early-matured girls compared to the later-matured.

5. Signs of more conduct disturbances (motor restlessness and concentration difficulties) among the earlier-developed girls could be found in early adolescence, according to teacher reports. The parents of the early-developed girls also reported somewhat more problems with their daughters in these ages than did the parents of the late-developed girls.

In addition to higher incidence of norm breaking among the early-developed girls, another main finding was a more negative educational situation for these girls:

1. After controlling for intelligence differences, the early-developed females still scored lower on the standardized achievement tests in Swedish at the ages of 10 and 15 and lower in mathematics (academic preparatory course) at age 15. Lower grade-point averages among the early developers, albeit at no age significantly lower, were found at all ages examined.

2. Significant differences with regard to register data on school absenteeism appeared at age 15, with the early developers being more frequently absent from school than were the later-developed girls.

3. Data on self-reported school interest showed no relationship to pubertal timing of girls at age 10. Significant differences for a few items, with early developers having a more negative view of school, were obtained at age 13. Profound differences among the four menarcheal groups were found at age 15 and at age 16. The early developers showed least school interest.

4. A considerably higher proportion of the early-developed girls than of the other, later-matured girls reported having cheated and played truant from school at the age of 14.

It should be noted that few differences with regard to self-reported school adjustment among the menarcheal groups appeared at the early grades but that these differences were more pronounced in the upper grades. Teachers' reports of the girls' school motivation at the ages of 10 and 13 showed the same tendency over time for increasing school difficulties among the early-developed girls.

The differences among the menarcheal groups of girls with respect to academic achievement were found utilizing different types of data: objective data on school achievement and absenteeism, and self-reported school interest. However, one exception was parents' reports. The parents of the early-developed girls thought that their daughters had about equally good school adjustment as that reported by the parents of the late-developed girls.

Adolescence is often considered a period in which the affective life of individuals is more labile and more expressive than in other phases. Depressive reactions increase from early adolescence, particularly for females, to encompass perhaps as many as one out of five females in adolescence (Rutter, Graham, Chadwick & Yule, 1976). Such feelings of unhappiness and misery on the part of teenage girls, to a considerable extent, tends to go unnoticed by parents.

In agreement with the hypothesis that early-developed girls had a more emotionally provoking life situation than did other girls in mid-adolescence, the early-developers were found to report more psychosomatic reactions and more depressive symptoms than did the other, later-developed girls.

A problematic adolescent life situation typically involves depressive and somatic reactions, parental conflicts, school problems, social maladaptation, and running away from home (Friedman & Sarles, 1980). Mood swings are more common among teenagers who spend their leisure time largely in the context of peers and less common among adolescents who spend more of their time with their parents. In a Swedish study dealing with peer influences on social behavior, it was reported that the girls who were most sexually active much more often reported weariness with life, compared to girls who were less sexually active, in contrast to boys for whom no such relationship was found (Marklund, 1985). Many of these typical features of a problematic life situation for teenage girls precisely characterized the present early-developed girls. They were less interested in school, were more involved in norm-breaking activities, and they perceived themselves to have more conflicting relations with their parents and with their teachers. One out of seven of the earliest-developed girls had once or at several occasions undergone an abortion by the age of 16.

The early-matured girls were clearly overrepresented among the girls with problem behavior. They also had more documented psychiatric disturbances of an extrinsic nature, such as antisocial behavior, compared with the later-developed girls. Antisocial problems among girls have previously been reported to be related to signs of depression (Duke, 1978, Duke & Duke, 1978), and a connection between delinquency, on the one hand, and depression and suicidal behavior, on the other, has been obtained rather consistently across studies. Gibbs (1980) reported that delinquent girls referred to parental and peer conflicts, problems connected with opposite-sex relations, and achievement difficulties, as reasons for their depressive states. For clinical samples, antisocial behavior in the teenage years has been found, for females, to be more connected with psychosomatic and other internalizing problems at adult age than with antisocial behavior at adult age, whereas teenage antisociality in males is found to predict adult antisociality primarily (Robins, 1986). The higher norm violations among early-developed girls, relative to later-developed girls, and their reports of more depressive and psychosomatic reactions in the present research agrees with this earlier pattern of findings on social and emotional adaptation for females generally and strengthens the argument of a more problematic adolescent transition for the early developers.

To the extent that problem behavior is part of a general pattern of social maladaptation, one could eventually expect that early-developing girls would express the same symptom patterns characteristic of socially deviant boys. As has been reviewed by Loeber (1982) and Loeber and Dishion (1983), the later delinquent is often a boy who, at an early age, shows signs of conduct problems, social problems, and troublesomeness. That aggressiveness is a major component in this socially maladapted behavior pattern has repeatedly been found in longitudinal studies on the early prognostic factors for crime (Feldhusen, Thurston, & Benning, 1973; Havighurst, Bowman, Liddle, Matthews, & Pierce, 1962; Kirkegaard-Sorensen & Mednick, 1977; McCord, 1983; Mitchell & Rosa, 1981; Mulligan, Douglas, Hammond, & Tizard, 1963; Stattin & Magnusson, 1984; West & Farrington, 1973). In the light of findings obtained in these studies it should be recalled that no differences with respect to aggressiveness were present among the four menarcheal groups prior to mid-adolescence, according to teachers.

To summarize, the present chapter has provided ample empirical evidence of more social and emotional problems among early-developed girls in mid-adolescence than among later-matured girls. The results are supported by different types of data. It must be admitted that the causality is largely unknown, both with regard to the association between antisocial behavior and depressive reactions among adolescent females generally, and with regard to the role of pubertal development in this context. Antisocial activities might be interpreted as a defense against intrapsychic problems. Psychosomatic reactions and depression might emerge as a consequence of social adjustment problems in everyday life. Both social and emotional problems might stem from some common third source, such as earlier conflicts with parents. They might be related to girls' current peer contacts or there might be feedback loops reinforcing the system of interconnecting factors, and so on. Given the present state of knowledge, we can only speculate about the causal relations. We return to this issue of the impact of pubertal timing on the association between social and emotional adjustment when hypotheses regarding the mediating factors of social and emotional behavior, between early- and later-developed girls, is explicitly tested.

7

The Short-Term Consequences

INTRODUCTION

The rapid psychological and social changes occurring in the adolescent years make it hazardous to develop a stable framework for the study of the impact of physiological maturity. What might be true at one point in time in this interval might be overridden at another. Partly as a way of validating the main empirical findings presented earlier, and partly to determine the course of development in the mid-adolescent years, it was decided to pick out domains where pubertal timing of girls had been found to play a central role, and to examine the relationship between menarcheal age and variations in these respective domains at a later point in time.

Obviously, all specific variables and scales that have been dealt with in the earlier chapters cannot be covered. The follow-up data constitute a small selection of the core items in certain domains that were available to use for a follow-up. The specific domains covered are *self-conception* (perceived maturity, perception of oneself as different from others), *social behaviors* (sexual intercourse, norm violations), *psychosomatic reactions*, and *interpersonal relations* (parent and peer relations).

A great deal of the data reported in previous chapters were collected at the age of 14 years. The norm inventory was administered at 14.5 years and the Adjustment Screening Test was given at 14.10 years. The follow-up data that are presented here refer to age 15.10 (i.e., 17 and 12 months, respectively, after these two self-report instruments were given). The data bearing on 15.10 years consist of a pupils' questionnaire administered at school. The research population at 15.10 years differed just marginally from the population of girls at 14.5 and 14.10 years. For most of the items more than 90% of the girls who were tested at 14 years were also tested at 15 years. This maximizes the conditions for an effective follow-up.

SELF-CONCEPT

Perceived Maturity

One of our basic claims for the understanding of how psychical maturation operates in development assumes an intimate link between actual pubertal timing in girls and their perceived maturity relative to peers. On the question: "Do you feel more or less mature than your classmates?" at 14.5 years, considerably more of the early-matured girls than of the late-matured thought of themselves as more mature, and the relationship between age at menarche and self-experienced maturation reached statistical significance at a high level (cf. Table 4.1).

Seventeen months after this item was given, the same question was re-administered. For the total research group the coefficient of correlation between self-perceived maturity at 14.5 and 15.10 years was .41 ($n = 449$, $p<.001$). The percentage of girls in the four menarcheal groups who considered themselves as more mature and less mature than their classmates at 14.5 and 15.10 years is depicted in Fig. 7.1.

The relationship between menarcheal age and perceived maturity was significant at a high level at the age of 15.10 years [χ^2 (6, $N = 433$) = 27.91, $p<.001$)]. The results were similar to age 14.5 years. A larger percentage of

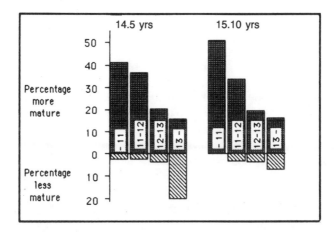

FIG. 7.1. The percentage of girls in four menarcheal groups who considered themselves more, or less, mature than classmates at 14.5 and 15.10 years.

the early-developed girls perceived themselves as more mature compared with late-maturing girls, and fewer of them perceived themselves as less mature.

The results from age 15.10 differed in two important respects from the same data at 14.5 years. As might be expected, a lower proportion of the late-developed girls stated themselves as less mature at 15.10 years compared with 14.5 years (6.7% vs. 19.6%). This illustrates a catch-up effect among the late developers over time. More unexpected was that a higher proportion of the earliest-developed girls rated themselves as more mature at 15.10 years (51.3%) than at age 14.5 years (41.7%). From an expected narrowing of the gap over time, the reversed situation would be expected.

The data presented in Fig. 7.1 clearly indicate that girls' conceptions of their maturity in mid-adolescence were connected with their physical developmental timing, and that differences among the menarcheal groups of girls as to their experience of feeling more or less mature compared with classmates were evident both at 14.5 years and 17 months thereafter. The apparent differences in perceived maturity between early- and late-maturing girls at the age of 15.10 years are quite noteworthy. At this time only a handful of the girls in the research group had not yet had their first menstruation, and most had reached adult physical status.

Perceived Difference From Peers

At 14.5 years the girls were asked whether they felt different from their same-age peers. Almost three times as many girls of the earliest-matured as of the latest-developed stated that they felt that way (Table 4.2). The same general tendency, with early-developed girls feeling more different from their peers was also evident at 15.10 years, [χ^2 (3, N = 430)= 14.87, $p<.001$]. The percentage figures were about the same of the menarcheal groups at both points in time, except that a lower proportion of the earliest-developed girls reported that they felt differently at the latter occasion than at the former (−11 years: 40.0%, 11–12 years: 23.5%, 12–13 years, 14.2%, 13 years: 17.6%).

For the research group as a whole, there was a phi coefficient of .33 ($n=$, 443, $p<.001$; gamma = .70) between the issue of perceiving oneself different from peers at 14.5 and 15.10 years.

On a follow-up item to the question of feeling different at 14.5 years, a number of different reasons were specified: "more religious," "other leisure time activities," "more romantic," "another political view," "another way to look at life," and "something else." The girls were asked to state which one of these alternatives primary explained their feeling at being different from others. As was reported in Table 4.2 in chapter 4, a larger proportion of the early-matured girls than of the other girls marked the option "more romantic" at 14.5 years. The statements reported by the other girls were more varied.

At 15.10 years, the same alternatives were given, but now the girls were asked to indicate for each of the alternatives whether it applied to them or not. A comparison among the four menarcheal groups for the alternatives is presented in Table 7.1.

Of the five alternatives, one differed significantly among the four menarcheal groups of girls. In agreement with the results from age 14.5 years, the early-matured girls stated more often than the later developed that they felt more romantic. There was also a tendency ($p<.10$) that more of the early-matured girls, particularly the earliest-matured, experienced themselves to have another way to look at life. The results overall tend to imply a stronger identification as a mature female among the early-developed compared with the later-developed girls at this later age.

SOCIAL BEHAVIOR

In this section, trajectories of certain social behaviors are examined over the mid-adolescent years. One of the issues to be addressed is the question of "catch-up" of the late-matured girls. Was the higher prevalence of advanced forms of social behaviors among the early maturers at the age of 14.5 years a temporary phenomenon to be followed by a greater net increase over time of such behaviors by the somewhat later-developed girls, and so on (the case of a general maturational impact), or did the differences that were observed early in mid-adolescence tend to prevail (the case of a differential maturational impact)? The former case would indicate the existence of social behaviors in adolescence being timed by processes of physical maturity. Entry into more advanced social behaviors are accelerated for the early-developers, but with the passage of time such behaviors will be matched by the later-developed

TABLE 7.1
Reasons Given for Feeling Different from Peers Among Girls in Four Menarcheal Groups at 15.10 Years

	Perceived difference from peers				
Age at menarche	More religious	Other leisure-time activities	More romantic	Another political view	Another way to look at life
−11 yrs	25.0	41.2	47.4	10.5	47.4
11–12 yrs	21.9	55.9	51.7	22.6	33.3
12–13 yrs	14.3	37.0	18.2	11.1	17.1
13– yrs	22.9	30.6	17.1	11.4	22.9
$\chi^2(3, N = 345)$	1.36	5.05	14.95	2.61	6.93
p	ns	ns	.01	ns	ns

girls. The latter case would indicate that the early-matured girls indeed had a unique progression of social behaviors in adolescence.

The specific behaviors to be investigated are sexual intercourse, reflecting the advancement in contacts with boys, running away from home, smoking hashish, and drunkenness, the latter behaviors reflecting norm-violating activities in this period of life.

Sexual Intercourse

The coefficient of correlation between sexual intercourse at 14.10 and 15.10 years for the total research group was a high as .57 ($n=424$), significant at the .001 level. Figure 7.2a graphically depicts prevalence data on sexual intercourse for girls in the four menarcheal groups as they relate to age 14.10 years and 15.10 years. The same type of information for frequent sexual intercourse at the two occasions is shown in Fig. 7.2b.

At the age of 14.10 years, 2 out of 10 girls in the research group had had sexual intercourse with boys. One year later the prevalence rate was almost the double: 4 out of 10.

One particular interesting issue concerns the comparison of these prevalence rates of sexual intercourse with samples from other countries. Sweden is generally recognized as a liberal society with regard to sexual issues. Education about sexuality, about living together, about contraceptive use, and so on, is obligatory in the elementary school. The prevalence rate of intercourse among the present group of girls, born in 1955, is also higher than has been obtained for comparable American samples of girls. From Rice's (1978) summary of three large surveys conducted during the same years as the testings of our subjects (Sorensen, 1973; Vener & Stewart, 1974; Zelnik & Kantner, 1972) it can be calculated that 13%–14% of 15-year-old American girls in the beginning of the 1970s had had sexual relations with boys and that 21%–23% of the 16-year-old girls had had sexual experiences with boys (Jessor & Jessor, 1975, reported a prevalence rate for sexual intercourse of 26% among 16-year-old girls). These figures should be compared with the prevalence rate of 22% among our sample of females at the age of 14.10 years and 39.5% at 15.10 years. Even if a comparison should be made with older American females, the sexual activity among the Swedish girls at 15.10 years would be comparatively high. For example, Zelnik, Kantner, and Ford (1981) reported prevalence figures of 26.3% among 15- to 19-year-old White American females surveyed in 1971 and 37.2% of same-age females surveyed in 1976. Also, comparing the sexual activity of our Swedish sample of girls, surveyed in 1970 and 1971, with those of Vener and Stewart's (1974) 1973 sample of American girls (24% sexually active at the age of 15 and 31% by the age of 16), it appears that the Swedish girls had a more advanced sexual behavior than that of the American females surveyed at a later point in time.

(a)

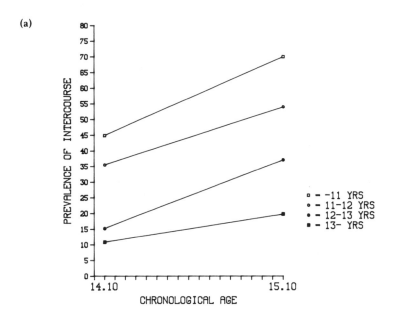

(b)

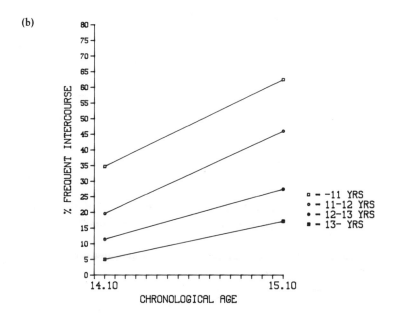

FIG. 7.2. (a) Prevalence of sexual intercourse among girls in four menarcheal groups at 14.10 and 15.10 years. (b) The percentage of girls in four menarcheal groups who reported frequent sexual intercourse at 14.10 and 15.10 years.

As shown in Fig. 7.2a, there were marked differences among the menarcheal groups of girls with respect to sexual experiences, with the early-matured girls having been engaged in such behaviors more frequently than were their later-developed counterparts both at 14.10 and 15.10 years. Among the earliest-matured group of girls an increase from 44.9% to 70% can be observed in the prevalence of sexual intercourse over the 1-year period. The same prevalence figures for the latest group of girls was 10.8% and 19.8%. Thus, the differences for sexual intercourse, related to pubertal development, which were present at 14.10 years were even *greater* 1 year thereafter. The increased differentiation over time was unexpected. It was thought that the age for the maximal differences between the early- and the late-matured girls would occur around 14 years and that thereafter a greater net increase would occur among the later-matured girls, resulting in equalization in late adolescence. Because we have no data on sexual behaviors at later ages in adolescence, we cannot determine the age for the maximal differences, nor can we say anything more definite about if and when the later-matured girls catch up with the earlier-matured. What is clear is that the differences between the menarcheal groups of girls widened up to the age of 16.

As to frequent engagement in sexual behavior (Fig. 7.2b), 14.1% of the girls in the research group reported having had sexual intercourse several times at the age of 14.10 years. At 15.10 years, more than twice as many, 32.4%, reported frequent intercourse. The same tendency with the early-matured girls being more advanced in their contacts with boys could also be observed for frequent sexual intercourse at both test occasions. The same was true for the observation that the increase over time for frequent sexual experiences was greater for the early-matured girls than for the later-matured.

Runaway Behavior

Running away from home is an adolescent phenomenon that has been found to occur as often or even more often among girls than among boys (Kratcoski & Kratcoski, 1975). Adolescents who run away have been characterized as having problems at home and at school, preceding the leaving of home. The other reasons for running away, apart from family conflicts and school-related problems, tend to be quite individualistic. However, girls who run away have been reported to do so because of punitive and controlling parents more often than do boys (Adams & Munro, 1979).

Running away from home was a type of behavior that did not occur often between 14.10 and 15.10 years. At the former occasion 7.9% of all girls reported having run away at least once, and a slightly higher percentage, 8.3% declared that they had done so up to 15.10 years. The phi coefficient between 14.10 and 15.10 years for running away from home was .52 for the total research group ($n = 423$, $p<.001$). At both test occasions a greater percentage of the earliest-matured girls reported that they had run away from home than among the later-developed girls. At the age of 14.10 years, 27.5% of the girls who had had their menarche before 11 years had run away from home once or at several occasions, in comparison with 5.9% of the girls who had had their menarche between 11 and 12 years, 8.2% of the girls who had had their menarche between 12 and 13 years, and 4.1% of those had who started to menstruate at the age of 13 or later. The differences for frequency of running away from home among the menarcheal groups were highly significant ($F = 7.00$, $df = 3,440$, $p<.001$; post hoc, $1>2,3,4$). The corresponding figures for age 15.10 were 23.9%, 8.6%, 6.7%, and 2.7%. Again there were strongly significant differences among the menarcheal groups of girls ($F=8.45$, $df = 3,430$, $p<.001$; post hoc, $1>2,3,4$). From these data we may conclude that runaway behavior was associated with menarcheal age in girls both at the age of 14.10 years and 1 year later.

Drug Use

Figure 7.3a graphically depicts the changes in prevalence of using hashish for the girls in the four menarcheal groups across three points in time: 14.5, 14.10, and 15.10 years. The same type of data for frequent use of hashish is shown in Fig. 7.3b.

Smoking hashish was rather infrequent at the age of 14.5 years but increased in frequency up to 15.10 years. At 14.5 years, 3% of the girls in the research group had smoked hashish. The prevalence rate increased to 5% at 14.10 years and to 12.2% at the age of 15.10 years. The stability over time, expressed as phi coefficients for those who had used hashish versus those who had not was .41 ($n = 437$, $p<.001$) between age 14.5 and 14.10, and .49 ($n = 435$, $p<.001$) between age 14.10 and 15.10 years.

At all three age points the prevalence rates were higher among the early-matured girls than among the later-matured. As indicated by chi-square tests, the menarcheal groups differed at the .001 level of statistical significance at each of the three age points, and the differences that occurred initially at 14.5 years among the four groups of girls were even more pronounced at the two latter occasions. Thus, for this type of behavior also, there was no evidence of a catch-up among the later-developed girls.

The girls who had used hashish more frequently (four time or more) were 0.2% at 14.5 years, 0.4% at age 14.10, and 4.6% of the girls in the research group at age 15.10 years. At all ages the frequent users were overrepresented among the early maturers, particularly the earliest-matured girls. Of the 20 girls who reported frequent use at 15.10 years, 12 belonged to the two most early-developed groups of girls.

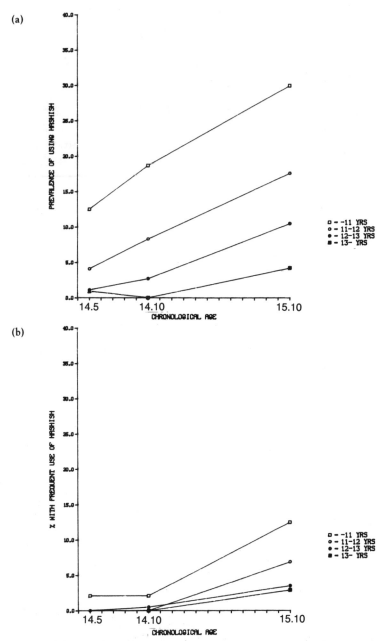

FIG. 7.3. (a) Prevalence of smoking hashish at 14.5, 14.10, and 15.10 years for girls in four menarcheal groups. (b) Percentage of girls in four menarcheal groups who reported having smoked hashish at several instances at 14.5, 14.10 and 15.10 years.

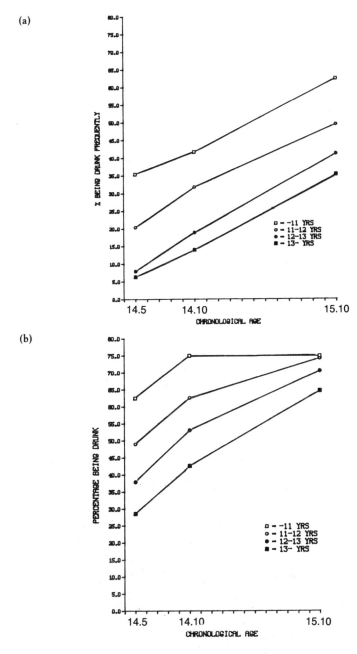

FIG. 7.4. (a) Prevalence of drunkenness at 14.5, 14.10, and 15.10 years for girls in four menarcheal groups. (b) Percentage of girls in four menarcheal groups who reported having been drunk at several instances at 14.5, 14.10, and 15.10 years.

Alcohol Drinking

Data on adolescent drinking, in the present case the frequency of drunkenness, were obtained at the average age of 14.5 years, at 14.10 years, and at 15.10 years. An overview of the data on the prevalence of drunkenness, and of frequent drunkenness (four times or more), at the respective ages for girls in the four menarcheal groups is presented in Fig. 7.4a and 7.4b.

The correlation coefficient between drunkenness frequency at 14.5 and a 14.10 years was as high as .77 ($n = 436$, $p<.001$) for the total sample. The correlation between drunkenness frequency at 14.10 and 15.10 years was .73 ($n = 429$, $p<.001$). The prevalence rate for the research group as to drunkenness rose from 40.7% at 14.5 years, to 54.9% at 14.10 years, and to 70.2% at 15.10 years (Fig. 7.4a). At 14.5 years 13.3% of the girls had had frequent experiences of drunkenness. The percentage figure rose to 23% at 14.10 years, and to 43.5% at 15.10 years (Fig. 7.4b).

There can be discerned a steady increase in the prevalence of drunkenness over time for all four menarcheal groups of girls (Fig. 7.4a). At the age of 14.5 years, 62.4% of the earliest-matured girls stated that they had been drunk at least once. This percentage figure increased to 75% at 14.10 years, and it remained on that level at 15.10 years. For the early developers, then, we seem to be witnessing a ceiling effect. Among the latest developed group of girls, 28.6% reported having been drunk at 14.5 years. There was a higher net increase over time for these girls compared to the earlier-matured. At 14.10 years, 42.6% of these girls reported having experienced drunkenness, and at 15.10 years 64.7% reported having been drunk at least once. The "spurt" or the catch-up for the later-matured girls is reflected in the statistical tests of significance in drunkenness experiences. At age 14.5 and at age 14.10 years these differences among the four menarcheal groups of girls were significant at the promille level (14.5 years: $\chi^2 = 19.37$, $df = 3$, $p<.001$; 14.10 years: $\chi^2 = 20.75$, $df = 3$, $p<.001$). However, at 15.10 years no statistically significant difference was obtained among the four menarcheal groups of girls ($\chi^2 = 3.26$, $df = 3$, ns).

The impact of biological maturation on alcohol habits, as determined by the prevalence of drunkenness at the three ages in mid-adolescence, seem to be temporary and restricted to a relatively short period. As a complement to the prevalence figures, data also were analyzed for frequent drunkenness over the same ages. As depicted in Fig. 7.4b, the catch-up effect among the late developers, observed for drunkenness per se, was not apparent in these analyses. There were significant differences among the four menarcheal groups of girls at all three age points ($p<.001$ at all three occasions), and the lead that the early-matured girls had at 14.5 years was not reduced over time.

PSYCHOSOMATIC REACTIONS

That the early-matured girls had more emotional difficulties in the mid-adolescent years than did other girls was found at the age of 14.10 years for reports of depressive reactions and psychosomatic reactions. In the present section a comparison among the four menarcheal girls with respect to self-reported psychosomatic reactions is made for the same type of data collected at 15.10 years. The percentages of girls in the four menarcheal groups reporting several times per week for the respective psychosomatic indices, nervous stomach, difficulty in falling asleep, and disturbed sleep, are presented in Table 7.2

The coefficients of correlations, expressing the stability over time for psychosomatic reactions in the research group, were .53 ($n = 432$, $p<.001$) for stomach problems, .51 ($n = 433$, $p<.001$) for difficulty in falling asleep, and .47 ($n = 433$, $p<.001$) for disturbed sleep.

As shown in Table 7.2, a higher percentage of the earliest matured girls reported the three types of psychosomatic problems than did the other, later-matured girls both at age 14.10 and 15.10 years. For example, frequent problems in falling asleep were about three times as common among the two early-matured groups of girls than among the two later-developed groups of girls at 14.10 years. At 15.10 years it was particularly the earliest-matured girls who manifested such problems. At 15.10 years disturbed sleep was also particularly characteristic of the earliest-matured group of girls.

In terms of observed p-values, there was some evidence that the differences

TABLE 7.2
Frequent Occurrences of Psychosomatic Reactions Among Girls in
Four Menarcheal Groups at 14.10 and 15.10 Years

	Problems several times per week					
	Nervous stomach		Difficulty in falling asleep		Disturbed sleep	
Age at menarche	14.10 yrs	15.10 yrs	14.10 yrs	15.10 yrs	14.10 yrs	15.10 yrs
−11 yrs	8.2	12.5	20.8	27.5	4.1	12.5
11–12 yrs	8.6	8.7	18.7	14.9	6.6	4.9
12–13 yrs	3.8	2.3	5.4	7.6	2.2	1.8
13– yrs	2.5	6.7	8.3	13.4	1.7	2.5
Total	5.0	6.0	10.9	16.5	3.3	3.7
F	3.70	3.07	5.00	3.61	6.94	2.18
df	3,454	3,429	3,456	3,427	3,453	3,426
p	.05	.05	.01	.05	.001	ns

among the four menarcheal groups decreased from 14.10 to 15.10 years. However, a closer inspection of the results showed that whereas the two early-developed groups of girls differed from the two later-matured girls at 14.10 years, it was primarily the earliest-developed girls who were differentiated from the other girls at 15.10 years.

INTERPERSONAL RELATIONS

Parental Relations

When they were 14.10 years, most of the girls in the research group reported satisfactory relations with their parents. The earliest-matured girls experienced that they had less satisfying parental relations in comparison with the later-developed girls. Conflicts were expressed especially with their mothers.

Table 7.3 presents data for the four menarcheal groups of girls with respect to mother and father relations at the age of 14.10 and 15.10 years. Data consisted of three representative items from each of the father relations and mother relations scales given at 14.10 years. Also included is the critical item

TABLE 7.3
Differences Among Girls in Four Menarcheal Groups With Respect to Parental Relations at 14.10 and 15.10 Years

	Age	F	df	p
		Mother relations		
Do you and your mother understand each other?	14.10	5.25	3,452	.01
	15.10	2.75	3,425	.05
Do you wish your mother to be different from the way she is?	14.10	1.81	3,448	ns
	15.10	3.51	3,424	.05
Do you and your mother quarrel often?	14.10	5.43	3,442	.01
	15.10	1.74	3,424	ns
		Father relations		
Do you and your father understand each other?	14.10	0.55	3,435	ns
	15.10	1.04	3,410	ns
Do you wish your father to be different from the way he is?	14.10	0.67	3,442	ns
	15.10	3.07	3,418	.05
Do you and your father quarrel often?	14.10	2.52	3,417	(.10)
	15.10	1.48	3,412	ns
		Parent relations		
Do you often do the opposite of what your parents want?	14.5	2.26	3,425	.01
	14.10	6.06	3,452	.001
	15.10	0.47	3,432	ns

picturing the girl's oppositional behavior relative to her parents. (This measure was also obtained at 14.5 years and is shown in the table.)

The correlation coefficients between the father and mother relations items at age 14.10 and 15.10 years, indicating the temporal stability of such relationships for girls in the research group, were around .5 to .6 (Mother: understanding $r = .53$, $n = 432$, $p<.001$, wished different $r=.51$, $n=433$, $p<.001$, quarreling, $r = .47$, $n = 433$, $p<.001$; Father: understanding, $r=.61$, $n = 413$, $p<.001$, wished different, $r = .55$, $n = 419$, $p<.001$, quarreling, $r = .60$, $n = 396$, $p<.001$). For the item of doing the opposite of what parents wanted, there was a correlation of .54 ($n = 430$, $p<.001$) between age 14.5 and 14.10 years and a coefficient of .50 ($n = 436$, $p<.001$) between age 14.10 and 15.10 years.

The overall impression of Table 7.3 is that the time of overt conflicts between the early-matured girls and their parents had diminished considerably at 15.10 years. At the age of 14.10 years, there were rather pronounced differences among the four menarcheal groups of girls with respect to relations to their mothers, with the early-matured girls quarreling more often ($p<.01$), and perceiving less of mutual mother–daughter understanding ($p<.01$). These differences had become considerably weaker at the age of 15.10 years. No significant differences among the menarcheal groups were obtained for the item of quarreling, and, albeit still significant, the former differences at 14.10 years with respect to mutual understanding were less pronounced at 15.10 years. The reverse trend over time occurred for the issue of wishing mother to be different from the way she was. No differences among the menarcheal groups were obtained at 14.5 years. However, at 15.10 years early-developing girls expressed this wish to a greater extent than did late-maturing girls. This result might indicate that weakened conflicts over time were accompanied by a disillusioned attitude on the part of the early-matured girls towards their mothers.

At the age of 14.10 years, the only item pertaining to father relations that differed among the four groups dealt with how often the girls quarreled with their fathers ($p<.10$). The earlier matured the girl was, the more often she reported quarreling. Neither for quarreling nor for mutual father–daughter understanding were there any differences among the four menarcheal groups of girls at 15.10 years. However, in agreement with the results on mother–girl relations at 15.10 years, the early-matured girls stated that they more often wished their father to be different.

When data on parent relations at 14.10 years were analyzed, to elucidate if, to what extent, and for which types of relational features the four menarcheal groups of girls differed from each other, there were rather strong suggestions that the main differences between the early and the late-matured girls occurred for those items that pictured the girl herself as the major instigator of the conflicts with her parents. The particular item that most significantly differed among the four menarcheal girls was "Do you often do the opposite of what your parents want?". This item was administered to the research group of girls at three occasions: at 14.5 years in the Norm inventory, at 14.10 years in the Adjustment Screening Test, and at 15.10 years in the Pupils' Questionnaire. Marked differences were obtained among the four menarcheal groups at 14.5 years for this particular item ($p<.01$). As shown in Table 7.3, the differences were stronger ($p<.001$) at 14.10 years. One year thereafter there were no significant difference among the menarcheal groups of girls. The change over time from 14.10 to 15.10 years was attributed both to a percentage decrease among the early maturers and a percentage increase among the late maturers with respect to behaving in opposition to their parents' wishes.

A more detailed examination of the topics on which girls perceived themselves to disagree with their parents was performed. Data were analyzed for four areas at 14.5 and 15.10 years: evening curfew time, alcohol use, peers, and schoolwork. The girls were asked to mark in which one or which ones of these areas they had different opinions from their parents. The results are presented in Table 7.4.

In contrast to the moderate to high stability over time reported for self-concept, social behavior, and parent relations items, the phi coefficients for

TABLE 7.4
Percentage of Girls in Four Menarcheal Groups who Reported Disagreements in Opinion With Their Parents on Different Issues at 14.5 and 15.10 Years

	Disagreement in opinion with parents on			
Age at menarche	Evening curfew time	Alcohol use	Peers	Schoolwork
		Age 14.5 years		
−11 yrs	62.2	42.2	35.6	35.6
11–12 yrs	61.5	40.0	23.2	36.8
12–13 yrs	67.8	28.8	25.1	25.4
13– yrs	55.5	25.9	22.4	33.6
Total	62.6	32.1	25.1	31.2
p	ns	ns	ns	ns
		Age 15.10		
−11 yrs	50.0	37.5	37.8	45.0
11–12 yrs	35.6	37.8	27.8	31.3
12–13 yrs	44.8	42.9	20.5	29.7
13– yrs	42.0	34.7	25.5	21.2
Total	42.4	39.0	25.3	29.1
p	ns	ns	ns	$p<.05$

the disagreement items between 14.5 and 15.10 were of a more modest magnitude (evening curfew time, phi = .23, n = 438, $p<.001$; drinking, phi = .37, n = 428, $p<.001$; peers, phi = .26, n = 388, $p<.001$; schoolwork, phi = .23, n = 435, $p<.001$). The lower temporal stabilities for these variables, compared to the previous ones are to some extent a function of the skewed distribution of the present dichotomized measures.

Evening curfew time was the issue that caused most of the disagreement between the girls and their parents at the age of 14.5 and also at 15.10 years. The lower percentage of girls reporting disagreement with parents over this issue at the latter occasion is likely to be due to an increasingly lenient attitude on the part of their parents over time. By virtue of being 1 year older at the second occasion, the girls experienced more permissiveness on the part of their parents with respect to staying out late. For the issue of evening curfew time no significant differences among the menarcheal groups were obtained either at age 14.10 nor at 15.10 years.

With regard to drinking alcohol, at the age of 14.5 there was a tendency ($p<.10$) that more of the early-matured than of the later-matured girls' opinions differed from their parents'. This difference between the early- and the later-developed girls at 14.5 years can be expected in view of the fact that the early-developed girls at this time were considerably more engaged in alcohol drinking than were the later-matured girls. As was described earlier in this chapter, a larger proportion of the later-matured than of the early-developed girls changed from a non-drinking to a drinking status from age 14.5 to 15.10 years. Consequently, one would expect an increase in reports of problems with parents connected with the change in drinking behavior to occur among the late-matured, as well as an attenuation of the differences between the early- and the late-matured. This was also the case. Whereas 27.7% of the two latest-matured groups of girls stated disagreement with parents' opinions at 14.5 years, 39.6% of these girls said that they had different opinions from their parents at 15.10 years. At this age, very few differences overall were observed between the early- and the late-matured girls. Thus, disagreement with parents over alcohol drinking was highest at 14.5 years among the early-matured girls, when these girls were the ones particularly engaged in alcohol drinking, but increased among the late-matured up to 15.10 years, by which time these girls had also begun to drink.

About as many girls at age 15.10 years as at age 14.5 years, one out of four girls in the research group, declared that they had different opinions with their parents on issues connected wit their peers. Given the earlier-documented more frequent contact with unconventional peers reported by the early-matured girls, it was assumed that a significantly higher proportion of the early-matured girls would report problems of this kind with their parents. However, although more of the earliest-matured girls stated that they had differences in opinions with their parents on peer-related issues than did the rest of the girls, the differences among the four menarcheal groups were not significant at any age.

About as many girls at both test occasions, around 30%, declared that they and their parents disagreed in areas connected with their schoolwork. No significant differences were obtained among the menarcheal groups of girls at 14.5 years. However, at 15.10 years more girls among the earliest-matured reported parental disagreement on this issue compared to the other, later-developed girls.

A final comment should be made on the interpretation of the results just given. Due to the fact that different standards of behavior, interests and attitudes might be systematically connected with menarcheal age, we should be careful not to interpret obtained frequency figures absolutely when comparing early- and later-developed girls. An example might be girls' reports of difference in opinions with their parents concerning evening curfew time. When x percent of the early-matured girls declare differences in opinions with their parents, and a similar x percent of the later-matured girls also report disagreement in opinions, this does not necessarily mean that the early and late developers perceive similar attitudes toward curfew times in their parents nor that the girls have a similar hour in mind for the issue at question. What we compare might be a group of early-matured girls who are allowed to stay out to 10 p.m. X percent of the early-matured girls want to stay out later because their older peers are allowed to stay out until around eleven. A similar x percent of the later-matured are allowed to be out until 9 p.m., but they think that 9 p.m. is too early and wish to be out to 10 p.m. at least. In this case, the permissiveness level among the parents, as well as the actual time when the girls should be at home, differs between the early- and the later-matured girls. In common is a group of x percent who wish to stay out until a later time in the evenings.

Peer Relations

Some of the items that dealt with the satisfaction experienced in contacts with peers at 14.10 years (Table 5.6) showed a curvilinear relationship to pubertal timing. Both the earliest- and the latest-developed girls reported less satisfying peer relations. Six items in the Norm Inventory administered at a 14.5 years dealt with similar peer issues, and these items were available for follow-up analyses at the age of 15.10 years. The coefficients of stability for the peer relations items between 14.5 and 15.10 years were in the range of .3 to .5, being lowest for the best friend item ("Is it important to you to be popular?", $r = .52$, $n = 452$; $p<.001$; "Do you feel popular among comrades?", $r = .46$, $n = 451$, $p<.001$; "Do you have a best friend?", $r = .21$, $n = 431$, $p<.001$; "Do you usually have the same opinion on things as your comrades?", $r = .31$, $n = 452$, $p<.001$; "Has it happened that you have gone along with things just to have the opportunity to be with your friends?", $r = .37$, $n = 438$, $p<.001$; "Do you feel shy and unsecure among same-age peers?", $r = .44$, $n = 439$, $p<.001$).

Table 7.5 summarizes a comparison among the menarcheal groups with respect to the emotional quality in relations to peers for the six items in the Norm Inventory at 14.5 years that were re-administered at age 15.10 years.

The curvilinear relationship with the earliest- and the latest-matured girl reporting less favorable responses were found for the two popularity items and the best-friend item at 14.5 years ($p<.10$, $p<.05$, and $p<.001$, respectively). However, at age 15.10 there were no significant differences among the menarcheal groups for these three peer variables. To take one example, the proportion of girls in the four menarcheal groups who thought themselves not particularly popular or hardly popular at all were 27.1%, 9.3%, 13.6%, and 25%, respectively, at 14.5 years. Seventeen months later the corresponding percentage figures were 15%, 10.8%, 10.4%, and 15%.

Two items pictured peer conformity. On the question of whether the girls usually had the same opinion as her comrades, few differences among the menarcheal groups were present at 14.5 years. At age 15.10 years, by contrast, highly significant differences appeared. The earlier matured the girl was, the less frequently she reported having the same opinion as comrades (post hoc, 1<3,4). At 14.5 years the proportion of girls in the four menarcheal groups who declared that they seldom had similar opinions on things were 12.5%, 7.1%, 1.7%, and 1.8%, respectively. The same percentages at 15.10 years were 17.5%, 4.9%, 1.2%, and 0.8%. The differences between the early- and the later-matured girls for this variable probably highlight the different attitudes that the early-developing girls had on everyday issues vis-à-vis their

TABLE 7.5
Differences Among Girls in Four Menarcheal Groups With Respect to Peer Relations at 14.5 and 15.10 years

	Age	F	df	p
Is it important for you to be popular among your comrades?	14.5	2.42	3,430	(.10)
	15.10	0.35	3,432	ns
Do you feel popular among your comrades?	14.5	3.41	3,430	.05
	15.10	0.25	3,431	ns
Do you have a best friend?	14.5	5.86	3,425	.001
	15.10	1.62	3,431	ns
Do you usually have the same opinion on things as your comrades?	14.5	1.71	3,430	ns
	15.10	6.21	3,432	.001
Has it happened that you have gone along with things just to have the opportunity to be with your friends?	14.5	0.61	3,430	ns
	15.10	0.40	3,431	ns
Do you feel shy and unsecure among same-age peers?	14.5	0.60	3,430	ns
	15.10	0.68	3,431	ns

same-age later developed peers. As to the second peer conformity item, conforming to peers in order to have the opportunity to be together with them, at none of the two age points were there any significant differences among the four menarcheal groups.

Finally, for the question on shyness and unsecurity in contacts with peers, no significant differences among the four menarcheal groups of girls were obtained at either 14.5 or 15.10 years.

Parent Versus Peer Orientation

In order to examine the parent versus peer orientation as related to the pubertal timing of the girls, the strategy used was to address issues where parents were contrasted with peers. One question asked the girls at 14.5 years what they would do in a situation where peers and parents wanted them to do different things (see Table 5.14). Another question asked was who usually understood them best—parent or peers (see Table 5.15).

For the first question, most of the girls reported an indecisive attitude, neither favoring their parents nor their peers at the age of 14.5 years. There was some tendency for the early-matured girls more than the other groups to report that they would do as their peers wanted but no overall significant differences among the four groups of girls was obtained. As to who understood them best, the alternative "peers" and the alternative "parents *and* peers" were more often reported among the girls in the research group than were "parents" exclusively. For this question also there was a tendency for the earliest-matured girls to report peers more often, but the overall test of differences among the menarcheal groups was nonsignificant.

The two questions were re-administered in the Pupils Questionnaire at 15.10 years. The results from a comparison among the four menarcheal groups at this age are presented in Table 7.6.

A comparison of the data for 14.5 years (Table 5.14) and 15.10 years, with respect to the first question, revealed strikingly similar results. At both occasions most girls reported an indecisive attitude, a minority stated that they would do as their parents wanted, and a smaller minority said that they would probably do what their peers wanted of them.

At neither of the test occasions did there appear any significant differences among the four menarcheal groups. However, the change over time for the earliest-matured groups of girls is worth observing. At 14.5 years, a greater part of these girls reported that they would rather follow their peers (28.3%) than their parents (23.9%). At 15.10 years the situation was reversed. Now 37.5% of the earliest-developed girls stated that they would rather do as their parents wanted (actually the proportion of girls favoring parents was highest among the earliest-developed girls), rather than as their peers wanted (12.5%). Tentatively, this indicates that the earliest-matured group of girls

TABLE 7.6
Comparison of Parent Versus Peer Issues Among Girls in
Four Menarcheal Groups at 15.10 Years

	If your parents want you to do one thing while your peers want you to do another—how does it usually end?		
Age at menarche	As my parents want	I am not sure	As my peers want
−11 yrs	37.5	50.0	12.5
11–12 yrs	19.6	59.8	20.6
12–13 yrs	20.0	69.4	10.6
13– yrs	22.7	68.9	8.4
Total	22.3	65.2	12.5
	Who usually understands you best—parents or peers?		
	Parents	Both	Peers
−11 yrs	15.9	52.5	30.0
11–12 yrs	18.8	45.5	35.6
12–13 yrs	21.4	39.9	38.7
13– yrs	24.1	39.2	36.7
Total	21.2	42.2	36.6

at the latter point in time was involved in a process of merging toward a recovery in their relations to parents. The results converge with earlier data presented in this chapter showing less parent–daughter conflicts among the early developers at 15.10 than 17 months earlier. One other circumstance in Table 7.6 is of interest. At 15.10 years it was the *second* menarcheal group (menarche between 11 and 12 years) that was the group of girls favoring peers most. Tentatively, it might be suggested that it is around this time, at 15.10 years, that this next-matured group of girls are involved in their "period of disequilibrium" (Steinberg, 1985) as far as relations to parents are concerned.

There is another feature of the parent versus peer comparison between early- and late-matured girls that should be observed. The proportion of girls who were indecisive as to whether to follow their parents or their peers tended to grow with pubertal timing at both test occasions. Later-matured girls were more indecisive than were the earlier developed. The total proportion of girls who stated an indecisive attitude was somewhat higher at the latter compared to the former test occasion.

Table 7.6 also presents results from the comparison among the menarcheal groups at age 15.10 as to whether parents or peers were perceived to understand the girl best. Compared to age 14.5 years (Table 5.15), a somewhat higher proportion of girls at 15.10 years stated that their parents understood them best, and a somewhat smaller proportion reported that their peers understood them best. Even though there was a tendency over time toward higher parent orientation for girls in the research group generally, we should not lose sight of the fact that even at 15.10 years a higher proportion of girls favored their peers exclusively rather than their parents. However, the most common response from the girls was that both their parents and their peers understood them best.

No significant overall differences among the four menarcheal groups were observed at the age of 15.10. Still, the changes from 14.5 to 15.10 years for the respective groups of girls are quite interesting. A considerably lower proportion of the earliest-developed girls stated that their peers understood them best at 15.10 years as compared with age 14.5 years. The decrease of perceiving that their peers understood them best over time among the earliest-matured girls was not, however, matched by an expected increase of reports that their parents understood them best. What seems to have happened from 14.5 to 15.10 years for the early maturers is a reconciliation between these two major socializing groups over time where now both parents and peers occupies important supporting influences for these girls.

SUMMARY AND CONCLUSIONS

Against the background of a generally more problematic adolescent situation—interpersonally, socially, and emotionally—for the early-matured girls at the age of 14 years (14.5 and 14.10 years), we undertook to investigate whether there were any changes that took place during the mid-adolescent years, and in what direction they pointed. What evidences were there that the early-matured girls were "coming back" to a more conventional life at 15.10 years (and/or that the late-maturing girls were catching up with the early-developers over time)? When answering this question it seems pertinent to examine the impact of biological maturity on the interpersonal network separately from the impact on behavior.

There were some indications of less parental conflicts and a conciliation between the parent and the peer reference groups over time in mid-adolescence for the early-developed girls:

1. For the issues on quarreling with the mother and mother–daughter understanding, there were rather great differences among the menarcheal groups at 14.10 years, with the early-matured girls quarreling more often and perceiving less mutual mother–daughter understanding. These differences were reduced considerably 12 months later. In the place of overt conflicts with the mother, the early-matured girls reported more often than the later-matured girls that they wished their mother to be different at 15.10 years.

SUMMARY AND CONCLUSIONS 245

2. The only item at 14.10 years pertaining to father relations that differed among the menarcheal girls was the one on quarreling. At 15.10 years no differences for this item among the four menarcheal groups of girls were observed. Similar to the girls' relations to their mothers, the early-matured girls reported more often that they wished their father to be different from the way he was. Summarizing the results for mother and father relations, a decrease over time in overt conflicts seemed to have taken place, but for both parental figures the early-matured girls expressed a more disillusioned attitude on the second occasion.

3. Perhaps the most direct evidence of decreasing conflicts with the parents occurred for the item that most significantly differentiated the four menarcheal groups at 14.10 years: "Do you often do the opposite of what your parents want?".

4. Two questions were asked to elucidate the parent versus peer orientation among girls in the four menarcheal groups. At age 14.5 the earliest-matured girls answered "peers" more often than did other girls on the question of whether they would follow their peers or their parents in a situation when these two reference groups wanted the girl to do different things. By contrast, as many of the earliest-matured girls as of other girls declared peers at 15.10 years, and a higher proportion of the earliest-matured than of the other three groups of girls stated that they would do as their parents wanted.

The second question concerned whether their peers or their parents understood them best. At age 14.5 more of the earliest-matured girls than of the other girls answered peers exclusively. At age 15.10 a somewhat lower percentage of the earliest-matured than of other girls declared peers exclusively, and a greater proportion of these girls than of the three other menarcheal groups of girls answered both parents and peers.

The results, suggesting a re-establishment of the affective parental relations during this period between 14.5 and 15.10 years for the earliest-matured girls, were quite consistent over different types of measures, albeit the magnitudes of the tendencies were of a modest size. To the extent that there was a parallel development for parental relations and social behavior, we would expect that the greater social problems among the early-matured girls found at 14.5 years, would attenuate over the period of time investigated.

This, however, was not the case. The behavior differences among the four menarcheal groups of girls at 14.5 years, whether sexual intercourse, running away from home, use of hashish, drunkenness, or psychosomatic reactions were considered, were also profound at 15.10 years of age. In several respects, the differences between the early- and the later-developed girls were even more pronounced at the latter test occasion.

For sexual intercourse the differences among the four menarcheal groups of girls widened over the time period in midadolescence. For the behavior of running away from home very little change occurred between 14.10 years and 15.10 years. The earliest-matured girls were markedly overrepresented at both occasions among the girls who had engaged in this type of behavior at both ages. Smoking hashish showed greater differentiation among the four menarcheal groups of girls at 15.10 than at 14.5 and 14.10 years.

Alcohol seemed to hold an exceptional position. In contrast to the behaviors previously reported, drinking was the only behavior for which an attenuation among the menarcheal groups could be traced over time. For the drunkenness variable per se (i.e., being drunk at least once versus never having been drunk) there was a catch-up among the later-matured girls over time.

In this connection, it should be noted that the greater net increase in prevalence of drunkenness from 14.5 to 15.10 years among the later-developed girls had its counterpart in the data on the girls' reports regarding those issues on which they had different opinions from their parents. At the age of 14.5 years, when the early-matured girls had considerably more advanced alcohol habits than did the late-matured girls, more of the early-matured girls than the later-developed girls reported that they had different opinions with their parents with respect to alcohol use. The percentage of the early-developed who reported opinion differences decreased somewhat at the test occasion seventeen months later, whereas it increased among the later-developed girls. Consequently, about as many among the early- as among the later-developed girls reported at 15.10 years that they had different opinions than their parents on alcohol drinking.

When we analyzed the data for alcohol for frequent drunkenness (drunk 4 times or more), again no attenuation of the differences between the early- and the late-matured girls from 14.5 to 15.10 years was found. At all three age points there were highly significant differences among the four menarcheal groups of girls.

Thus, whether frequent alcohol use, smoking hashish, running away from home, or sexual intercourse were considered, the findings were conclusive. Either there was no evidence of reduced differences among the menarcheal groups of girls, or there was actually evidence of increasing differentiation over time. For different indices of psychosomatic reactions there was also little attenuation of the differences among the four menarcheal groups from 14.10 to 15.10 years.

In the light of prevailing differences in social behavior and emotional adjustment among the menarcheal groups of girls at 15.10 years and considering that little apparent difference in overt physical characteristics following pubertal development can be observed at this age, a most interesting result was the finding that dealt with self-perceived maturity for the menarcheal groups of girls. At 15.10 years, as well as 17 months earlier, a higher percentage of the early-matured girls, compared to the later-matured, thought of themselves as more mature than their classmates. As might be expected, a lower percentage of the later-developed girls reported feeling less mature at 15.10 than at 14.5 years. More unexpected was that a higher percentage of the earliest-matured girls reported that they felt more mature at age 15.10 years than they had at age 14.5 years. The latter finding should, perhaps, not be exaggerated in view of chance fluctuations in data.

In agreement with the results obtained at 14.5 years, a greater proportion of the early maturers than of the late maturers at 15.10 years stated that they felt different from their same-age peers. The percentage of subjects who reported feeling different from their peers among the earliest-developed girls decreased somewhat from 14.5 to 15.10 years, whereas the proportions of girls in the other three menarcheal groups who felt this way was about the same on both occasions.

On a subsequent question at 15.10 years, asking in what respect the girls who declared themselves different from peers perceived themselves as different, the early-matured girls were differentiated from the later-developed girls by more often reporting being "more romantic." Actually, the same type of differences between the early- and the later-developed girls were obtained when the question was given 17 months earlier.

Taken together, the main results in this chapter on the short-term consequences of pubertal development in mid-adolescence can be summarized as follows: Early pubertal development was connected with perceiving oneself as more mature than others both at 14.5 years and 15.10 years. Different types of problem behavior, which were more prevalent among early- than among late-developed girls at the age of 14, also tended to be more prevalent among the early developers 17 months later. For many of them, the differences that were observed among the menarcheal groups at the age 14 were even more pronounced later on in midadolescence. However, the higher prevalence of parental conflicts among the early-developed girls, observed at the age 14, seemed to decrease over time, as did the peer orientation of these girls.

8
Developing Girls in a Developing Environment

A "CO-DEVELOPING" ENVIRONMENT

This chapter focuses on the norm climate in the social ecology of early- and later-matured girls. As discussed earlier, a premise in the proposed model is that differences between early- and late-developed girls with respect to new social behavior in adolescence should be viewed in the light of the more advanced social behaviors and leisure-time activities among the friends of the early-developing girls, relative to the friends of their later developed counterparts. With empirical support for a differential association with peers based on a girl's biological stage of maturity, it then becomes important to try to illuminate the normative dimensions through which the peer influence is operating.

These analyses are not limited only to peers as norm transmitters. They also involve an examination of the parental support for new social behavior in adolescence. By their own behavior and standards of behavior, their sanctioning of the girl's social behavior, and through their explicit and implicit expectancies regarding behavior, parents communicate their own attitudes and beliefs of the appropriateness of new social behavior to their daughters. Our assumptions are that we cannot expect a similar norm climate in the social network of the early and the late maturers: with respect to either the peers or to the parents.

That there probably occurs some mutual reinforcement between girls and their parents with regard to social behavior, a reinforcement connected with the pubertal development of the girls, is the conclusion of previous research on attitudes and interests among adolescent girls, presented by Stone and Barker (1937, 1939). In their first study, the Pressey Interest-Attitude Test and the Sullivan Scale for Measuring Developmental Age in Girls were given to post- and premenarcheal girls aged 12 to 15. Differences in chronological age among girls were controlled through a matched-control design. Generally, more of the postmenarcheal than of the premenarcheal girls favored the mature responses of the items included in these inventories. For the items in the former inventory, the postmenarcheal girls were estimated to be about 15 months "older" than the premenarcheal girls. For the latter inventory the differences in responses favoring a more mature orientation were estimated at 11 months between the pre- and postmenarcheal girls. Thus, despite being chronologically of the same age, the postmenarcheal girls gave responses which were more common among girls chronologically 11 to 15 months older. Guided by the results from the first investigation, that the domains differentiating the pre- and postmenarcheal girls predominantly concerned physical appearance, heterosexual interests, avoidance of physical exertion, daydreaming, and home conflicts, additional items from these domains as well as from other theoretically interesting domains were gathered to form a new attitude and interest instrument (Stone & Barker, 1939).

Although about the same proportion of items in the "family adjustment" domain in the new study by Stone and Barker (1939) were favored by the early developers (17 items) as by the late developers (18 items), a closer inspection of the specific items in this area supports a mutual parent—daughter reinforcement interpretation of differences in social behavior between pre- and postmenarcheal 12- to 15-year-old girls. The postmenarcheal girls were allowed more freedom by their parents than were the premenarcheal girls. The parents gave the postmenarcheal girls more freedom and autonomy to develop their own personal identity and style than did the parents of the premenarcheal girls, who were not allowed as much freedom in deciding on everyday life activities. A greater proportion of the postmenarcheal girls than of the premenarcheal answered affirmatively that they were allowed to stay up late in the evenings, go riding with boys in the day time, go to parties, have dates and boyfriends, use their spending money as they wished, wear the clothes they liked, wear their hair in the way they themselves wanted, use make-up, and so forth. On the other hand, the postmenarcheal girls also had more responsibilities, and they reported more often that they had to do more housework than they liked. Indications of parental conflicts occurred more frequently among the early developers. More postmenarcheal girls than premenarcheal felt themselves as a burden to their family and declared more often that they would like to leave home. They complained that their parents teased them about their boy friends, and they felt embarrassed when their parents bragged about them to other people. They admitted more often than the premenarcheal girls that they concealed their worries and troubles from their parents, had secrets they did not wish adults to know, and were bothered if their parents wanted to know details of what had happened at parties. Finally, they claimed more freedom.

The premenarcheal girls, by contrast, seemed to have a more sincere and trusting relationship with their parents. A greater proportion of these girls than of the postmenarcheal answered affirmatively that their mother understood them well enough so that they could talk about their problems, that they were not bothered if their parents wanted a detailed account of what happened at parties they attended, that they did not conceal worries and troubles, that they had no secrets they did not want adults to know, that they did not find themselves teased by their parents, and that they did not think of themselves as being troublesome to their parents. Although they were not given as much freedom as the postmenarcheal girls, this did not seem to concern the premenarcheal girls greatly, as they answered more often than their postmenarcheal counterparts that they did not want more freedom and agreed that girls their age should obey their parents.

Stone and Barker's studies contribute to our understanding of the process of parent–daughter development as related to pubertal development in girls. Their results indicate that we cannot treat the parents of the early- and the later-developed girls as a homogeneous group with equal support for new social behaviors that emerge in adolescence. To the extent that we can talk about a faster social development among the early-matured girls, we have to acknowledge the possibilities of an analogous "co-development" among the parents of these girls. The postmenarcheal girls behaved more in a grown-up fashion, leading to a corresponding permissive attitude on the part of their parents. By contrast, the mothers of the premenarcheal girls set higher restrictions with regard to their daughters' engaging in more mature social behaviors, and these restrictions were not regarded as unjust by these girls.

Similar findings have been reported by others as well. In a study by Danza (1983), postmenarcheal girls, 11 to 14 years of age, were found to be allowed by parents to show more advanced interests and behavior to a greater extent than were the premenarcheal girls, and they were allowed more decision leeway in everyday matters ("more likely to wear make-up, a bra, to date, and such," "more limit setting on their own," "slept less on school nights," "more liberal time limits"). At the same time, the postmenarcheal girls experienced more parental conflicts than did their later maturing counterparts ("experienced less comfort in emotionally charged discussions with their mothers," "less comfortable in discussions with their fathers than their mothers," "more conflict with their parents," "evaluated their parents as getting along less well").

The empirical evidence that different permissiveness levels on the part of parents are connected with the pubertal timing of the girl herself is a thought-provoking finding that has not attracted much attention in the literature on social adjustment in adolescent females. That girls, through their level of biological development, may influence the attitudes of their parents toward greater permissiveness for mature forms of social behaviors seems quite possible in the light of the data presented by Stone and Barker (1937, 1939). The question, as to whether or not social behaviors generally recognized as problem behaviors among girls at a certain age, and consequently viewed negatively by adults, are more permitted by parents of early-developed girls, is a delicate one. Intuitively it seems quite reasonable that this might be the case, at least for certain types of behaviors that have a more grown-up connotation. Also, what is regarded as problem behavior among early-matured girls from a chronological age point of view, may be met with greater tolerance on part of their parents because such behaviors more frequently occur among the chronologically older peers whom they resemble and with whom they interact.

The literature of adolescent development and problem behavior has until quite recently been heavily biased toward recognizing the conflicts between networks—the "generation gap," parent versus peer orientation, and so on—to the neglect of the correspondences. Over time the insight has deepened that many of the developmental processes through which girls pass in the adolescent years have their counterparts in changes in their immediate environments. Parental permissiveness and privilege-giving and the adolescent's own striving toward adaptation and adoption of new social behavior are often complementary processes over time, rather than solely a source of strife (Bandura, 1972; Steinberg, 1985). Initially, the emancipation on the part of the adolescent girl leads to her stumbling; her parents respond by alternatively giving the girl greater privileges and withdrawing them in the fear that the girl is not mature enough to handle her new freedom effectively. The fragile equilibrium proceeds toward greater permissiveness and greater autonomy with the passage of time.

Viewed from this perspective of supporting networks, two questions arise. First, what types of behaviors are likely to be more tolerated among the parents (and peers) of the early maturers, and second, what features of the parents (and peers) as norm transmitters are likely to be involved?

For certain types of norm-violative activities connected with more absolute norms (such as those involving violence, threat, or property damage, etc.), one can expect a one-sided repudiation from parents. However, other more age-related behaviors (alcohol drinking, staying out late in evenings, and petty offenses) might be condemned but might still be more tolerated among the early-matured girls, with the excuse that such things happen in the teenage culture to which she seemingly belongs.

In the present case, the tolerance among parents is examined for norm violations at home, at school, and during leisure time, which are not necessarily legally enforced by the law, but which represent more or less the break of unwritten behavioral standards. The results from this analysis will not only yield information about which violations the parents of the early-matured girls are more likely to tolerate relative to the parents of the late maturers, they also may yield information about which behaviors are closer to the "normal" development in girls. The assumption here is that such behaviors that are more connected with permissiveness on part of the parents of the girls are the ones signifying the way to independence, the approach of adult status, or, generally, a more "matured" status on part of girls.

The second question concerns the normative factors that support or excuse violation of norms. The great importance attached to examining norm transmission for different attitudes or normative components has been emphasized repeatedly in research on attitudes (Fishbein, 1967; Fishbein & Ajzen, 1975; Katz & Stodtland, 1959; Triandis, 1967). Either or both of the following two conditions are possible. The higher norm-violative activity among early-matured girls might reflect the presence of higher acceptance of such behaviors in the reference groups of these girls. It might also reflect the absence of sanctions of such behaviors in the reference groups. Whether or not the main reference groups of girls in the mid-adolescent years actively support or tolerate violation of age-related behavior among girls in these years is a crucial issue in the interpretation of individual differences in social behavior in adolescent females.

From the perspective of supporting reference groups, an investigation was designed to examine what features of peers and parents, as norm transmitters, were involved in norm-governed behavior among early- and later-developed girls. Both the perceived *evaluation* of different types of problem behavior among parents and peers, and the perceived *sanctions* that girls expect would emanate from parents and peers, if they found out about the behavior, were investigated. In addition, a closer investigation was made of the *expectations* the parents were thought to have regarding the girls' behavior in above respects.

The evaluative element of norm transmission concerns the general opinion of the girl's social behavior: whether it is positively or negatively viewed. It reflects the general opinion of behavior rather independent of the particular motive forces that appear in concrete everyday life situations. As a distal factor relative to situational behavior, we expected that few differences would appear for this norm-transmitter component among the reference groups of the early- and the later-developed girls. The parents and the peers, respectively, of the early-developed girls would hold about the same evaluative opinions with respect to problem behavior as would the parents and the peers of the later-developed girls.

The sanctioning of problem behavior would be a normative component that more probably would differentiate the early- and the later-developed girls, with the parents and the peers of the early developers being more tolerant and reacting less harshly than the parents and the peers of the later-developed girls.

Finally, it was assumed that the expectations the parents had of their daughter's behavior would be another normative component that would differentiate between the early- and the later-matured girls. The expectancy component not only contains a future-directed prognosis based on earlier experiences with the girl, it also contains an element of comparing a girl with other girls in her developmental situation. With more experiences of advanced social behavior among their daughters, the parents of the early-matured girls were hypothesized to expect more problem behavior in their daughters than were the parents of the later-developed girls.

PARENTS AND PEERS AS NORM TRANSMITTERS

Data for these analyses were taken from the Norm Inventory administered at 14.5 years. The instrument contained eight types of norm violations common among teenagers—ignoring parents' prohibitions, staying out late without permission, cheating on an exam, playing truant, smoking hashish, getting drunk, loitering in town every evening, and pilfering from a shop. As was described in chapter 6, all of these eight items, except cheating at school, showed higher incidence among the early-matured girls than among the later-matured. For these items the girls who reported frequent rule breaking were also significantly overrepresented among the early-matured girls.

The girls were asked how they themselves, how their parents, and how their peers evaluated each of these norm-related behaviors. With respect to peers, it was stated in the instruction that "by peers we mean those whose opinion you care most about, whether they are in you class, your pals, or your best friend." The instructions for the *evaluative ratings* were: "Say here what you yourself (your parent, your peers) think about cheating (etc.) .. I (My parents, My peers) think it is. . . ." The answers were given on 7-point scales with the alternatives: (1) very silly, (2) silly, (3) rather silly, (4) not really OK, (5) rather OK, (6) OK, and (7) quite OK.

For each of the norm-breaking behaviors, concrete situations were formulated (see Magnusson, Dunér, & Zetterblom, 1975, pp. 101–103). After each norm-breaking situation that was depicted, the questions on sanctions and parental expectations were given. For expected *sanctions* the following questions were asked: "How do you think your parents (peers) would react, if they found out that you had cheated (etc.) in this situation?" Answers were given on 5-point scales with the alternatives: (1) they would certainly disapprove, (2) they would probably disapprove, (3) I am not sure they would react, (4) they would probably not care, and (5) they would certainly not care.

The important aspect for the developmental course of the parents' expectations about behavior is how they are perceived by the girl. This aspect of the *parents' expectations*, was assessed in the following way. For each of the situations the following question was asked: "What do you think your parents expect you to do in a situation like this?", with the alternative responses: "That I (1) would absolutely not, (2) would probably not, (3) would perhaps not, (4) I am not sure what they think, (5) would perhaps, (6) would probably, or (7) would most certainly pick up the notes and cheat (etc.)."

The testing of differences in norm-transmitter components among girls in the four menarcheal groups was performed using one-way analyses of variance. Table 8.1 presents the results of the comparison among the four groups of girls. To facilitate an overview of the main findings only the information concerning whether or not significant differences were obtained is presented.[1]

Evaluations. The assumption that few differences among the four menarcheal groups would occur for parents' and peers' evaluations of normbreaking was basically correct. Except for the item "get drunk," more accepted by the peers of the early-matured girls, the ANOVA tests did not produce any significant differences among the four menarcheal groups of girls, whether for parents' or peers' evaluations.

That few differences that related to pubertal development were obtained with respect to parents' and peers' evaluation of norm breaking by no means implies that there were no differences in acceptance of these behaviors between the two reference groups. Indeed these differences were strong. Figure

---

[1] Complementary analyses were made with the response alternatives dichotomized into accepting versus nonaccepting evaluations (scale value 1 to 4 vs. 5 to 7), sanctions versus no sanctions (scale value 1 to 3 vs. 4 and 5), and positive expectancies versus negative expectancies (scale value 1 to 4 vs. 5 to 7).

The results with regard to these dichotomized measures were as follows:

	Evaluations		Sanctions		Expectations
	Parents	Peers	Parents	Peers	Parents
Home					
Ignoring parents' prohibitions	—	—	—	.10	.01
Staying out late	—	—	—	.05	.05
School					
Cheating	—	—	.10	—	—
Playing truant	—	.10	.10	—	—
Leisure time					
Smoking hashish	—	.10	—	.001	—
Getting drunk	—	.01	.10	.001	.001
Loitering in town	—	—	.05	.10	—
Pilfering	—	—	—	.05	—

8.1 graphically pictures the average evaluative ratings for parents and peers respectively over the four menarcheal groups. For comparison, a profile across the four menarcheal groups, for how the girls in respective menarcheal groups themselves evaluated norm breaking of the type included in the Norm Inventory, is also shown. (The girls' own evaluation ratings were the same as the peer and the parent ratings.)

As depicted in Fig. 8.1, the parents of the girls were perceived to take exception to the norm-breaking behaviors included in the Norm Inventory. Peers were perceived to be more accepting with regard to norm breaking than were either the parents or the girls themselves. The girls' evaluations were in a middle position between the more restricted norm evaluations of the parents and the more accepting norm climate among peers. In contrast to the nonsignificant differences for parent and peer evaluations among the four menarcheal groups, the girls' own evaluations of the normbreaking types differed significantly among the four menarcheal groups ($F = 3.31$, $df = 3,423$, $p<.05$), such that the early-matured girls were more accepting of norm violations than were the later-matured girls.

The observation that teenagers often stand in the middle between the more conventional norms of parents and the more antisocial norms of peers has been reported earlier in the literature (Rommetveit, 1954). Parents, as

TABLE 8.1
Tests of Differences in Evaluations, Sanctions, and Expectations
of Norm Breaking of Parents and Peers among Girls
in Four Menarcheal
Groups: p-values

	Evaluations		Sanctions		Expectations
	Parents	Peers	Parents	Peers	Parents
Home					
Ignoring parents' prohibitions	—	—	—	.05	—
Staying out late	—	—	—	—	.05
School					
Cheating	—	—	—	—	—
Playing truant	—	—	—	—	—
Leisure time					
Smoking hashish	—	—	—	.001	—
Getting drunk	—	.05	.05	.05	.001
Loitering in town	—	—	—	—	—
Pilfering	—	—	—	—	—
Total Norm breaking	—	—	—	.01	.05

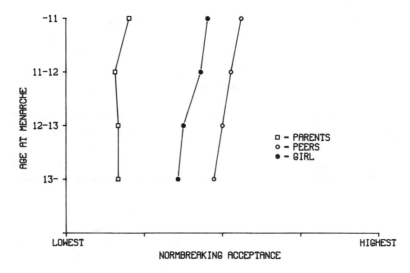

FIG. 8.1. The subjects' own evaluation of norm breaking and the evaluation of norm breaking by parents and peers, by menarcheal age.

representatives of the norm system of the society, are perceived as strict and as repudiating breaches of norms. With respect to peers, the concept that an adolescent's friend are more deviant than the adolescent himself or herself seems to be related to the proximity of the contact. When teenagers are asked about the evaluations of peers in general, or about the evaluative norm climate in the peer culture, they perceive more social deviance than when they are asked to rate the norm-breaking acceptance of their close friends (Hauge, 1971). However, in the latter case also, friends are thought to be more accepting of norm violations than is the actual respondent. This tendency, to experience peers as more prone to break rules than is oneself, is a fiction to a large extent. It was labeled "pluralistic ignorance" by Allport, in the 1930s.

Sanctions. It was assumed that the parents of the early-matured girls would be more tolerant of norm violations on the part of their daughters by sanctioning these girls' behavior less than would the parents of the later-developed girls. Contrary to these assumptions, few differences in parental sanctioning of the girls' behavior were observed among the four menarcheal groups. For one behavior only, drunkenness, did there appear significant differences, with the early-matured girls expecting milder parental sanctions after having been drunk than were expected by the later-maturing girls.

The sanctioning of the girls' social behavior by peers tended to be a norm transmitter component that was more strongly related to the pubertal development of the girls than were peers' evaluations. The early maturers reported consistently lower peer sanctions than did the later-matured girls. For specific behaviors, the peers of the early-developed girls were perceived to exercise fewer sanctions with regard to ignoring parents' prohibitions, getting drunk, and, in particular, smoking hashish (a tendency in the same direction, at the .10 level of significance, was obtained for "loiter in town").

Parents' Expectations. The last column in Table 8.1 refers to the parental expectations as perceived by the girls. Were the parents of the early-developed girls perceived to expect more norm violations in their daughters than expected by the parents of the later-developed girls? As judged by the comparison in Table 8.1, the early-matured girls reported that their parents were more likely to expect them to stay out late in the evenings and to get drunk, as compared to the parents of the later-matured girls (a tendency at the .10 level of significance also appeared for "ignore parents' prohibitions"). The behavior "get drunk" was the social behavior in particular where the parents' expectations regarding their daughters differed between parents of the early- and the later-matured girls.

Comments. What types of behaviors were more tolerated by the reference groups of the girls? Of the norm violation behaviors in Table 8.1, drinking behavior seemed to hold a unique position. The early-matured girls had peers who were more accepting of getting drunk and had parents and peers who were less sanctioning of this behavior. Moreover, the parents were perceived to expect the early-matured girls to engage in this type of behavior more than did the parents of the later-matured girls. Only with respect to parents' evaluation of getting drunk were there found no differences among the four menarcheal groups of girls. These results are important, and they suggest that the more advanced drinking habits of the early-matured girls were sanctioned in *both* the main reference groups of the girls. As far as can be judged from the comparison in Table 8.1, and the frequency figures for getting drunk presented in chapter 7 (Fig. 7a and 7b), established drinking behavior among the early-matured girls seems to be what is normatively expected of these girls at this age and does not have a particularly deviant connotation. Alcohol drinking as a "transition behavior," as conceived by Jessor and Jessor (1977), seems to be an appropriate designation.

A norm-violating behavior that tended to be closely bound to the peer reference group, but not to parents, was drug use. The peers of the early-matured were expected to exercise milder sanctions if the girl was found to use hashish than were the peers of the later-developed girls. There were also differences related to peer sanctions among the four menarcheal groups with respect to ignoring parents' prohibitions.

In conclusion, the present data indicated that the higher incidence of norm breaking among the early-matured girls to a certain extent was connected with a higher tolerance level for norm breaking among their peers. This was most obvious with regard to drinking behavior. The early-developed girls' more frequent use of hashish was also connected with weaker sanctioning in the peer group. From the perspective that the early developers largely had their peer circles localized outside the school context, it should be noted in Table 8.1 that the peers of the early-developed girls were not less sanctioning with regard to breaking norms at school—cheating or playing truant—but were perceived to express weaker sanctions for norm breaking during leisure time. These tendencies were particularly evident for the analysis utilizing a dichotomized measure of peer sanctions (see footnote 1), thus suggesting that the impact of peer association was differential, depending on the context in which breaches of norms occurred.

What features of the parents and the peers as norm transmitters were most related to the pubertal development of the girls? The last row in Table 8.1 presents overall tests of differences in evaluations, sanctions, and expectations of norm breaking among parents and peers among the four menarcheal groups of girls. Data were collapsed across the eight norm violation behaviors in the table.

As judged by these overall testings, there were no significant differences among the four menarcheal groups with respect to the perception of parents' evaluation of these behaviors. Neither were there any significant overall differences among the four groups with regard to parents' sanctioning. However, the early-matured girls experienced, to a significantly higher extent than later-maturing girls, that their parents expected that they should break norms.

If the peers of the early-matured girls actively motivated these girls to break norms, then we should find differences in the peer evaluative ratings among the menarcheal groups. This was not the case. With some exceptions for particular behaviors, the overall measure of peer evaluation did not significantly differentiate the early- from the later-matured girls. However, the expected sanctions after own normbreaking among peers were indeed related to the pubertal timing of the girls. The peers of the early-matured girls were perceived to exercise less overall sanctioning than were the peers of the later-matured girls. This result indicates that it was not the presence of active peer instigations that was connected with more frequent norm violations among the early-matured, but, primarily, the absence of sanctions once the girls had violated a norm.

CLASSMATES AND "OTHER PEERS" AS NORM TRANSMITTERS

Throughout this book is the message that the circle of peers among teenage girls include more than classmates and that the friends outside the framework of the classroom play an important role in girls' engagement in advanced forms of social behavior. The designation "peers" employed in the formulation of the questions in the Norm Inventory was carefully defined so as to allow the girls to chose the peers that they individually regarded as their *peer reference group*. These might be friends in the same class as well as other types of peers, such as pals or the girl's best friend.

The girls might have had different types of peers in mind when answering questions of this type. In order to be able to make a more detailed account of peer evaluations for different types of peers, an attempt was made to ascertain whether girls perceived any difference in norm-breaking acceptance among those in their class and other friends outside this classroom frame. After the questions on peer evaluations, a question was formulated addressing the issue of which peer reference group the girls had had in mind when they had answered the questions on peer evaluations. To the girls who had thought of their outside-classroom peers the question was asked: "Do you think that your classmates have another opinion about the situations above than do your other friends?". The results of the comparison among the four menarcheal groups for this question are presented in Table 8.2.

This analysis showed that the tolerance of norm breaking of the outside classroom peer frame versus that of classmates indeed was perceived differently by the early- and by the later-developed girls. As shown in Table 8.2, of those who had peers outside their class—pals or best-friend relations—almost twice as many among the earliest-developed group of girls perceived that their "outside classroom" peers had more accepting evaluations of breaking norms compared with those of the "outside classroom" friends of the later-matured girls.

SUMMARY AND CONCLUSIONS

The idea that pubertal development in girls coincided with development of the interpersonal ecology was investigated here. It was applied to norm-breaking behavior, and a hypothesis was advanced that early-developed girls,

TABLE 8.2
"Peers' " Evaluation of Norm Breaking Versus Classmates': By Menarcheal Age

Age at menarche	Peers versus classmates		
	Peers less accepting	Peers and classmates have equal opinion	Peers more accepting
−11 yrs	21.1	36.8	42.1
11–12 yrs	11.8	57.6	30.6
12–13 yrs	16.9	62.3	20.8
13– yrs	21.3	59.6	19.1
Total	17.3	58.0	24.8

$\chi^2(N = 371) = 14.13$, $p<.05$

having been found to break conventional norms to a higher extent than later-maturing girls in mid-adolescence, would face more tolerant attitudes toward their norm violations, from the two main socializing agents, than would their later-matured counterparts. The parents of the early-developed girls would be more apt to accept new social behaviors in their daughters, even of a problem behavior type. The more permissive attitude of the parents would have its counterpart in more permissiveness on part of the friends of these early-maturing girls. Thus, the early-matured girls would be surrounded by a social network that was more favorable to her social transition than were the later-maturing girls.

Results indicated that the perceived norm climate of parents and peers in some respects differed between early- and later-developed girls. The greater incidence of problem behavior among early-developed girls than among later-maturing girls had, in some respects, its counterpart in higher tolerance for engaging in this behavior among parents and peers.

Few differences in the evaluation of norm breaking among the parents of the early- and the later-developed girls were obtained. They had, from the perception of the girls, about the same nonaccepting opinions. However, for drinking behavior the parents of the early-developed girls were perceived to exercise less sanctioning than did the parents of the later-developing girls. For behaviors like staying out late, getting drunk, and ignoring parents' prohibitions, the parents of the early-developed girls were also perceived to expect their daughters to be engaged in these behaviors more than did the parents of the later-developed girls. To the extent that we can talk about social transition behaviors that are likely to be more tolerated by parents of the early- relative to the parents of the later-developed girls, it is these types of behaviors, which signify less parental restrictions on leisure time activities, which will differentiate the early- from the later-developed girls.

Alcohol drinking, in particular, was a social behavior that was bound by close ties to both the norm climate among parents and among peers. The parents of the early developers were perceived to sanction this behavior less severely than did the parents of the later-developed girls. The parents of the early-developed girls also were perceived to expect that their daughters should be involved in such behavior to a higher extent than did the parents of the later-developed girls. With respect to peer support, the peers of the early developers were perceived by them to have less negative evaluations of getting drunk than did the peers of the later-developed girls, and they were perceived to react less critically to this type of behavior than did the peers of the later-developed girls. The experience of more approval for drinking among the socialization agents of the girls who matured early biologically, supports the argument that alcohol drinking is a social facet of a normal maturational process.

Part of the investigation concerned the normative dimensions through which peer and parent influences operated. The results, for both socializing agents, showed that the evaluative dimension did not differentiate the early- from the later-developed girls. Rather, the sanctioning of norm breaking among parents and peers was the normative component that differentiated the early-developed girls from their later-developed counterparts. At the same time, when a comparison was made for the subsample of girls in the research group who had thought of their outside classroom friends when answering the questions on peer evaluations, these peers (pals or the best friend) were perceived to accept norm violations more than did the same type of peers among the later-developed.

Earlier chapters have dealt with the "conflict" view of the parent–daughter relations in mid-adolescence. Ample support was found for the proposition that earlier pubertal development was connected with more conflicting parent relations in mid-adolescence. The present chapter focused on the other of the two views about the trajectory of development of parent–daughter relations that was suggested at the outset: namely, that the accelerated social development among early developed girls also would co-occur with greater permissiveness and with reinforcement of social behaviors among the parents of these girls. As determined from the perception of the early-developed girls, their parents were less sanctioning and more expecting of the girls to be involved in behaviors normally connected with a more mature status than were the parents of the later-developed girls. Obviously, the behaviors that were examined here were just a small sample of social behaviors normally connected with the transition to more grown-up patterns of social life. Before more qualified conclusions can be reached about the extent of tolerance and support by parent for early- and late-developed girls' approaching mature behaviors, more research is needed to investigate parents' social reinforcement of various other facets of psychosocial transitions in their daughters.

One conclusion, however, that can be drawn from the results presented here is that the "conflict" and the "co-development" view should be recognized of as parts of the same developmental process. While seemingly contradictory, these perspectives enable us to understand in a more comprehensive way the features within the family typically connected with an earlier pubertal maturation in females.

9
Mediators of the Influence of Pubertal Timing

A DEVELOPMENTAL MODEL

The physical growth perspective taken in this book has shown that social and emotional problems in the mid-adolescent years among girls to some extent have their roots in timing of the girls' own development. It has been possible to identify for which girls whose behavior is likely to be affected by the maturity operant in the mid-adolescent years and for which domains this is true. An empirically documented relation between individual variations in pubertal maturation and individual differences in social and emotional adjustment in adolescence raises questions as to the nature of the factors that contribute to this relationship.

At the outset of this book it was emphasized that understanding of the role of physical maturity among girls requires a model for development: (a) recognizing the complex network of biological, psychological, and social influences; (b) delineating the domains that are likely to be affected by pubertal maturation, and (c) outlining the factors responsible for the relationship between pubertal maturation and these respective domains. The latter issue of how pubertal maturation affects behavior is to be addressed here.

Chapter 2 presented a general model incorporating the primary changes in self-conception and in the interpersonal area assumed to occur as a function of changes in the biological system. As a conceptual tool for systemizing which factors are the operating ones connecting pubertal development and behavior, this more general model is useful. Throughout the text, analyses have tapped the various areas delineated. For a direct test of the proposed relationship between individual differences in pubertal maturation and individual differences in behavior, a more circumscribed developmental model is investigated in this chapter. The particular role presumed to be played by *self-perceived maturity* and *peer network* for the influence of pubertal development on behavior motivates a simplification of the overall model, as depicted in Fig. 9.1.

The developmental progression is initiated by changes in the biological system. Individual differences in the timing of physical maturity are thought to have consequences primarily for (a) the self-system: resulting in differences in perceiving oneself as "psychologically," "socially," and "reproductive" mature; and (b) social network constellations: associations with peers who are congruent with one's biological stage of maturity and adopting behavior patterns common in these peer networks. Self-perception and peer constellations are proposed to connect individual differences in pubertal maturation with the individual differences in psychosocial functioning observed at a certain point in time in mid-adolescence. No direct effect of pubertal maturation on behavior is assumed. Rather, the impact is indirect, operating through self-perceived maturity and the peer network.

The utilization of the term *mediators* rather than moderators for the factors of self-perception and peer network is deliberate (cf. Baron & Kenny, 1986). If a systematic connection is found between pubertal maturation and psychosocial functioning in girls, it is expected that this relationship is indirect and mediated by self-perception and peer network factors. The intervening factors in the developmental model in Fig. 9.1 provides an answer to the question of *how* pubertal maturation expresses itself in behavior.

Existing models in research on physical growth typically involve three of the four systems depicted in Fig. 9.1—the biological, the psychological, and the behavioral. It is argued that by virtue of the fact that some girls enter puberty earlier and others later, different adult images and female orientations result, making certain types of behavior that are more mature in status more likely to emerge in the earlier-developed girls. By contrast, the model presented earlier proposes that the peer network system is a necessary piece in the developmental puzzle when examining individual differences in behavior among teenage girls. It is in the association with peers who are differentially

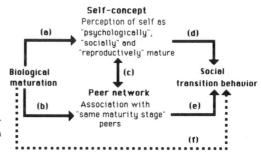

FIG. 9.1. A short model for the influence of biological maturation on social transition.

advanced socially that the key to the differences between early-developed and late-maturing girls in social transition issues is to be found.

In general, early-matured girls are assumed to develop a sense of being psychologically, socially, and reproductively more mature than later-developed girls. Consequently, they more readily adopt behaviors that are more mature in status and seek out the company of peers who match them with respect to maturity status and sexual maturation. In the association with these peer groups, which are more socially advanced than the same-sex and same-age peers of the later-developed girls, they encounter an interpersonal environment that reinforces more advanced and mature habits and leisure-time activities, reinforces the independence of parents, instills more tolerant attitudes toward norm breaking and reinforces breaking of conventional norms. In summary, many of the social transition problems expressed by the early-matured girls are to be seen as a way of identifying with and modeling their social behavior in the direction toward the socially more advanced older peers, working peers, and boyfriends. The association with the nonconventional types of peers also is thought to have negative impact on the early-matured girls' *school adjustment*. Finally, the strain caused by simultaneously adapting to two types of peer cultures, that of same-age peers and that of chronologically older peers, is assumed to affect the early-matured girls' *emotional adjustment* negatively. The more problematic life situation of being squeezed between two age strata of peers is assumed to be the psychological background for the early-developed girls' greater psychosomatic problems and greater expression of weariness of life in these years.

To the extent that the proposed developmental model for the effects on social behavior of pubertal maturation is valid, the following assumptions can be formulated:

a. *Pubertal maturation* and *self-concept*. Early-developed girls perceive themselves as more mature than others, compared to later-matured girls' perception.

b. *Pubertal maturation* and *peer network*. Early-matured girls engage more often with older peers and working peers, and they have established boyfriend relations to a greater extent than other girls.

c. *Self-concept* and *peer network*. Girls who perceive themselves as more mature than their classmates associate more often with older peers, working peers, and with boyfriends, and less with younger peers, than do girls who perceive themselves as less mature.

d. *Self-concept* and *social transition*. Girls who perceive themselves as more mature than others have more advanced social habits in mid-adolescence than do girls who perceive themselves as less mature.

e. *Peer network* and *social transition*. Girls who engage with older peers, with working peers, and with boys have more advanced social habits in mid-adolescence than do girls who lack these peer contacts.

f. *Pubertal maturation* and *social transition*. If the model is valid, little direct effect of pubertal maturation on social transition behavior should be present. The relationship between pubertal maturation and social transition issues is indirect and mediated, operating through the paths: (d) *self-concept* to *social transition*, and of (e) *peer network* to *social transition*.

In earlier chapters, ample evidence for the validity of paragraphs a (chapter 4) and b (chapter 5) have been presented. For the present purpose, before testing the mediation model, a more detailed account of data covering paragraphs c, d, and e are given.

Self-Concept and Peer Network

According to the developmental model presented in Fig. 9.1, early pubertal maturation is connected both with perceiving oneself as more mature than agemates and with more frequent association with older peers, working peers, and boys, but with less association with younger peers. The model implies a relationship between self-perceived maturity and peer interaction, such that girls who perceive themselves as more mature than agemates to a greater extent associate with older peers, working peers, and with members of the opposite sex; however they associate less frequently with younger peers. This refers to paragraph c in Fig. 9.1. Results showing the relationship between perceived maturity and the characteristics of peers outside the same-age–same-class frame is presented in Table 9.1.

Overall, the assumption of a relationship between self-perceived maturation and the characteristics of the peer network turned out to be valid. As outlined in Table 9.1, girls' conception of their maturational status was clearly connected with the types of peers in their circles of friends. The girls who felt

TABLE 9.1
Association With Different Types of Peers in Mid-adolescence as Related to Perceived Maturation

Perceived maturation	Characteristics of peers			
	Younger peers	*Older peers*	*Working peers*	*Heterosexual relations*
More mature	7.1	69.7	35.4	30.3
About as mature	15.1	46.4	18.1	21.6
Less mature	3.8	34.6	23.1	9.7
χ^2 ($df = 2$)	6.15	18.84	12.16	6.71
p	$\leq .05$	$\leq .001$	$\leq .01$	$\leq .05$

more mature than others engaged more often with older peers and with working peers, and a higher proportion of these girls had stable heterosexual relations than did the girls who felt less mature. About twice as many girls who declared that they felt more mature than their classmates had older friends and three times as many of these girls had stable relationships with boys at this age compared to the girls who felt themselves less mature than other girls. Thus, about the same relationship to nonconventional peers as was observed for actual pubertal maturation was the case for self-perceived maturation.

Also in agreement with the expectations, the earliest-developed girls had no extensive contacts with younger peers. Contrary to expectations, however, it was not the girls who conceived of themselves as less mature than their classmates who primarily associated with younger peers, but the girls who thought themselves about as mature as their classmates.

Self-Concept and Social Transition

To what extent are girls' conceptions of their maturation related to different aspects connected with the issue of social transition? This refers to paragraph d in Fig. 9.1. The question was investigated by relating differences in perceived maturation to the specific variables in mid-adolescence for which the menarcheal groups of girls differed significantly (at the .05 level at least). The relationships of perceived maturation to these indicators of psychosocial functioning in midadolescence are presented in Table 9.2.

Interpersonal Relations. A significant effect was obtained between perceived maturity and peer relations. Both the girls who felt themselves as more mature and those who felt themselves as less mature than their classmates perceived themselves to have less satisfying relations to their peers than did the girls who felt themselves about as mature as their classmates. A statistically significant effect also appeared for popularity among the opposite sex in the class, measured as the sociometric status in the class. The girls who felt themselves more mature than classmates were rated as more popular among boys than were other girls. Perceived maturity was not significantly related to any of the interpersonal scales measuring relations to adults.

Social Transition Issues. The effect of perceived maturity on the total norm-breaking scale was statistically significant at the .001 level, at least. For the respective indicators, significant differences were found for 6 of the 11 norm-breaking behaviors in Table 9.2. The directions of the relationships should be explicated. The relationships between perceived maturity and the norm-related behaviors were nonlinear. The girls who felt themselves more mature than their classmates were generally more involved in these types of

TABLE 9.2
The Relationship Between Perceived Maturation and Psychosocial Functioning in Mid-adolescence

	F	p	post hoc
Interpersonal relations			
Parent relations	1.15		
Mother relations	2.93		
Teacher relations	2.63		
Peer relations	3.27	*	
Sociometric status: own sex	0.71		
Sociometric status: opposite sex	4.38	*	1>2
Social transition issues			
Norm breaking: Total	8.75	***	1>2
Home:			
Ignoring parents' prohibitions	2.13		
Staying out late	5.18	**	1>2
Running away from home	3.02	*	
School:			
Playing truant	7.53	***	1>2
Leisure time:			
Smoking hashish	0.95		
Using harder narcotics	2.03		
Getting drunk	13.85	***	1>2,3
Stealing goods	0.20		
Forging signatures	3.16	*	
Loitering in town	3.37	*	
Pilfering	1.82		
Unwanted pregnancy	1.02		
School adjustment			
Grade-point average	4.56	*	1>3
Grade: math (academ prep)	2.36		
Stand. achievement (Swedish)	5.56	**	1>2,3
School motivation	0.30		
Hours absent from school	6.74	**	1>2
Affective reactions			
Psychosomatic reactions	4.04	*	1>2
Depressive reactions	4.97	**	1>2

*$p<.05$. **$p<.01$. ***$p<.001$.

behaviors relative to other girls. Moreover, the group of girls who felt less mature than their classmates had somewhat higher rate of norm breaking than did the girls who felt themselves about as mature as their classmates. Of the different types of norm-breaking behaviors, the impact of perceived maturity was greatest for drinking behavior. In this connection it can be mentioned that in a recent Swedish study on 4,500 pupils in the upper grades of the comprehensive school, it was found that the girls who judged themselves as "early-developed" drank more frequently than did other girls (Marklund, 1985).

Perceived maturity was not connected with unwanted pregnancy among girls in midadolescence.

School Adjustment. As can be seen in Table 9.2, an impact of perceived maturity occurred for three of the variables in the school adjustment area: grade-point average, standardized achievement results in Swedish, and school absenteeism. Girls who saw themselves as more mature than their female classmates had higher grade point averages (less mature: 2.9; about as mature, 3.2; more mature, 3.4) and higher mean standardized achievement results in Swedish (less mature, 49.9; about as mature, 54; more mature, 58.3). They also had higher absenteeism from school. They were absent an average of 62 hours in Grade 8, in comparison with 44 hours for the girls who felt about as mature as others, and 54 hours for the girls who felt less mature.

Affective Reactions. The girls who felt themselves more mature than agemates showed more psychosomatic and depressive reactions than the girls who experienced themselves about as mature as others in the class.

Comments. In view of the fact that an earlier pubertal development was thought to be connected with perceiving oneself as more mature than agemates, such self-perception would presumably be related to more conflicting adult relations, perception of less satisfactory peer relations, more social transition problems, less school interest, and more emotional problems. The results confirmed the expectations as far as norm-violating activities of different kinds and affective reactions were concerned. However, perceived maturity was not related to either parent or teacher relations. Contrary to what was expected, it was the girls who perceived themselves to be more mature than agemates who showed higher school achievement than other girls. This implies, by inference, the existence of a group of girls, whose perceived maturity is unrelated to actual pubertal maturation but is closely associated with superior school achievement.

Peer Network and Social Transition

To what extent are the type of peers with whom girls associate connected with social transition issues in midadolescence? This issue refers to paragraph e, and answers to the question can be found in Table 9.3. The table summarizes results of t tests between girls who engaged versus those who did not engage with the respective peer types outside the same-age–same-class frame in midadolescence.

Interpersonal Relations. To have stable heterosexual relations was generally connected with less satisfying adult relations but more satisfying relations to peers and higher popularity among boys. For all three scales measuring relations to adults—parent relations, mother relations, and teacher relations—girls who had stable opposite sex relations reported less satisfying relations than girls who had no such stable heterosexual relations. In addition, to associate with working peers was connected with worse relations to the mother. Generally, girls who associated with nonconventional peers reported more satisfying relations to their peers and they were more popular among boys in their class than were the girls who did not have nonconventional peers. Conversely, engaging with younger peers was connected with lower opposite-sex popularity.

Heterosexual relations was the peer type that was most significantly related to the girls' mother relations. This is in line with several previous studies in the literature that have shown that girls' contacts with members of the opposite sex might have a negative relationship with relations to their mothers (cf. Frisk, Tenhunen, Widholm, & Hortling, 1966; Lewis, 1963). The additional evidence that associating with working peers had almost the same negative impact is a further insight into the aspects of the peer network that might contribute to conflicting mother relations.

Girls with a steady boyfriend and those with working or older peers reported more satisfying peer relations than did girls who did not associate with boys steadily and did not engage with working peers or with older peers. Having each of the peer types was also positively related to popularity among boys, whereas having younger peers was negatively related to opposite-sex popularity. These results support the conclusions that others have drawn, for example Faust (1960) and Kandel and Lesser (1972): namely, that opposite-sex relations and older peer contacts are status-heightening qualities for teenage girls. Status in the peer group during this period seems to concern properties that have developmentally important values at this age—establishing relations with the opposite sex and acquiring more mature and advanced social life patterns.

Social Transition Issues. A rather conclusive pattern was found for the

TABLE 9.3
Relation of Mid-adolescent Peer Types to Psychosocial Functioning[a]

	Younger peers		Older peers		Working peers		Heterosexual relations	
	t	p	t	p	t	p	t	p
Interpersonal relations								
Parent relations	0.67		0.42		−1.72		−3.14	**
Mother relations	1.60		−0.44		−2.31	*	−2.48	*
Teacher relations	1.51		−0.23		−1.29		−2.21	*
Peer relations	−1.96		2.20	*	2.70	**	3.94	***
Sociometric status: own sex	−1.38		1.19		−0.33		1.01	
Sociometric status: opposite sex	−3.50	***	4.45	***	2.13	*	5.16	***
Social transition issues								
Norm breaking: Total	3.06	**	−2.79	**	−3.76	***	−4.46	***
Home:								
Ignoring parents' prohibitions	0.50		−0.27		−2.42	*	−3.51	***
Staying out late	2.10	*	−3.74	***	−3.62	***	−3.94	***
Running away from home	1.10		0.75		−0.03		−2.75	**
School:								
Playing truant	3.98	***	−2.68	**	−2.85	**	−3.42	***
Leisure time:								
Smoking hashish	0.54		−2.13	*	−1.25		−0.19	
Using harder narcotics	1.21		−0.22		0.41		−1.05	
Getting drunk	5.07	***	−4.26	***	−3.85	***	−5.49	***
Stealing goods	3.74	***	−0.15		−2.02	*	−2.79	**
Forging signature	0.75		−0.92		−1.87		−2.02	*
Loitering in town	0.27		−1.33		−0.41		−1.62	
Pilfering	2.07	*	−0.47		−2.25	*	−2.11	*
Unwanted pregnancy	2.99	**	−0.13		−1.27		−2.20	*
School adjustment								
Grade-point average	1.20		−2.01	*	−3.23	***	−2.18	*
Grade: math (academ prep)	0.60		−2.40	*	−3.28	***	−3.01	**
Stand. achievement (Swedish)	0.05		−0.14		−0.90		−1.28	
School motivation	1.14		−1.55		−2.18	*	−1.42	
Hours absent from school	1.89		−4.55	***	−3.41	***	−0.80	
Affective reactions								
Psychosomatic reactions	1.52		−1.53		−0.60		−1.42	
Depressive reactions	2.65	**	−0.80		−1.69		−0.23	

*p<.05. **p<.01. ***p<.001.

[a] a positive *t* value refers to *better* interpersonal relations, *less* social transition problems, *higher* school adjustment, and *less* psychosomatic and depressive reactions.

connection between peer-type associations and norm breaking, such that girls who engaged with nonconventional types of peers were more involved and girls who engaged with younger peers were less involved in various types of norm breaking. Again, except for smoking hashish, to have stable opposite-sex relations tended to have the greatest implications for behavior of this type. Girls with steady relations to boys reported higher rates of violating norms at home, at school, and during leisure time than did girls without such steady relations. Significant differences between girls who had stable boyfriend relations and girls who did not were found for 8 of the 11 behaviors in Table 9.3. The results have important implications for how to view and understand the appearance of problem behavior among teenage girls and which factors in the social ecology of the girls in these years are responsible. Although not having its roots precisely in male norm breaking, female norm breaking tends to be closely connected with associations with the opposite sex and the norm climate prevailing among males.

To engage with working peers also implied a higher risk for norm violations than not having such friendship contacts. Girls with employed peers in their circle of friends had a higher rate of norm breaking generally than did girls without such peers. The effect of this peer type was significant for 6 of the 11 norm-related behaviors.

Older peers was the third peer-related characteristic that had a strong positive impact on norm violations in mid-adolescence. For one particular norm-breaking activity, smoking hashish, it was the only one of the four peer characteristic variables that showed a significant effect. Girls with older peers reported having smoked hashish more often than did girls without older peers.

The impact of engaging with younger peers on norm breaking was significant for the total norm-breaking scale and for five of the specific behaviors in the scale. Girls who engaged with friends who were younger than themselves had a lower rate of norm breaking generally than did the girls who lacked such friends. The effect of associating with younger peers was particularly associated with drinking. Overall, younger peers tended, as expected, to function as a protective factor against engaging in norm-violating activities.

It comes as no surprise that heterosexual relations was the peer type that was significantly related to unwanted pregnancy. Of 26 girls who had complete data for the peer network variables and had undergone an abortion, 46.2% had stable boyfriends in comparison with 23.1% among girls in the total research group. A more unexpected finding was that having younger peers also was significantly negatively related to pregnancy in mid-adolescence. Unwanted pregnancies thus occurred more seldom among girls who had younger friends than among girls who did not associate with such peers. Of the girls who had had one or more unwanted pregnancies, only 1 girl of the 26 (3.8%) had younger peers in her circle of friends in comparison with 13.7% of the girls in the total research group.

School Adjustment. Lower marks were connected with associating with all three types of nonconventional peers. Lower school motivation was expressed by the girls who associated with working peers. Girls who engaged with older or working peers were more frequently absent from school.

Of the four peer types, working peers tended to have the greatest overall effect on school adjustment. Girls who associated with peers who were already employed received lower marks, were less interested in schoolwork, and were more frequently absent from school, compared to girls who did not engage with such peers.

Affective Reactions. When the girls in mid-adolescence reported on different aspects of their emotional adjustment, two of the domains, psychosomatic reactions and depressive reactions, differed considerably among the menarcheal groups, with the early maturers reporting greater difficulties. These differences were interpreted in terms of more stress among the early-matured girls to adjust to different peer environments.

As shown in Table 9.3 there is little justification for such an assumption as far as relations to nonconventional peers are concerned. Only one significant difference appeared, and it concerned differences with respect to depressive reactions between girls who associated with younger peers and those who did not. To associate with younger peers was connected with reports of less severe depressive reactions.

INVESTIGATING THE PROPOSED MEDIATING MODEL

What remains now is to apply a systematic design to examining the issue of the role of perceived maturity and the peer network as mediators for the association of pubertal maturation to psychosocial functioning. If the hypothesis is correct, that perceived maturity and peer contacts are the principal factors mediating the association between pubertal maturation and actual behavior in mid-adolescence, then the original differences in psychosocial functioning among the menarcheal groups of girls should disappear or decrease considerably once differences among the groups of girls, with regard to these mediating factors, are corrected for.

Two questions arise. The first one is: Does there remain any significant effect of menarcheal age on the respective psychosocial indicators once the variation in the dependent variables due to the proposed mediators is eliminated? This is analogous to asking whether there is a hypothesized nonsignificant pathway f in Fig. 9.1. Our expectations are that much of the original effects of pubertal maturation on various forms of behavior are reduced once corrections have been made for the self-concept and the peer-type measures.

The second question is: Which of the self-concept and the peer-type variables is the most potent mediator for the relationships of menarcheal age to the respective dependent measures involving diverse aspects of interpersonal relations, social behavior, school adjustment, and emotions?

Methodological Considerations

Different measurement options are available for the analysis of the mediating function of perceived maturity and peer network. The most common would be some form of regression or covariate analysis. It is clear from earlier analyses that nonlinear relationships between menarcheal age and the variables within the domains of interpersonal relations, social and emotional adjustment were common. For the most part the effects were concentrated to the early-matured girls, particularly to the earliest-matured group of girls. For other variables, like peer status, the effects were concentrated to both the extreme ends of the pubertal-timing dimension. Therefore, ordinary linear techniques for data treatment (multiple regression, LISREL, etc.) were not appropriate. The principal treatment technique adopted had to treat the menarcheal age variable as a categorical or nonmetric variable.

The present issue is to partial out the influence on the dependent variables accounted for by self-perceived maturity and the peer-type variables, to see if the effect of menarcheal age on the dependent variables are reduced after these intervening factors are brought under control. The general question of how to remove the extraneous variance of certain intervening variables from the association between the independent and the dependent variable is a difficult one. As has been repeatedly emphasized (Glass, Peckham, & Sanders, 1972; Snedecor & Cochran, 1967), there are no ideal techniques for controlling for pre-existing differences between groups in nonexperimental settings. Results from such analyses always have to be interpreted with caution.

Given the nonlinear relationships, an analysis of covariance was applied as the most appropriate method for conducting the control of perceived maturity and the peer-type variables. Several assumptions imposed by the ANCOVA on the data should be met. As with the ordinary analysis of variance, the ANCOVA ideally requires a randomized experiment (e.g., that the allocation of the subjects to the categories of the menarcheal age variable should be done by randomization). In our case the menarcheal variable was a nonrandom event. This situation in nonexperimental research, of utilizing nonmanipulative independent variables with analysis of variance, we share with many other investigations both in longitudinal and cross-sectional research. Other requirements of the ANCOVA are parallel within-group regression slopes, error-free measure, and no correlation between the independent variable and the covariate. The last requirement is the most problematic. Such relationships occur in our case, and it will, to some extent, affect the precision of the F test. Therefore, the level of significance should be interpreted cautiously.

A final comment should be made with regard to the correcting for the covariates. The ANCOVA controls for the linear effects of the covariates on the dependent variables. To the extent that nonlinear relationships exist (as was seen previously, such was the case for self-perceived maturity at several instances), the impact of the mediating factors will be underestimated and, thus, bias the results *against* confirmation of our hypothesis.

The specific calculations were performed in the following way:

1. Ordinary ANOVAs were initially computed for the subjects, with menarcheal age as the independent variable and each of the variables investigated in mid-adolescence serving, in turn, as the dependent variable. The calculations were performed for all the variables in mid-adolescence that had differed significantly (at the .05 level at least) among the menarcheal groups of girls. The calculations were done for the subjects in the research group with complete data for self-perceived maturity and the four peer-type variables. This was done in order that all calculations would refer to the same individuals so as to have a basis for comparing the F values in the subsequent covariance analyses.

Of the 466 girls in the research group, 381 girls or 81.8% had complete information with respect to perceived maturation and peer-type characteristics. The percentage of girls with complete data was somewhat greater for the later-developed girls than for the early developers (menarcheal group 1: 79.6%; group 2: 75%; group 3: 84.9%; group 4: 83.7%). The greater loss of subjects among the early-developed girls, among whom the effects were primarily located in previous analyses, has the consequence of somewhat attenuating the original relationship between menarcheal age and the dependent variables under study. As is seen here, a few of the earlier-reported significant relationships between menarcheal age and the respective interpersonal, social and emotional variables were not significant for the more circumscribed group with complete data for the mediating factors.

2. Next, a series of ANCOVAs were computed. They tested the extent to which menarcheal age had a statistically significant effect on the respective dependent variables after adjustments had been made for the covariates, self-perceived maturity and the peer-type variables.[1]

---

[1] The fact that the covariates were not measured on the same occasion should be noted. The peer-type variables—younger peers, older peers, and working peers, were measured at the age of 14.7 years in the Peer Contact Inventory. The covariate Heterosexual Relations (see the question "Do you have or have you had a steady relationship with a boy?" in the Opposite-Sex Relations Scale in Table 5.7 was administered at 14.10 yrs in the Adjustment Screening test. However, the possible bias, caused by having data on peer relations from two occasions, with an interval of 4 months, can be regarded as quite marginal.

The following items subsumed under broader domains were investigated. As described, the criterion for inclusion was that a particular dependent variable had previously shown significant relationships with the menarcheal age variable:

- <u>Interpersonal relations</u>
  Parent relations
  Mother relations
  Teacher relations
  Peer relations
- <u>Social transition problems</u>
  Normbreaking: Total
   Ignoring parents' prohibitions
   Staying out late in evenings
   Playing truant
   Smoking hashish
   Getting drunk
   Pilfering
   Stealing goods
   Running away from home
  Unwanted pregnancy[2]
- <u>School adjustment</u>
  Grade-point average
  Grade: Mathematics (academic preparatory)
  Standardized achievement: Swedish
  School motivation
- <u>Emotions</u>
  Psychosomatic reactions
  Depressive reactions

To maximize the initial differences, only the scale items that earlier had shown significant differences among the menarcheal groups of girls were saved for the analysis of covariance. This means, for example, that six items of the nine in the scale Parent Relations (Table 5.12) were subjected to the ANCOVA.

Excluded from further investigations, due to an insignificant effect of

---

[2] Even though the differences for unwanted pregnancy in the teenage years among the menarcheal groups for the more selected sample did not reach the .05 level (the statistical effect of menarcheal age was at the .08 level), due to the theoretical importance of knowing what factors were connected with teenage pregnancies, it was decided to include this variable in the ANCOVA.

menarcheal age, were measures on social status among the same sex as well as among the opposite sex, three items on norm-related behavior (use harder narcotics, forge signatures, loiter in town every evening), and the measure of school absenteeism.

The Effect of the Mediating Variables

The first issue to be examined is to what extent the original effects of the menarcheal age variable on the dependent measures decreased after adjusting for group differences in perceived maturity and associating with the four peer types. This is a critical issue. If a substantial impact of menarcheal age were to prevail after the corrections for the presumed mediating factors, the results would not support the validity of the proposed model. The results covering the issue are presented in Table 9.4.

The results in Table 9.4, which concern the present issue, are the unadjusted effects of menarcheal age and the adjusted effects of the independent variable, controlling for all covariates.

Interpersonal Relations. When the effect of menarcheal age was adjusted for differences among the menarcheal groups of girls in perceived maturity and peer network, there were no statistically significant efforts remaining for three of the four interpersonal scales; those measuring relations to adults: parent relations, mother relations, and teacher relations. A significant effect still prevailed for the peer relations measure.

Social Transition Issues. For the total norm-breaking scale the original difference among the menarcheal groups of girls decreased considerably after the introduction of the covariate terms. It was still significant after correcting for the covariates, but at a much lower magnitude. For specific norm-violating behaviors, the differences found initially for "ignoring parents' prohibitions," "pilfering from a shop," "stealing goods," and "running away from home" were reduced to insignificance after correcting for the covariates. The differences among the menarcheal groups for "staying out late without permission," "playing truant," and "getting drunk" were also much reduced after introducing the covariates, but they still remained significant at the .05 or at the .01 level. The norm violation behavior "smoking hashish" seemed to be the behavior least affected by the covariate correction. Differences among the menarcheal groups of girls for this behavior have to be explained by factors other than those proposed.

One important result concerns the data on unwanted pregnancy. As can be seen in Table 9.4, the original effect of menarcheal age on abortions decreased considerably after the covariance control.

TABLE 9.4
The Relationship of Menarcheal Age to Various Forms of Psychosocial Functioning With and Without Control for all Covariates, and for the Strongest Covariate, Respectively

	Control for all covariates				Control for the strongest covariate			
	Unadjusted effect		Adjusted effect					
	F	p	F	p	Covariate	F	p	n
Interpersonal relations								
Parent relations	3.85	**	1.68		heterosex rel	2.25		381
Mother relations	3.87	**	1.62		heterosex rel	2.62		368
Teacher relations	3.22	*	1.51		heterosex rel	1.88		381
Peer relations	3.53	*	2.78	*	heterosex rel	2.61		380
Social transition issues								
Normbreaking: Total	11.11	***	3.27	*	heterosex rel	6.07	***	372
Home:								
Ignoring parent's prohib.	3.06	*	0.59		heterosex rel	1.42		378
Staying out late	9.17	***	2.90		heterosex rel	5.30	***	381
Running away from home	4.92	***	2.33		heterosex rel	3.32	*	364
School:								
Playing truant	11.61	***	5.08	**	heterosex rel	7.91	***	380
Leisure time:								
Smoking hashish	4.75	**	4.21	**	older peers	3.42	*	381
Getting drunk	12.56	***	2.88	*	heterosex rel	6.73	***	380
Stealing goods	4.64	**	1.94		heterosex rel	2.91	*	379
Pilfering	3.70	*	0.96		heterosex rel	1.95		381
Unwanted pregnancy	2.30	.08	0.78		heterosex rel	1.32		381
School adjustment								
Grade-point average	2.73	*	1.13		working peers	1.44		380
Grade: math (academ prep)	6.23	***	3.09	*	working peers	4.17	**	247
Stand. achievement (Swedish)	7.06	***	5.94	***	perc maturity	6.09	***	343
School motivation	4.92	**	2.09		working peers	2.57		366
Affective reactions								
Psychosomatic relations	6.24	***	4.37	**	perc maturity	5.23	**	362
Depressive reactions	7.03	***	5.05	**	perc maturity	5.94	***	368

*$p<.05$. **$p<.01$. ***$p<.001$.

School Adjustment. Differences among the four menarcheal groups with respect to grade-point average and girls' self-reported school interest (containing the significant variables in the school adjustment scale in Grade 8), were both reduced to insignificance once the covariates were introduced into the equation. For the academic preparatory course in mathematics, substantial reduction of the impact of menarcheal age was observed, albeit there was a remaining, statistically significant effect of menarcheal age after the covariate correction. The introduction of the covariates had little impact on standardized achievement results in Swedish.

Affective-Reactions. As can be seen in Table 9.4, the expectations that perceived maturity and peer network factors operated as the prime mediating factors for emotional differences among the menarcheal groups of girls did not turn out to be valid. The initial differences among the menarcheal groups for psychosomatic reactions and depressive reactions, which approached significance at a high level, were little affected by the covariance adjustment. To explain differences among the menarcheal groups of girls with respect to psychosomative and depressive reactions, factors other than perceived maturity and peer network must be considered.

To summarize, the present results indicate that the mediation model fit the data well as far as interpersonal relations and social behavior are concerned, and for measures of school satisfaction and marks within the school adjustment domain. However, another picture emerged for the differences in affective reactions between early- and later-developed girls. Only slight reductions were found in the original effect of menarcheal age on psychosomatic reactions and depressive reactions after controlling for the mediating factors.

The Strength of the Mediators

The model proposed was investigated with regard to the total effect of the self-concept and peer network variables as mediators. All covariates were entered simultaneously into the equations. We now turn to the second question: namely, which of the covariate terms was the most effective mediator for the respective dependent measures?

For every dependent variable, each one of the five covariates was introduced separately into the equation, with menarcheal age as the independent variable. To identify the covariate that had the greatest mediating power, the covariate was selected that, after its effect was partialed out, yielded the lowest F value for the effect of menarcheal age on the dependent variable under study. By comparing the adjusted effect of menarcheal age when all covariates are corrected for with the adjusted effect after correction only for the strongest covariate, it is possible to ascertain the extent to which this covariate constituted a powerful mediator. The results, pertaining to which of the five covariates produced the greatest drop in the relationship of menarcheal age to the dependent variables under consideration, and the resulting F values, are shown in the right columns in Table 9.4.

Interpersonal Relations. Heterosexual relations seemed to be the essential mediator between menarcheal age and the dependent measures in the interpersonal relations area. When this peer-type covariate was entered as a single covariate, no significant effect of menarcheal age remained on the scales measuring parent relations, mother relations, teacher relations, and peer relations. In fact, it did not seem to matter much whether all the covariates or just the heterosexual relations variable was introduced as the single covariate. These results suggest that relations to boys in mid-adolescence contributed much to the influence of menarcheal age on relations to adults and peers.

Social Transition Issues. For the variables in the social adjustment domain, association with boys again seemed to be the most important mediator. For all but one of the norm-breaking variables, "smoking hashish," heterosexual relations was the covariate that produced the largest drops in the F values for the effect of menarcheal age on the respective norm-related behaviors. With regard to "smoking hashish," older peers was the only covariate that demonstrated a significant contribution.

For the measure of unwanted pregnancy, heterosexual relations, as expected, had the greatest impact as a mediating factor. No significant differences among the four menarcheal groups of girls with respect to abortions in mid-adolescence were obtained once this peer-type variable was corrected for.

School Adjustment. Whereas heterosexual relations was the dominant mediator within the areas of interpersonal relations and social transition issues, working peers was the important mediator in the school adjustment domain. No significant differences among the menarcheal groups of girls were obtained for grade point average and school motivation after correction for the single covariate working peers. Moreover, a modest decrease in the effect of menarcheal age on grades in mathematics was found after control of this peer type.

With respect to results on standardized achievement in Swedish, perceived maturity was the only significant covariate. However, little reduction in the effect of menarcheal age on standardized achievement was obtained after correction for this covariate.

Affective Reactions. Perceived maturity was the covariate that produced the greatest drop in the F values for the effects of menarcheal age on psychosomatic reactions and depressive reactions. Its contribution as a mediating factor, however, was limited.

To summarize, the results suggest that the mid-adolescent peer types had *differential* impact on the domains investigated. Relations to boys was the essential mediator for the effect of menarcheal age on various indicators in the interpersonal relations and in the social adjustment domains. Working peers showed the highest contribution as a mediating factor within the school adjustment area. Finally, the differences among the menarcheal girls with respect to psychosomatic and depressive reactions were but marginally reduced by the covariate perceived maturity.

DIFFERENTIAL SUSCEPTIBILITY TO PEER INFLUENCES

Two main arguments have been proposed for an assumed earlier display of social transition behavior among early-developed girls. The first deals with the *opportunity* to encounter advanced social behavior patterns. Teenage girls, based on their timing of physical development, encounter differential opportunities to have contact with more "advanced" forms of social behavior. Earlier pubertal maturation in girls is connected with associating with peers who are chronologically older than themselves and with peers outside the school context. An earlier sexual development in these girls makes them more likely to establish relations with older boys. Overall, these nonconventional peer contacts yield more opportunities for the early-developed girls to encounter and practice more mature forms of social behavior than for the later-developing girls. One implication of this differential opportunity argument is that we do not expect much difference with regard to social transition issues between early- and later-developed girls who do not associate with nonconventional peers.

The second argument concerns *susceptibility* to peer influences. An earlier pubertal maturation is thought to be connected with perception of oneself as more mature than other same-age peers, with the consequent implication of being more open to change in the direction of more mature forms of behavior. Of two girls who both have older peers in their circle of friends, the girl who has come further in her pubertal maturation is assumed to be more sensitive to the social lifestyle among her older peers and more psychologically "ready" to change as a result of this peer influence. So, the hypothesis of differential susceptibility suggests that the peer influence for the girls who associate with nonconventional peers will be relatively stronger for the early-maturing than for the later developed girl.

The two hypotheses on issues connected with social transition for early- and later-developed girls are investigated in the following. They have different implications for (a) the girls who did not associate with peers outside the same-age–same-sex group—no effect of menarcheal age on the dependent variables are expected, and (b) the girls who had peer contacts outside the same-age–same-sex group—strong differences among the four menarcheal groups of girls are expected for the dependent measures. Because a total presentation would be too lengthy, a representative sample of the findings is shown here.

Interpersonal Relations. As was shown in Table 9.4, heterosexual relations was the covariate that had the greatest impact as a mediator for mother relations. Consider now the differences in mother relations among the menarcheal groups of girls separately for the girls with and for the girls without stable heterosexual relations. As shown in Fig. 9.2, there were little differences in conflict level among the four menarcheal groups for these girls who had no stable relation to boys. A one-way ANOVA revealed no statistically significant differences among the four groups of girls ($F = 1.48$, $df = 3,278$, ns). This result supports the hypothesis of an absence of impact of pubertal maturation on adult relations for girls who are outside the influence circle of "critical" peer groups. In contrast, significant differences among the menarcheal groups were indeed obtained for the subsample of girls with established boyfriend relations ($F = 2.84$, $df = 3,83$, $p < .05$). In accordance with the second hypothesis, earlier-developed girls in this subsample reported higher conflict level than did the later-maturing girls (observe in Fig. 9.2 that the later-developed girls with stable opposite-sex relations who had reached their menarche at age 13 or later had in fact more satisfying relations to their

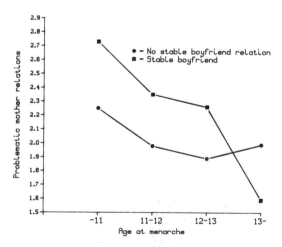

FIG. 9.2. The impact on mother relations of having versus not having a steady boyfriend among girls in four menarcheal groups.

mothers than did the later-developed girls who did not have such stable boy relations).

The question may be framed similarly for the comparison of early- and later-developed girls, with and without younger peers. A graphical presentation of this comparison is shown in Fig. 9.3 Because only two girls among the most early-developed had associated with younger peers, the two earliest-developed groups of girls were merged into one group (girls with menarche before age 12).

Consider first the majority of girls who did not associate with younger peers. Among these girls the earliest-maturing girls reported greater mother conflicts than did the two later-developed groups of girls. The other interesting comparison is the one comparing mother relations, within each of the three menarcheal groups of girls, between girls who had versus did not have younger peers in their circles of friends. As shown in Fig. 9.3, a greater impact of younger peers was obtained for the latest-developed girls than for the earlier-maturing groups of girls. Thus, analogous to the greater impact of opposite-sex contacts among the earliest-maturing girls, there was greater effect of younger peers on the latest-maturing girls.

The results reported here emanated from hypotheses that were specified beforehand: namely, that girls would be differentially susceptible to the influence of peers who were congruent with their level of pubertal maturation and that this peer contact affected their parental relations. However, it must be acknowledged that the causal effects could possibly have operated in the opposite direction. For example, it might be the case that the early-maturing girls who got along well with their mothers preferred to associate with conventional types of peers in order not to endanger their mother relations.

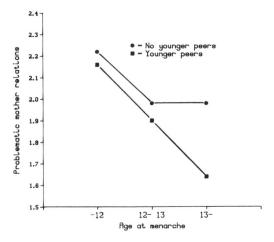

FIG. 9.3. The impact on mother relations of having versus not having younger friends among girls in three menarcheal groups.

Social Transition Issues. Figures 9.4a to 9.4d present graphical accounts of differences in total norm breaking (scale) among the menarcheal groups of girls, separately for the girls with and for the girls without younger friends, older friends, working friends, and stable boy relations, respectively.

Differences in norm violation among the menarcheal groups of girls without stable heterosexual relations in Fig. 9.4d were significant at the .05 level ($F = 3.26$, $df = 3,282$, $p < .05$), but more clearly so for the girls with stable opposite-sex relations ($F = 6.39$, $df = 3,82$, $p < .001$). Similar results were obtained for working peers (Fig. 9.4c). Among the girls who had no working friends, the differences among the menarcheal groups were significant at the .01 level ($F = 3.85$, $df = 3,283$, $p < .01$). Stronger differences among the four menarcheal groups were found for the subsample of girls who had friends who were employed ($F = 8.94$, $df = 3,81$, $p < .001$). A similar pattern was found for associating with older peers (Fig. 9.4c). There were no significant differences among the menarcheal groups of girls who had no older peer in their circles of friends ($F = 2.15$, $df = 3,180$, ns). By contrast, for the subsample of girls who had friends who were older than themselves, there were highly significant differences in norm violations among the menarcheal groups ($F = 9.56$, $df = 3,184$, $p < .001$).

Overall, for girls who declared that they had no friends of the nonconventional peer type (neither older peers, working peers, or stable boyfriend relations), no significant differences in norm violation related to menarcheal age was found ($F = 1.04$, $df = 3,153$, ns). By comparison, among the girls who associated with at least one nonconventional friend, profound differences in norm violation were found between the early-maturing and the later-maturing girls ($F = 9.24$, $df = 3,211$, $p < .001$).

Consider next the impact of having younger peers (Fig. 9.4a). More limited impact of pubertal maturation was found among the girls who had no younger peers ($F = 3.15$, $df = 3,43$, $p < .05$) than among the girls who associated with younger friends ($F = 8.29$, $df = 3,321$, $p < .001$). Again, among the later-developed girls, to have younger peer contacts was connected with less norm breaking than to have no such friends. Among the early-developed girls, small differences with regard to norm breaking frequency were present between those who had younger peers and those who did not engage with younger friends.

Altogether, for the peer types proposed to facilitate violation of conventional norms—older peers, working peers, and heterosexual relations—the peer effect was greater within the group of early- than within the group of later-developed girls. In contrast, for association with younger peers, proposed to inhibit social maladaptive behavior, the positive effect of such peer impact was stronger among the later-developed girls than among the early developers.

Alcohol drinking and the developing of alcohol habits among teenagers has received much attention in the teenage literature. Investigations of the

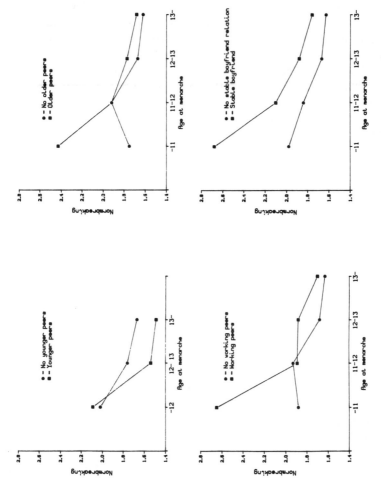

FIG. 9.4. The impact on norm breaking in mid-adolescence among girls in four menarcheal groups: (a) of associating versus not associating with younger peers, (b) of associating versus not associating with older peers, (c) of associating versus not associating with working peers, (d) of having versus not having a steady boyfriend relation.

initiation into drinking habits among females, however, have not been conducted from a bisocial point of view as presented here. Therefore, illustrations of the menarcheal impact on drunkenness frequency, when the total subject sample is broken down in subgroups with and without different types of peer contacts, are graphically depicted in Figs. 9.5a to 9.5d.

For each one of the peer types, older peers, working peers, and heterosexual relations, the menarcheal impact on drunkenness frequency in girls was greater among the girls who engaged with these types of peers than among the girls who did not have the respective peer associations. For the subsample that had no nonconventional friends at all in their circles of friends, there were no significant differences in frequency of drunkenness among the menarcheal groups ($F = 2.17$ $df = 3,156$, ns). Pronounced differences, with early-developed girls reporting more drunkenness occasions were found in the subsample of girls who had such peers ($F = 8.05$, $df = 3,216$, $p < .001$). As can be seen in Fig. 9.5b to 9.5d, the impact of associating versus not associating with nonconventional peers was relatively stronger among the early-developed than among the later-developed girls. The results strongly favor the hypotheses that (a) the impact of pubertal maturational is limited among girls who have no nonconventional peers, and (b) that nonconventional peers have more negative impact on a girl if she is early than if she is later-developed.

The opposite situation occurred for associating with younger peers (Fig. 9.5a). Such peer association had, relatively speaking, more effect on the later-developed girls than on the early-maturing. In fact, there was not one single girl of the later-developed girls who associated with younger peers who reported ever having been drunk!

School Adjustment. It was shown in Table 9.4 that, within the realm of school adjustment, working peers was the essential mediator for the effect of menarcheal age on grades and school motivation. Figure 9.6a shows a graphic representation of the grade-point averages for early- and later-developed girls with and without working peers. The same comparison of the influence of employed peers with regard to self-reported school motivation is presented in Fig. 9.6b.

For both grade-point average and for school motivation there was a menarcheal impact at the .05 level, at least, for the subgroup of girls who associated with employed peers (grades: $F = 2.79$, $df = 3,81$, $p < .05$; school motivation: $F = 3.20$, $df = 3,77$, $p < .05$). No significant impact of menarcheal age on grade point average nor on school motivation was obtained among the girls who did not have such peer association (grades: $F = 1, 67$, $df = 3, 291$, ns; school motivation: $F = 1.24$, $df = 3, 281$, ns). Both Figs. 9.6a and 9.6b show that the impact of working peer association was stronger for the early- than for the later-developed girls who engaged with peers already employed full time.

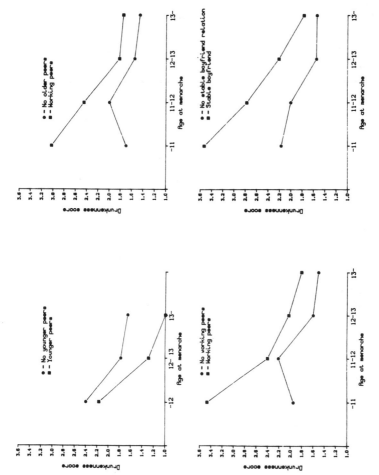

FIG. 9.5. The impact of drunkenness frequency in mid-adolescence among girls in different menarcheal groups: (a) of associating versus not associating with younger peers, (b) of associating versus not associating with older peers, (c) of associating versus not associating with working peers, (d) of having versus not having a steady boyfriend relation.

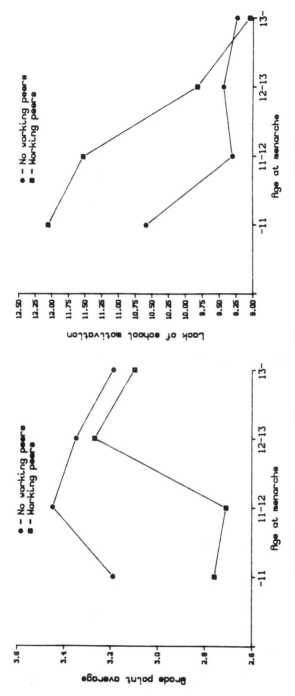

FIG. 9.6. (a) The impact on grade-point average of associating versus not associating with working peers among girls in four menarcheal groups. (b) The impact on school motivation of associating versus not associating with working peers among girls in four menarcheal groups.

A summary of the results now can be made. Whether relations to parents, norm-breaking behavior, or school adjustment is considered, a similar tendency appears. To associate with nonconventional peers had more impact on the behavior of the early- than the later-developed girls. Given two early-developed girls, the girl who engaged with nonconventional peers tended to have more conflicting mother relations, a higher rate of norm breaking, and lower school adjustment as reflected in marks and reported school disinterest. In contrast, a similar comparison between girls who had nonconventional peers and girls who did not, among the later-developed girls, did not reveal much difference in the previous respects. This supports the hypothesis of greater susceptibility to the influence of nonconventional types of peers among early-developed girls compared with later-maturing girls.

No statistically significant effects of menarcheal age on mother relations, norm breaking, or school adjustment, were found for the subsample of girls who did not associate with nonconventional peers. The results overall underline the importance of the peer contact in mid-adolescence for the understanding of differences in social adjustment between early- and later-developed girls.

Moreover, it is of particular interest to recognize the impact on behavior for associating with younger peers. For mother relations and norm breaking, the impact of this type of friends was greater for the later-developed than for the early-developing girls. For the later-developed girls, having younger peers was connected with more satisfying mother relations and less norm breaking, relative to those same issues among late-maturing girls who did not associate with younger peers.

DISCUSSION

The issue of the mediating factors for the impact of pubertal timing on psychosocial functioning in mid-adolescence was empirically approached in this chapter. It was advocated that a great part of the differences observed between early- and later-matured girls, with regard to social adjustment in a broad sense, to relations with parents, and to relations with teachers and peers, as well as with respect to the emotional adjustment of the girls in this period, had to be seen in the light of the role of girls' concept of their maturity and their interpersonal ecology in the mid-adolescent years. The hypotheses were built on a developmental model hypothesizing that individual differences in the timing of physical maturity primarily resulted in individual differences in perceiving oneself as socially and psychologically mature and in associating with peers who were congruent with one's biological stage of maturity. The self-system and the peer network system of factors were assumed to be the major mediators responsible for the link between the menarcheal age of girls

DISCUSSION

and their psychosocial adjustment in mid-adolescence. In terms of measurement, few differences in central social and emotional variables were presumed to appear between early- and later-matured girls once the factor of self-perceived maturity and the peer-characteristics factors were controlled for.

The first question raised dealt with the overall validity of the model and concerned the direct effect of menarcheal age on various indicators of satisfaction in interpersonal relations, social transition, school adjustment, and emotional reactions, independent of its mediated impact through perceived maturity and peer network. No such significant impact was expected.

When the differences in self-conceived maturity and associating with older, working, or younger peers, and heterosexual relations, were controlled for, much of the difference with respect to interpersonal relations, social adjustment, and school adjustment among the four menarcheal girls was eliminated. However, the differences between early- and later-developed girls with respect to affective reactions—psychosomatic reactions and depressive reactions—remained strong after corrections for the girls' self-concept and peer association. Thus, the results supported the mediation model as far as interpersonal relations, social behavior, and school adjustment were concerned, but not with regard to affective reactions.

A second question concerned which of the covariates mediated the effects of menarcheal age on the indicators within the areas under investigation. In the domains of interpersonal relations, social transition issues and school adjustment, the peer-type variables rather than perceived maturity were powerful as mediating factors, whereas perceived maturity was the only one of the covariates that was significantly related to measures in the domain of affective reactions (albeit its mediating impact was not strong).

The results revealed differential effects of types of peers. For interpersonal relations and for social adjustment, association with members of the opposite sex overshadowed in importance the other covariates. When the heterosexual relations variable was introduced as a single covariate, the former significant differences among the menarcheal groups with respect to parent relations, teacher relations, and peer relations, were all insignificant. Substantial reductions with respect to norm breaking was also found when introducing heterosexual relations as a single covariate. Finally, the differences among the menarcheal groups of girls with regard to unwanted pregnancy was largely mediated by stable opposite-sex contacts.

For the school achievement variables, on the other hand, association with working peers played the greatest role as a mediator. No statistically significant effects of menarcheal age on grade-point average and school motivation appeared after the covariate working peers was entered as a single covariate.

Perceived maturity was the only covariate that showed a significant relationship with the psychosomatic reactions and depressive reactions. However, its mediating role was quite marginal.

Taken together, stated in terms of the proposed development model, peer-type characteristics, particularly, constituted strong mediating factors. They accounted for a large part of the observed effect of pubertal timing of girls on their interpersonal and social adjustment in the mid-adolescent years. What the results tell us is that the impact of menarcheal age on psychosocial functioning is shaped largely by the interpersonal ecology of the girls. For interpersonal relations and norm-breaking behavior, the primary mediating factor is opposite sex relations, while it is contacts with working peers for school adjustment.

The Role of Heterosexual Relations

Perhaps the most conspicuous result in the examination of the mediating factors for pubertal timing is the role that girls' relations to the opposite sex plays for their social adjustment. Across the areas investigated—interpersonal relations, social adjustment, and also school adjustment—opposite-sex relations were consistently related to the dependent variables. Girls who dated steadily had more problematic parental relations, were more involved in drug and alcohol use, as well as in other norm-breaking activities, did worse in school, and had poorer relations to their teachers, compared to the girls who did not have established relations to boys.

Norm Breaking. Most notable in the present investigation was the role of heterosexual relations as a mediating factor for norm-violating activities. For all except one of the norm-breaking behaviors, heterosexual relations appeared as the primary mediating factor. The important role of opposite-sex contacts for social adjustment issues among adolescent females have been emphasized also in earlier research. In a retrospective study by Brown (1982), peer pressures in high school and in college for behaviors like alcohol/drug use and cigarette smoking were found to be related to have a boyfriend and to have sex among females. Females who had had intercourse in high school reported more peer pressure to use alcohol and drugs in these years than did females who remained virgins.

Support for the finding that engaging in norm breaking is connected with relations to the opposite sex was also reported by Jessor and Jessor (1975). They compared 16- to 18-year-old adolescent girls who had sexual experiences with a comparable group of girls who lacked such experiences for a range of social behaviors commonly occurring in the adolescent years. Their theoretical position made them attend to features in the girls that departed from the expected age-appropriate pattern:

> Not only sexual intercourse but other behaviors that mark transitions during the course of development—beginning to drink, for example, or taking a full-

time job—are normatively age-graded. . ... Engaging in such behavior at earlier stages constitutes a departure from regulatory norms, and it is precisely in this context that a social psychology of deviance or problem behavior has its logical applicability. (p. 474)

Engaging in sexual behavior at an earlier age than is socially normative, they argued, is exactly the type of behavior which denotes a transition "toward a more mature status." Thus, the comparison between girls who had sexual experiences with boys and girls who lacked such experiences would isolate girls who generally engaged in behaviors that departed from age-normative social behaviors from the girls who mainly engaged in social behaviors that followed the conventional age-appropriate pattern.

With longitudinal data, Jessor and Jessor (1975) confirmed these theoretically derived hypotheses. The nonvirgins reported placing lower value on achievement than did virgin females. They expressed lower expectations for achievement and higher expectations for independence, a more tolerant attitude toward deviance, lower religiosity, more peer orientation, less church attendance, and higher general deviance. Finally, their friends approved more of deviance and were more models for deviance than were the friends of the virgin females. In an attempt to predict socially deviant behavior, a comparison was made between girls who, over a 1-year period, changed from virgin to nonvirgin status and girls who stayed as virgins over this period. This comparison revealed that the former group of girls, at the second test occasion, placed a lower value on achievement, were less achievement- than independence-oriented, showed less expectation for achievement, had more positive attitudes toward deviance, showed less religiosity, were more peer- than parent-oriented, had friends who approved more of deviance and functioned as models for deviance. The girls who changed sexual status were also more involved in deviant activities by the end of the year than were the girls who remained virgin.

The theoretical framework of Jessor and Jessor (1975) shares much of the same perspective found in our research. In common is the outlook that deviance proneness among girls in adolescence must be viewed against what is the normative and socially expected behavior at a particular age. We also concur that an earlier display of person features that characterize a more mature status implies a higher risk for engaging in behaviors of a problem type. Moreover, we share the viewpoint that relations to the opposite sex are involved in the process toward maturity. However, a question remains in the Jessor and Jessor (1975) model: What causes girls to go from a virgin to a nonvirgin status? This issue is where the pubertal maturation is introduced in our model.

Jessor and Jessor (1975) made sexual behavior their starting point, and one is left to guess what made some but not other girls engage in sexual intercourse with boys. It is an apparent circularity in their argumentation. A relationship between sexual and social transition may work both ways. Either changes in the social area may give raise to changes in the sexual domain, or the other way around. Or both types of transition might be caused by another, unknown instigator. From our point of view, sexual intercourse is intimately connected with physical maturity, and it is another way of stating that some girls, by virtue of an earlier pubertal maturation, are within the influence circle of more nonconventional peers than are later-physically developed girls. Problem behavior does not occur as a function of becoming nonvirgin. It occurs as a consequence of entering peer circles where the prevalence of problem behavior is higher than among the usual same-age–same-sex friends of other girls.

Our claim that sexual intercourse is closely connected with nonconventional peer influences is supported by data presented by Jessor and Jessor (1975). Friends' approval of deviant behavior, and friends as models for deviance, were the only two factors that, both when comparing the virgins with the nonvirgins, and when comparing girls who stayed virgin from one year to another with girls who changed sexual status over the year, were significantly discriminative at the .001 level at least. Of particular interest in the Jessor study was also the finding that the two peer influence variables were more discriminative for girls than for boys. This is in agreement with other empirical findings that characteristics of peers and their social behaviors are more important information for understanding female norm-violative activity than for accounting for male norm breaking. Numerous studies on alcohol drinking show a clear correspondence between adolescents' drinking patterns and their friends' alcohol drinking, an association that in several investigations has been reported to be stronger among females (cf. Downs, 1985 for studies in this field). Pulkkinen (1983a), to take one example, found that girls who used alcohol frequently also had peers who had similar alcohol patterns, whereas the correspondence between boys and their peers' drinking patterns were lower. She interpreted this finding as suggesting more dependence on peers as models for drinking behavior among girls than among boys. Downs (1985) similarly found that peer influences on alcohol use were more prominent among girls than among boys and, for females, that drinking patterns among close friends were causally prior to the girl herself drinking.

The results on the role of heterosexual contacts for female norm breaking reported in this chapter also have their counterpart in studies examining the etiology of female delinquency. Campbell (1980) reported that female delinquents had more friends than their nondelinquent counterparts and that the interaction took place mainly outside school. In particular, the delinquent girls engaged more often with older males who had left school and were employed. These males were often delinquents.

With regard to norm breaking, a pattern now begins to take form, and isolated findings in the adolescent literature become more understandable, in the light of the triad, early physical maturation—heterosexual relations—norm-violating activities, found in the present study. On the one side, there is ample empirical evidence in the physical growth literature of stronger heterosexual orientation in the teenage years among early-developed girls than among later-matured (Dornbusch et al., 1981; Stone & Barker, 1937, 1939; Garwood & Allen, 1979; Simmons, Blyth, & McKinney, 1983, Simmons, Blyth, Van Cleave, & Bush, 1979; Crocket & Petersen, 1987; Jones & Mussen, 1958). Data also have been presented in the literature showing a more advanced social life among teenage girls with more stable relations to boys than among girls with less advanced heterosexual relations (Jessor & Jessor, 1975; Schofield, 1965; Simmons, Carlton-Ford, & Blyth, 1987). Girls who have stable heterosexual relations tend to show more independence relative to parents (Jessor & Jessor, 1975; Schofield, 1965; Westney, Jenkins, & Benjamin, 1983), and there are much data accumulated that suggests that heterosexual relations and engagement in premarital coitus is linked with a nonconventional behavior pattern (cf. Billy, Rodgers, & Udry, 1984; Schofield, 1965). Miller and Simon (1980) summarized these data in the following manner:

> Strong correlations between coital experience and other forms of deviant behavior (e.g., drug use, general delinquency, etc.) are reported in each study. Moreover, adolescents who have engaged in (or are about to engage in) coitus are more likely to be estranged from those institutions—the family, the church, the school—that serve to monitor and maintain commitments to conventional values. This sexual experimentation during adolescence apparently continues to be embedded in the social context of deviance. (p. 398)

The present study brings together these two lines of research findings. There seems to be a maturational basis both for engaging with boys and for norm breaking. Early maturation is connected with earlier in time establishing stable relations with boys, and it is in these boyfriend contacts that norm-breaking activities develop. Data were presented in chapter 5 showing more stable heterosexual relations and more advanced sexual behavior among early- than among later-developed girls. Chapter 6 demonstrated strong empirical evidence of more norm violation among the early- than among the later-developed girls. The results from the testing of the mediating model in this chapter lend further clarification to this point, demonstrating the mediating role of heterosexual relations for the influence of pubertal timing on norm-breaking.

Interpersonal Relations. Another area in which heterosexual relations played an important role as mediator concerned *interpersonal relations*, particularly with respect to the girls' relations to adults. When heterosexual relations was entered as a single covariate, it reduced the original significant differences among the four menarcheal groups with regard to parental relations to insignificance, and the same was true for mother relations. Thus, the more problematic parental relations among the early-developed girls seemed to be closely related to these girls more advanced heterosexual relations in mid-adolescence. These results are not only of interest for the issue of pubertal development and relations to adults, but have implications for our understanding of parent–daughter conflicts generally in adolescence.

The present findings, attributing parent–girl conflicts to the girls' heterosexual relations agree with previous studies of parent–adolescent relations generally and with earlier findings in the physical growth literature.

Schofield (1965) reported that what in particular differentiated sexually experienced girls from their sexually inexperienced counterparts was that they were less influenced by and were more hostile toward their parents. In screenings of normal adolescent populations, teenage girls' heterosexual relations have been found to constitute one of the greater conflict areas with parents. Ellis-Schwabe and Thornburg (1986) reported a study conducted in the United States showing that dating was among the most common conflict area with parents for girls 10 to 14 years of age. Asked to report the three most conflicting areas with mothers and fathers, respectively, 30% of the girls stated dating as a conflict with mother and about the similar proportion reported such conflicts with father in these years. As to the issue of the impact of pubertal development, one main reason behind the more conflicting mother–girl relations for the early-matured compared with the late-developed girls in the clinical sample investigated by Frisk, Tenhunen, Widholm, and Hortling (1966) was these early-developed girls' relations to boys. The anticipated trouble the girl might encounter in her intimate contacts with boys evoked anxiety in the mother and caused her to place restrictions on the girl's contacts with the opposite sex. A spiral of increased conflicts was evoked, with heightened anxiety and greater demands on the part of the mother which, ultimately, had the effect that the girl sought even more and was forced to satisfy her interpersonal contacts outside the family.

In this connection it should be recalled that our measure of the girls' affective relations to peers was positively correlated with their parental relations while the composite opposite-sex measure was negatively associated with parental relations. Thus, the distinction between relations to peers and relations to members of the opposite sex is an important one when studying peer culture influences on parent–teenager relations. A differential role for social behavior of peers was also reported by Simmons, Carlton-Ford, and Blyth (1987). They showed that peers and boyfriends had opposite implications for the girls' behavior in Grades 6 and 7, such that positive behavior was associated with peer popularity whereas a mixed impact was obtained for opposite-sex popularity. Girls who perceived themselves as popular among peers showed less problem behavior and higher self-esteem, whereas girls who perceived themselves as popular among boys had greater problem behavior and lower school achievement (but, not unexpectedly, higher perceived attractiveness). The results raised the old questions of whether or not the girls who associated with boys at this age were too emotionally immature for such contacts.

Other interpretations of the parent–daughter disagreements in adolescence, which places less responsibility on peer influences, have also been advanced in earlier research. One such common interpretation is that family conflicts are particularly related to the strivings for autonomy in the girls. Relatedness to parents and the development of own autonomy are closely intertwined processes (cf. Simmons, Blyth, & McKinney, 1983; Stone & Barker, 1939). In the normal course of development, the teenager is to assume responsibility for his or her actions and decisions, and to achieve autonomy, at the same time as he or she is to maintain a positive relation with parents, who, up to the mid-adolescent years have had the socializing power (Murphey, Silber, Coelho, Hamburg, & Greenberg, 1963). From this perspective, the higher conflict level of the parent–daughter relation among the early-developed girls can be interpreted as an attempt to gain independence by the early developers at a time when this level of independence is not customary among other teenagers of the same age. If this interpretation is correct, then we would expect to find that the early-developed girls, to a higher extent than their later-matured counterparts, would seek to achieve a more independent functioning by wanting, more than later-developed girls, to handle things in everyday life more by their own than later developed girls.

Data available in the present longitudinal program directly or indirectly connected with the issue of the amount of autonomy wanted by the females (in contrast to autonomy given by parents) show little association with pubertal timing. A prototypical item is a question covering independence from parental supervision and own handling of everyday life activities, administered to the girls at 14.10 years. The girls were directly asked: "How much to do you want your parents to decide for you: for example leisure time activities, choice of friends, what to do at the summer vacation, vocational plans, and so on?". A comparison among girls in the four menarcheal groups revealed no differences whatsoever. About as many of the early-developers as of the later-developed wanted to decide on their own. Two out of three girls in each of the four menarcheal groups stated that their parents should decide "very little" or "rather little," and only 1 out of 20 thought that their parents should decide "very much" or "rather much." To the extent that this question covers the core of striving for independence in daily life, it suggests that this factor is not the one primarily responsible for the obtained differences in parental relations between the early- and the later-matured girls. In fact, the similar finding was reported by Dornbusch, Gross, Duncan, and Ritter (1987), using data from the National Health Examination Study. They found little relationship between decision making in the family and sexual maturation among females, but more so among males.

School Adjustment. A third area on which heterosexual relations showed significant impact concerned *school adjustment*. Girls with stable heterosexual relations had lower grades than girls who had no such stable boyfriend relations. Also in earlier research has opposite-sex relations been shown to have an adverse effect on girls' educational aspirations (Schofield, 1965; Simmons, Blyth, & McKinney, 1983). Schofield (1965) reported that the girls who engaged in sexual behavior early left school earlier than did the girls whose sexual contacts with boys started later. The sexually experienced girls also expressed more dislike of school and more school problems than did the sexually more inexperienced girls. That educational achievement is connected both with menarcheal age and with dating was reported by Simmons, Blyth, Van Cleave, and Bush (1979). They found lower grade-point averages and lower scores on standardized achievement tests among sixth- and seventh-grade girls who had begun menstruating, as well as lower achievement results for girls who had dated most.

Having observed the crucial role of opposite-sex contacts for girls' adjustment in different areas, and its role as a mediating factor for the impact of pubertal development, the question obviously arises as to what in the contact with members of the opposite sex reinforces parental conflicts, more nonconformist behavior, and less interest in school.

The role played by opposite-sex relations for girls' social adjustment is open to different interpretations. Miller (1979) reviewed the traditional view on the specific matter of females and delinquency. This viewpoint is one where females primarily tend to commit sex-related offences. Her limited success in opposite-sex contacts or achieving the feminine goal of family building makes her engage in activities—sexual misconduct, property crime, running away from home, and the like—commensurate with this underlying motive:

> The girl who is anxious about her accomplishments in this regard or who has experienced conspicuous failure in her attempts to attract or maintain the attention of boys may offer them her sexual favors. . . . Or, to enhance her attractiveness, she may steal clothing or cosmetics and in this way attract the attention of the authorities. . . . Accidents of nature, such as acne, obesity or late physical development, social disadvantages such as minority group status or poor economic circumstances, and interpersonal inadequacies represent various constraints that conceivably limit or preclude success in dating. Offences like sexual misconduct, running away from home and shoplifting become understandable in the context of such disadvantages in conjunction with the strategies available to adolescent girls for improving their competitive positions in the rating and dating games of adolescence. (pp. 124–125)

DISCUSSION

The argument of limited success in opposite-sex contacts' leading to antisocial behavior in order to improve one's status among boys does not seem to be the chief initiating reason for the early-matured group of girls in the present sample to engage in norm-breaking activities. Among our subjects, the early developers expressed the least anxiety about not being able to attract the attention of boys (see chapter 4).

Other interpretations are more probable. The role of opposite-sex relations as a mediating factor may be interpreted as a direct person–person influence: modeling of the boyfriend's behavior or social reinforcement from the boyfriend. For drug and alcohol use, and for delinquency, the mere number of friends who themselves use drugs or are involved in delinquent activities have been found to constitute a forceful predictor of a subject's initiation into these behaviors (Kandel, Kessler, & Margulies, 1978). This is true also for the perception of one's friends' attitudes toward norm breaking (Jessor & Jessor, 1975). As earlier reported, engaging in alcohol drinking is closely related to drinking patterns of friends among girls (Downs, 1985; Pulkkinen, 1983a), perhaps more so than among boys.

The influence of members of the opposite sex on early-maturing girls' social behavior also could be interpreted in environmental terms: these girls have entered the male territory where availability of alcohol and drugs are greater, and exposure to a more norm breaking-accepting climate is higher than in the peer culture of girls who are not going steady in mid-adolescence. The consequent changes in behavior, and in attitudes toward such behavior, eventually result in open family conflicts when these girls are faced with the more restricted value system of their parents.

Simmons and co-workers have reported an impact of opposite-sex contacts both on problem behavior and on scholastic achievement. In their longitudinal study of girls aged 12 to 16, Simmons, Blyth, and McKinney (1983) noted more active and intense opposite-sex relations among the early-matured girls and also more problem behavior at school among these girls compared to later-maturing girls. They interpreted their results such that the early-developing girls' more extensive contacts with boys in these years reduced the amount of time for studying and for other activities that are associated with advanced education. They suggested that the early maturers were distracted from achieving in school and also were tempted to problem behavior due to this more close relations to boys, and they proposed a causal four-step sequence with (a) pubertal change leading to (b) change in physical appearance and endocrine level, which in the next step affected (c) *adults'* reactions (giving the girl more independence), *own* reactions to the physical change (body image and interest in the opposite sex), and *boys'* reactions (popularity, dating). Finally, the girls' relations to boys were assumed to (d) distract school performance and lower the girls' educational aspirations.

An alternative, but related, interpretation is that the boys the early-developing girls are dating in most cases are chronologically older teenagers, presumably with lower educational aspirations than the educational aspirations of same-age peers. The role of boyfriend relations also might be interpreted in terms of the consequential social role or future lifestyle to which the girl aspire. Steady contacts with boys might make the girls more willing to engage in activities such as family life than to aspire to higher education.

Altogether, a multitude of reasons can be given for the influence of heterosexual relations on girls' social adjustment and its role as mediating the pubertal impact. Our data do not permit further specifications on the nature of this influence. However, we want to address an aspect of this influence that directly concerns the mediating role of heterosexual relations for the impact of pubertal maturation on norm breaking: namely, the value system of the girls' boyfriends. It was hypothesized here that because more of the early-developed girls than of the late developers had stable boyfriend relations, more of the early developers would subsequently be socialized to "boy norms," and their more mature status would make them identify and model this behavior to a greater extent than would the later-developed girls. However, it is possible, and we have no data to confirm or reject this, that girls, and perhaps more often the early-developed girls, selected boys who were more norm breaking prone than other boys. To the extent that these boys were more likely to be norm breakers than were other boys, other aspects, in addition to differential opportunity and differential susceptibility, are at work.

The Role of Working and Older Peers

That *working* peers constituted the primary mediating factor for the effect of menarcheal age on school achievement is an important finding of the present study. School fatigue among the early-developed girls to a large extent seemed to be fostered in a peer environment which did not value educational achievement highly. To associate with peers who are already employed was connected with low achievement in school, low educational aspirations, and high school absenteeism. Working peer associations was also significantly related to a number of different indicators within the domains of interpersonal relations and to various norm-breaking behaviors. Girls with employed peers in their circles of friends had poorer relations with their mothers and higher norm-breaking activity.

Older peers also had significant impact on some of the indicators within the areas of interpersonal relations, norm breaking, and school adjustment. Having older peers in one's circle of friends was connected with more frequent norm violations, lower grade-point average, and more school absenteeism, but, on the other hand, with more comfortable peer relations and higher opposite-sex popularity. More beneficial peer group relations were also the case for the working peers variable.

Overall, the directions of the relationships with the dependent measures for the working peers and older peers were similar, and also similar to boyfriend relations, thus implying that these three nonconventional peer types had homogeneous behavior implications.

Differential Susceptibility

In order to make more precise statements on the efficacy of peers to change the behavior of the girls, the term *differential susceptibility* was introduced. The term relates the level of pubertal maturation in the girl to her psychological preparedness to adjust to the behavior patterns of her peers. To be early-developed was assumed to be connected with being psychologically ready to change one's behavior in the direction toward the social life among those peers who were congruent with the girl's advanced level of pubertal maturation. The impact of such peers was expected to be more powerful on the early- than on the later-developed girls.

The hypothesis was investigated for different interpersonal and social measures. About the same patterns of results were obtained across areas. In accordance with the hypothesis, little differences for measures within the interpersonal relations, social transition, and school adjustment areas were found between early- and later-developed girls among the subsample of girls who had no nonconventional peers in their circles of friends. By contrast, and in accordance with the differential susceptibility hypothesis, profound differences were found in these respects among the menarcheal groups of girls for the subsample of girls who engaged with such peers. The earlier matured the girl was, the higher the conflict level with her parents, the higher the norm-breaking activity, and the lower her school adjustment.

The idea of differential susceptibility was found to work both ways, not only for associating with nonconventional peer groups but also for associating with younger peers. If a girl was early-developed and engaged with younger peers, her interpersonal relations and social behavior did not differ from other early-developed girls who lacked such peer contacts. However, if the girl was later-developed and associated with younger peers, her parental relations were more comfortable and her normbreaking activity was lower than was the case for those later-developing girls who did not associate with younger peers. In summary, nonconventional peers, being in a developmental sense more like the early-developing girls, seemed to have more impact on the behavior of the early-maturing girls compared with the later-developed, while the opposite situation occurred for associating with chronologically younger peers.

One merit of the susceptibility hypothesis is that it offers the possibility of circumscribing the range within which maturational influences can be expected to appear. We are therefore in a somewhat better position than has hitherto been the case to judge when maturational influences are likely to be manifested and when they are not likely to emerge, and for which girls this will be true.

Concluding Comments

One message of the present findings is: different types of peers have different implications for behavior. To associate with nonconventional peer groups—older peers, working peers, and boyfriends—was connected with problematic relations to adults and more social adjustment problems in midadolescence, whereas the opposite was the case for associating with younger peers. Moreover, although nonconventional peer contacts had about similar behavior implications across various areas, the strength of their influence differed for different areas. Boyfriend relations was particularly disadvantageous in terms of relations to adults and normbreaking behavior, whereas associations with working peers had a more negative impact on girls' school adjustment.

The insight of the role of the peer culture for differences between early- and later-developed girls, with regard to social transition issues in various respects, is an important outcome of preceeding analyses. Contrary to common notions, our a prior hypotheses, which tests confirmed, demonstrated that it was not the impact of peer per se that was the crucial influence for the type of behavior girls expressed. It was the specific constellation of peers, the engagement with which was partly governed by the girls' own maturational level. The impact of peers on the social behavior of early- and later-developed girls was found to be a function of the match of the girl's maturational feature and corresponding maturational properties of her peers. The impact of a certain peer type was more prominent if a girl's maturational level matched that of the particular peer type.

10

The Long-Term Consequences

THE LONG-TERM PERSPECTIVE

Up to the present point, data for our research group have been analyzed and the literature in the field has been reviewed, for the teenage years. Some data have covered late adolescence (see, for example, Jones & Mussen, 1958, for TAT data on 17-year-old females; Stone & Barker, 1934, for personality data among females, 18–19 years of age; and Weatherley, 1964, for personality descriptions of 19-year-old college girls).

When we now raise the question of the long-range consequences of early versus late physical maturation for females, we enter a field of research that is largely unknown. Not only is there a scarcity of systematic studies addressing the issue of the long-range impact of biological maturation, there are only a few attempts at delineating the domains in which such influences might manifest themselves, and there are virtually no discussions on the intervening factors between menarcheal age and adult functioning. A few investigations have been conducted comparing adult family formation and childbirth among females who were early and late developers. More rare are the consequences in other areas—self-concept, gender role, emotionality, social adjustment, personality, social network, support systems, and so on.

As is true of other domains in psychology, answers to the question regarding the impact on behavior of individual differences in physical growth cannot be finally settled before both the short- and the long-term consequences are taken into account. Cross-sectional studies might be useful to solve some questions. However, a comprehensive knowledge of the maturational processes involved in individual development presupposes following the same subjects over extended periods of time. Many of the ideas and prevailing conceptualizations of the impact of physical growth have their base in the longitudinal studies initiated in the end of the 1920s and the beginning of the 1930s at the Institute for Human Development at Berkeley, California: the Berkeley Growth Study, the Guidance Study, and the Oakland Growth Study. The presence of marked differences related to physical growth at adult age for males reported in these longitudinal projects, but the remarkable absence of such studies reported for females in these longitudinal projects, indicates either (a) the consequences that exit for adult personal and social functioning for females are too complex and incomprehensible to be easily settled within the limited format of ordinary scientific journals, or otherwise are too complex to be explained from our present knowledge of the impact of pubertal development on behavior, or (b) that there actually are very few long-term implications of differential maturational timing among females. The latter explanation of why long-range studies on the role of biological maturation have not been reported in these longitudinal programs is supported by a striking absence of such studies elsewhere in the literature.

Which, then, are the reasons, if any, for expecting differences in behavior in adult life between females who were early or later maturing? From data showing that early-maturing females in late adolescence had undergone more personal crises than had the late-maturing females, Berzonsky (1981) argued that to the extent these conflicts were coped with effectively, more satisfying personal adaptation would be found among these early-developers at adult age. Elaborating on Peskin's (1973) hypothesis on limited preparations for early developers in early adolescence, Simmons, Blyth, and McKinney (1983) advanced the hypothesis that late maturation might cause a distressing transition time into adult life for some females, due to their limited adolescent experiences. However, their hypothesis was not particularly well substantiated. Shipman (1964) argued that such differences were most likely to appear if the psychosocial conditions that were associated with physical growth in the adolescent years would create distinctive future-time orientations: social and personal styles that would extend into adult life. Discussing the possibility for specific events in adolescence to foretell certain life orientations, Bronfenbrenner (1984) gave the example of teenage pregnancy from the developmental literature (Furstenberg, 1976) as one such event that has been shown in the literature to forecast a peculiar and definite life course: "In contemporary America, when a teenager becomes pregnant before marriage, much of the rest of her life is foreordained and, indeed foreclosed, in terms of future education, work opportunities, income, marriage, and family life" (p. 33).

There is much that speaks in favor of the hypothesis that the early-matured girls' stronger adult image, stronger interest in adult-type behavior, stronger heterosexual interests, and their more advanced contacts with the opposite sex, together with less interest in an educational career in mid-adolescence, set the stage for their future life conditions. This conclusion is reached by bringing together quite disparate empirical data.

Early-matured girls have been characterized as more romantic than late developers in adolescence (Jones & Mussen, 1958), as showing more interest and contact with boys (Crocket & Peterson, 1984; Faust, 1960; Garwood & Allen, 1979; Stone & Barker, 1937, 1939), as being more frequently involved in sex-appropriate behaviors (Simmons, Blyth, & McKinney, 1983), such as displaying interests in infants (Goldberg, Blumberg, & Kriger, 1982). School data indicate lower satisfaction and lower educational aspirations (Andersson, Dunér, & Magnusson, 1980; Davies, 1977; Frisk, Tenhunen, Widholm, & Hortling, 1966; Simmons, Blyth, & McKinney, 1983; Simmons, Blyth, Van Cleave, & Bush, 1979). Altogether, the more developed maturity status connected with a greater concern for the biological sex role and heterosexual concerns in early-developed girls might have the long-term implication of making marriage, household, own children, and a traditional family life more welcomed by these girls, relative to a professional orientation and an occupational role.

Our own results lend support to the assumption of a primary adult sex-role commitment toward homemaking among the early-developed females. The early-maturing females were clearly overrepresented among the girls in mid-adolescence who had stable relations to boys. Whereas the later-matured girls reported difficulties in establishing contact with boys, the concerns of the early-matured girls were that their boyfriends would get tired of them. In responding to the question concerning the way that they differed from other girls, more of the early-matured girls mentioned that they were romantic. The early developers also had more frequent sexual experiences, as was reflected in official data on unwanted pregnancies. A considerably greater proportion of the early maturers than of the other, later-maturing girls had undergone one or several abortions up to the age of 16. This circumstance, that heterosexual contacts intervened early in the lives of the early-developed girls, is one indication for an expected prolonged influence on childrearing and marriage in future life.

In addition, the early-developed girls were not as motivated for school work in mid-adolescence as were the other, later-developed girls. They reported more conflicts with their teachers, and they showed more school discipline problems. School records also showed that they were more frequently absent from school. More of the early-developed girls than of the other girls reported having differences in opinion with parents on schoolwork at 15.10 years. The early-developed girls, in addition, had, more often than other girls, friends who had quit school and were working full time. This is suggestive evidence that long-range effects of pubertal development might occur in the educational domain. To summarize, data from the present investigation suggest that early maturation tends to carry an impetus for a future traditional family life orientation connected with lower adult education: a sequence ranging from early maturation—a general sensitivity to aspects (behavior, attitudes, etc.) that connote a more mature stance—early heterosexual orientation, early sexual intercourse, low academic aspirations, early childbearing and a family life orientation, to low adult educational level and low job status.

Not unexpectedly, most of the studies in the literature that have examined the long-term implications of variations in biological maturity among females have concentrated their efforts on investigating the impact of pubertal timing on fertility and the future family circumstances of the girls. That early-matured girls reach sexual maturation at an earlier time would suggest that they would also marry and give birth to own children earlier in time. Kiernan (1977) found no relationship between age at menarche and age at marriage, except for a significantly later marriage age for the girls with more pronounced late menarche (14 years or later). However, an analysis directed to the majority of females with no premarital conceptions or births yielded a significant relationship at the .001 level between menarcheal age and age at marriage. No significant association between menarcheal age and age at motherhood was reported. The sample of females was a nationwide British longitudinal cohort born in the first week of March 1946, and marriage and parenthood data were collected up to age 26.

In a sample of child-bearing women at fertile age, Buck and Stavraky (1967) reported a highly significant positive relationship between age at menarche and age at marriage. Of the women with a menarcheal age under 11 years, 46.6% married before their 20th birthday in comparison with 25% of the women who had their first menstruation at the age of 15 or later. Circumscribing their analysis to those women who married at 20 or later, 76.8% of the females with early menarche married at the age of 20–24 years. The same percentage figure for the late-developed females was 66.3%.

Presser (1978) interviewed a sample of women, 15 to 29 years of age, who had given birth to their first child. She found a correlation of .19 ($p < .001$) between age at birth of the child and menarcheal age for Black mothers but a nonsignificant correlation of .09 for White. The same correlations for noncontraceptive users were .16 ($p < .05$) for the Black mothers and .22 ($p < .05$) for the White. Among the Black mothers, early dating was associated with early menarche ($r = .29$, $p < .001$), whereas there was no such association among White mothers ($r = .01$, ns).

In a study of 3,000 30-year-old Black and White women in low income neighborhoods, Udry (1979) reported that earlier age at menarche was connected with earlier first sexual intercourse. Both Black and White women with earlier intercourse had their first pregnancy at an earlier point in time. In addition, it was demonstrated that the earlier-developed females had their first pregnancy at an earlier time. The results were suggestive of a sequence ranging from early biological maturation, through early sexual intercourse, to early pregnancies. Sandler, Wilcox, and Horney (1984) reported data from a large-scale study, the Menstruation and Reproduction History Study, of women followed with a prospective design over 20 years. Subjects were subdivided into two cohorts, one consisting of 1,003 women born before 1925 and one consisting of 1,059 women born in 1925 or after. It was found that early-developed women married earlier and had their first conception at an earlier age than did the later developers in both cohorts. The differences were fully 1 year between the subjects who had their menarche before age 12 and those who had their first menstruation at age 15 or later. It was concluded that the earlier age at conception could be explained by the younger age at marriage among the early developers. Although the timing of marriage and childbirth was connected with menarcheal age, the number of pregnancies was not.

In order the broaden the scope of earlier studies of the relationship between the age at menarche, marriage, and the first birth of a child, Udry and Cliquet (1982) reported results from a cross-cultural study including several adult samples of subjects for which the age at marriage and the age at first birth of a child differed widely (U.S. White and Black females, and Belgian, Pakistan, and Malaysian samples of females). Rather consistently across the different samples, there was found evidences for relationships among menarcheal age, age at first birth, and age at marriage.

Overall these studies are indicative of an earlier timing of sexual and sociosexual issues for females who develop early: early age at menarche, early heterosexual orientation, early sexual intercourse, early marriage or other forms of stable partner relations, early pregnancy. One of the complications in the studies reviewed is that the age at birth of the first child is not necessarily the same as the age at which the first pregnancy occurs (cf. Sandler, Wilcox, & Horney, 1984). Another complication is that age at marriage may not be a sensitive measure of living in a stable relationship with a partner. Moreover, Buck and Stavraky and Presser did not include an unselected sample of females but restricted their analysis to child-bearing mothers. Also, Sandler et al.'s and Udry's investigations employed samples that had somewhat circumscribed ecological generalizability.

Much of the data that are presented in this chapter are based on a follow-up assessment when subjects in our sample were, on the average, 25.10 years old (i.e., about 10 years after our main investigation of the subjects' life situation in the mid-adolescent period). Data were collected by a mailed questionnaire covering educational–vocational career, family life, working conditions, social networks, and leisure-time activities (Andersson, Magnusson, & Dunér, 1983). Of the 466 females in our sample, 417 (89.5%) answered the postal questionnaire at adult age. There was no systematic relationship between age at menarche and answer rate (−11 years: 87.8%, 11–12 years: 94.4%; 12–13 years: 89.2% 13– years: 85.4%).

An interview and psychological testings were performed at the same age on a selected subsample (see chapter 3 for a full description of the medical sample). As a complement, to provide more detailed information on specific topics, data from this sample are reported here. Participating in the interview were 101 females. An additional 10 females could not be located, and 19 females refused to be interviewed or tested. Of the 101 females, 71 had menarcheal information and were born in 1955.

The adult life situation is also discussed, as it can be measured by information from official registers. A description of these official data are given in the text.

The areas of interest for the present examination of the long-term effects of differential menarcheal age in females are:

1. the *social life* situation, including work, education and the family situation;
2. availability of *social support* systems;
3. adult *personality* and *life values*; and
4. *social adjustment* as it can be measured in terms of delinquency, alcohol and drug use, and smoking).

THE SOCIAL LIFE SITUATION AT ADULT AGE

To make a broad description of the social living conditions at adult age, three areas are investigated: family life, work, and education. In addition, the issue to what extent the females who were early- and later-developed perceived that they had control over their life at adult age, is examined.

Family Life

Questions in the postal questionnaire were used to investigate the relationship of menarcheal age to the adult family life situation. The proportion of the females in the four menarcheal groups, who at the age of 25.10 years were married or lived together in a stable relation with a male, and the proportions who had own children at the time, are shown in Table 10.1.

Two females out of three were married or lived together in a stable relationship with a man. There was no systematic relationship between menarcheal age and living under stable family circumstances.

About 4 out of 10 females had given birth to children by 25.10 years. More of the women who were early-developed had own children than did the later-developed women. The earliest-matured women also had more children per person than did the later-matured. The latter relationship between menarcheal age and the number of children, circumscribed to those who had

TABLE 10.1
The Family Life Situation at 25.10 Years for Females Who Were Early or Late Developed. Percentage of Subjects in Four Menarcheal Groups Who Were Married, Had Own Children, and the Mean Number of Children

	Married—living together	Own children	Mean number of children[a]
–11 yrs	67.4	60.5	1.65
11–12 yrs	70.6	51.5	1.38
12–13 yrs	64.5	33.7	1.34
13– yrs	73.3	35.2	1.38
Total	68.5*	41.2**	1.41***

[a] Circumscribed to those Ss who had own children at 25.10 yrs.
* $\chi^2(3, N = 416) = 2.62$, ns
** $\chi^2(3, N = 415) = 16.36, p<.001$
*** $F(3,167) = 1.57$, ns

children at this age, however, did not reach statistical significance. The relationship between age at menarche and number of children for the total research group of subjects was significant at a high level ($F = 6.97$, $df = 3,411$, $p < .001$).

To summarize the results in Table 10.1, biological maturation had an effect in adult age on the childbirth of the females, but was not connected with whether the females lived under stable family circumstances or not. Observing the relationship between menarcheal age and childbirth, it will be recalled that more of the early-maturing subjects than of the late-maturing had undergone one or several abortions in the adolescent years.

Work Conditions

At the time in adult life when the postal questionnaire was administered about 9 out of 10 females were employed. About as many of the early developers as of the late held employment. For more than 8 out of 10 this employment was of permanent kind. There was no relationship between menarcheal age and permanent versus part-time employment [x^2 (3, $N = 354$) = 5.60, ns].

The women had had on the average 3.4 employments since they finished school. The number of employments were about as many among the early- as among the later-developed [F (3, 404) = 1.87, ns]. Neither did the number of months employed—on the average 72 months—differ among the early- and the later-matured [F (3, 319) = 0.67, ns]. One third of the sample had once or at several occasions been unemployed, but the proportion who had been out of work was about the same for the early- as for the later-matured females [x^2 (3, $N = 399$) = 0.30, ns]. There were no differences among the menarcheal groups with respect to the number of months unemployed [$F(3,395) = 0.30$, ns]. The average female had been unemployed 1.3 months.

Seven women out of eight (83.3%) thought that they had chosen the right occupation. About as many of the early-matured as of the late stated that this was the case. The postal questionnaire included a work satisfaction scale covering satisfaction with work assignment, work environment, colleagues, and superiors. No significant relationship with the menarcheal age of the women was found. At 25.10 years, 15.6% of the females reported thinking about changing occupation, and 27.2% thought about changing place of work. No relationships with the menarcheal age of the subjects for these work aspects were discerned.

In summary, the actual work history of the women—length of time employed and unemployed, whether or not they were presently employed, and the specific conditions surrounding employment—did not differ among females who were early- or late-matured. Neither did their present work satisfaction, as indicated by several measures. Now, let us make a more detailed examination of the type of occupation that the females had at the time the postal questionnaire was administered.

An open question allowed the subjects to describe the kind of employment they had at the present time. They were instructed to give as detailed information as possible. Of the 417 females, it was possible to classify 352 descriptions (84.4%) according to type of occupation and 349 descriptions (83.7%) according to the minimum education that normally is required for this occupation. The type of occupation was coded in eight categories. These categories and the percentage of subjects who held an occupation in each of these, are: (a) technical and scientific (3.7%), (b) industry and trade (8.5%), (c) office and organization (25.3%) (d) commerce and economy (7.7%), (e) communication and military (1.1%), (f) open-air and farming (1.1%), (g) service (6.8%), (h) social service (44.3%), and (i) artistic (1.4%). As expected, most of the females held employment within the social welfare sphere or in an office. As can be seen in Table 10.2 there were no indications of a systematic relationship between occupational type and being early- or later-developed.

The employment the subjects held were categorized according to the minimum education required. The four requirement levels were (a) compulsory school education only, (b) vocational training, (c) theoretical education at the secondary school level, and (d) academic education or college education. Although few differences among the menarcheal groups of females appeared above for the occupational domain, quite strong differences were obtained with regard to the position in the hierarchy within the respective domain. Early-developed females systematically held lower position employments than did later-matured females. A detailed presentation of job position as related to menarcheal age is presented in Table 10.3.

As shown in Table 10.3 it was quite common that the early-matured girls

TABLE 10.2
Percentage of Females Who Were Early Developed (Menarche Before Age 12) and Late Developed (Menarche at or After Age 12) Who Held Employment in Different Occupational Domains at 25.10 Years

Occupational domains	Early developed	Late developed
Technical and scientific	4.2	3.4
Industry and trade	11.8	6.9
Office and organization	23.5	26.2
Commerce and economy	9.2	6.9
Communication and military	1.7	0.9
Open-air and farming	0.8	1.3
Service	6.7	6.9
Social service	40.3	46.4
Artistic	1.7	1.3

$\chi^2(8, N = 352) = 4.48$, ns

TABLE 10.3
The Minimum of Education Normally Required for the Specific Employment Females in the Research Group Held at 25.10 Years. Broken Down by Age at Menarche of the Subjects

Age at menarche	Academic or college education	Higher secondary school, theoretical	Higher secondary school, practical	Compulsory school education
−11 yrs	1 (3.0%)	3 (9.1%)	17 (51.5%)	12 (36.4%)
11–12 yrs	11 (12.9%)	18 (21.2%)	38 (44.7%)	18 (21.2%)
12–13 yrs	8 (5.4%)	38 (25.7%)	82 (55.4%)	20 (13.5%)
13– yrs	11 (13.3%)	20 (24.1%)	44 (53%)	8 (9.6%)
N	31	79	181	58
%	8.9	22.6	51.9	16.6

$\chi^2 (9, N = 349) = 23.40$, $p<.01$

were employed in work that did not require as much education beyond the obligatory compulsory schooling. Perhaps most noteworthy was that 36.4% of the earliest-matured females had job positions that required only compulsory schooling, in comparison with 14.6% of the rest of the females. Whereas only 12.1% of the earliest-developed females were employed in jobs which required at least theoretical education at the secondary school level, almost three times as many of the other females, 33.5%, held such employment.

Thus, whereas early- and later-matured subjects held employment within similar occupational domains, the job positions within these domains differed among the menarcheal groups, with early-matured subjects holding jobs that did not require so much theoretical education.

This difference in job position is open to different interpretations. It might be the case that the early-matured girls, due to their domestic situation with more child-care and more home and household responsibilities, were prevented from taking a job that matched their education. Another possibility, not necessarily contradicting the first, is that the early-matured females had, in fact, a lower education, and that their lower job positions pictured their true lower educational situation at adult age. This issue is examined more closely in the next section.

Adult Education

The educational situation at adult age among the females in the four menarcheal groups is of particular interest, not only because of the difference in job positions found at adult age between early- and later-matured subjects, but also in view of the differences in school achievement and motivation between early- and later-developed females in adolescence.

In order to examine the possible long-range differences with regard to adult education among early- and later-matured females, it was of interest to compare not only the subjects' current level of education at adult age, but also their views of their schooling in the past as well as how they looked at the possibilities for future education.

As to the retrospective view of school experiences, two questions in the postal questionnaire addressed this issue. The subjects were asked whether they had enjoyed their time in the compulsory school, and whether they had had clear plans after the compulsory school. The great majority of girls stated that they had liked their time for the most part in the compulsory school and most reported that they had had somewhat unclear or rather clear plans for their future after the compulsory school. Recognizing the retrospective bias that might occur for these reports on prior school experiences, no differences among the four menarcheal groups of subjects were found either with respect to liking school $[F\ (3,411) = 1.16$, ns], or with respect to plans after the compulsory school $[F\ (3,411) = 1.05$, ns].

Questions on the current educational level at 25.10 years were also asked in the postal questionnaire. The level of education was coded into four categories:

1. compulsory school education (compulsory school only, compulsory school education together with manual training less than 2 years),
2. higher secondary school education, practical (higher secondary

school—2 years of vocational preparation, higher secondary school—2 years of direct work preparation courses),
3. higher secondary school education, theoretical (higher secondary school—3 years of theoretical education), and
4. academic and college education.

Results are shown in Table 10.4, which presents the educational status at age 25.10 years for females grouped according to menarcheal age.

As can be seen in Table 10.4, there were marked differences among the four menarcheal groups of girls in the level of education at adult age. The difference among the menarcheal groups was significant at a high level [$F(3, 411) = 5.43\ p = .0011$]. Post hoc analyses showed that the earliest-matured group of girls differed significantly from each one of the other maturational groups. Most striking was that a minority of the most early-developed females (27.9%) had some form of theoretical education above the 9-year compulsory schooling, whereas a majority of the latest-developing girls (60%) had such an education. Only 2.3% of the earliest-matured females had entered college or university compared to 12% to 15% of the rest of the girls. About 4 out of 10 of the most early-developing females left school after the obligatory 9 years in the compulsory school system. Of the latest-developing females this happened for only 2 out of 10.

Having observed considerable differences in job position among the menarcheal groups of subjects in the previous section, and now substantial differences in level of education, a control analysis was made examining the impact of menarcheal age on job position after covariate correction for actual level

TABLE 10.4
Level of Education at Adult Age for Females in Four Menarcheal Groups

Age at menarche	Academic or college education	Higher secondary school, theoretical	Higher secondary school, practical	Compulsory school education	N	%
−11 yrs	1	11	13	18	43	10.4
	2.3%	25.6%	30.2%	41.9%		
11–12 yrs	13	39	20	28	100	24.2
	13.0%	39%	20%	28.0%		
12–13 yrs	21	86	25	34	166	40.1
	12.7%	51.8%	15.1%	20.5%		
13– yrs	16	47	19	23	105	25.4
	15.2%	44.8%	18.1%	21.9%		
N	51	183	77	103	414	
%	12.3	44.2	18.6	24.9		

$\chi^2\ (9, N = 414) = 21.83,\ p<.009$

of education. Most of the differences among the menarcheal groups disappeared after adjusting for educational differences among the females [$F(3, 342) = 2.58$, ns]. Thus, it may be concluded that the difference in position in the occupational hierarchy between early- and later-matured females to a large extent could be explained by their educational differences.

As to the issue of plans for future education, the subjects were asked whether they thought about further education and the obstacles that were connected with it. Altogether 31% of the females reported reflecting on additional education at 25.10 years. About as many of the early as of the late developers were interested in further education.

A number of different reasons were specified that might interfere with the future educational plans. These eight reasons, rank-ordered according to how frequently they were mentioned, were:

1. cannot afford (11.8%)
2. problems with day care (8.4%)
3. no such education where I live (4.8%)
4. unsure of managing the education (3.8%)
5. tired of studying (3.4%)
6. my employment makes it difficult (2.4%)
7. does not result in better work (1.7%)
8. my family does not want it (0.5%)

Differences among females in the four menarcheal groups were found for two of the categories: day-care problems and difficulties due to employment. More of the earliest-matured females stated that problems with day care for their children interfered with future educational plans. This was also the case for the alternative specifying that their present employment made it difficult to return to studying. A detailed account of the differences among subjects in the menarcheal groups for these two educational obstacles is presented in Fig. 10.1a and 10.1b.

To summarize, great differences in level of education at adult age between the early- and the later-matured females were found. The early-matured females thought, when they were looking back, that they had had as good time in the compulsory school as had the later-developed and that their future educational plans after the compulsory school were as decisive. There were no differences among the menarcheal groups of females with regard to plans for further education at adult age. However, directly asked which type of obstacles they experienced for further education, a greater proportion of the earliest-developed group of subjects stated that day care and things connected with their present employment interfered with such educational plans.

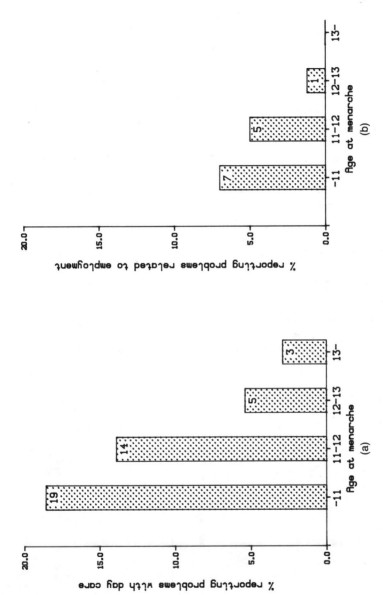

FIG. 10.1. (a) The percentage of females in four menarcheal groups reporting problems with day care as an obstacle for further education. (b) The percentage of females in four menarcheal groups reporting that their present employment causes difficulties for further education.

Control Over Life Conditions

A central concept in research on individuals' lifestyle and on their ways of coping with environmental obstacles is the concept of control. We were interested as to whether the subjects' sense of control over their lives—in the past, at present, and for the future—were connected with their maturational timing. Therefore, direct questions concerning perceived control, as given in the postal questionnaire, were analyzed. The subjects were asked whether they saw themselves as the determining agents when they chose their education and occupation after the compulsory school ("Have you yourself decided what education and occupation to choose?"). They were also asked whether they experienced control over their present life situation ("Do you yourself control your present life situation?"), as well as over their future life situation ("Do you yourself control your future life situation?").

Significant differences among the menarcheal groups were obtained for control over past experiences but not for present or future conditions. The percentage of females in the four menarcheal groups who thought that they had not decided their education and occupation themselves is shown in Fig. 10.2.

As can be seen in Fig. 10.2, a higher proportion of the earliest-matured group of subjects reported that their choice of education and occupation were out of their control [x^2 (3, $N = 409$) = 9.18, $p < .05$]. In fact, almost three times as many of the earliest-developed females as of the other women did not perceive their educational and occupational decisions to be within their control.

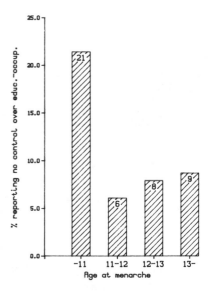

FIG. 10.2. The percentage of females in four menarcheal groups who stated at 25.10 years that their past educational and vocational choices had been out of their control.

SOCIAL SUPPORT

The peer network of the average- and later-developing girls in mid-adolescence was to a large extent made up of conventional peers—friends in the same class and same-age peers. The peer network of the early-matured girls, by contrast, was more often localized outside the conventional peer group of teenage girls: to older peers, working peers, and boys.

The status in the peer group was highest among the average-developing girls throughout the years in the compulsory school. In early school years it was lowest among the early-developed girls. Their peer status improved over time.

In mid-adolescence, the most early- and the most late-developed girls reported less satisfactory peer relations than did average-developing girls. However, the early-matured girls reported having the highest number of friends generally.

The relations to adults differed among the early- and the later-matured girls in adolescence. The early-matured girls reported more conflicting parental relations, particularly with their mothers, and they expressed more conflicting teacher relations.

The question as to whether the pattern of interpersonal relations peculiar for the early- and later-developing girls in mid-adolescence has any resemblance to their future social network conditions has not hitherto been investigated, except as it relates to living single versus married. Of particular interest is the future relations with parents. In chapter 7 there were some indications that the early-matured girls' parental conflicts in mid-adolescence were improving over time. Now, are there any reminiscents of the parental conflicts at adult age, or were these conflicts circumscribed to the adolescent years for the early developers? This is one of the questions that we investigate here. It will be remembered that Hill and Holmbeck (1987) anticipated prolonged parental conflicts among the earliest-developed group of girls.

In the postal questionnaire at 25.10 years, an attempt was made to assess the social support of the subjects at this age. Based on earlier research on the type of persons associated with affection and trust, the subjects' relations to seven categories of people was examined. For each one of these person categories they were asked to respond whether or not they (a) were on visiting terms, (b) could talk openly about their problems, and (c) could trust that they would receive help if they were in a difficult situation. A comparison among subjects in the four menarcheal groups for the seven person categories, each one measured by three social support dimensions, is presented in Table 10.5.

An inspection of Table 10.5 shows that most of the females had a great many friends they saw often, friends with whom they could talk, and whom they could trust in a difficult situation. The social support system was primarily

TABLE 10.5
Differences in Social Support Dimensions among Females
in Four Menarcheal Groups ($N = 414$)

	Be on visiting terms[a]	Talk openly with	Trust
Parents	ns (80.9)	ns (68.6)	.05 (91.8)
Siblings	ns (64.5)	ns (47.3)	.05 (60.1)
Other relatives	ns (31.9)	ns (8.9)	ns (24.4)
Friends of one's childhood	ns (33.3)	ns (23.2)	ns (15.9)
Fellow student / Fellow worker	ns (54.6)	ns (34.3)	ns (24.2)
People in clubs, associations, etc.	.05 (20.8)	ns (7.2)	ns (6.3)
Neighbors	ns (32.1)	ns (9.7)	ns (7)

[a] Figures within parentheses refer to the percentage of females in the total group who marked respective person category.

localized among family members. More than 9 out of 10 subjects stated that they could confide their personal problems to their parents, and all but 3.4% had some relatives (parents, siblings, or other relatives) they could trust in a difficult situation. Interestingly, the subjects regarded parents more as the persons they could trust and the persons who would help them if they got into some difficulty than they did the persons they talked with about their problems. Thus, parents tended primarily to have back-up functions for the females. Overall, also taking into account outside family relations, the great majority of females seemed to have an adequate social support network.

The persons the females met often also tended to be the same persons with whom they could talk openly and trust. There were some exceptions. People in clubs, associations, and neighbors were persons with whom the subjects often were on visiting terms, but with whom they more seldom talked openly or shared confidence. "Other relatives" were often visited and trusted. However, the subjects more seldom talked openly with them. Of persons outside the family, fellow students or fellow workers were the persons in particular who had social support functions.

Perhaps the most interesting finding from our perspective was the differences among the menarcheal groups of girls with respect to parental relations. There was a tendency for early-matured subjects to report less often that they met their parents frequently ($p < .10$), and less often that they could trust their parents when getting in a difficult situation ($p < .05$), thus reflecting that the conflicting parental relations among early-matured females in the adolescent years to some extent prevailed up into adulthood. Figure 10.3a and 10.3b present the percentage of females in the four menarcheal groups who were on visiting terms with their parents and the percentage who trusted their parents for help in a difficult situation.

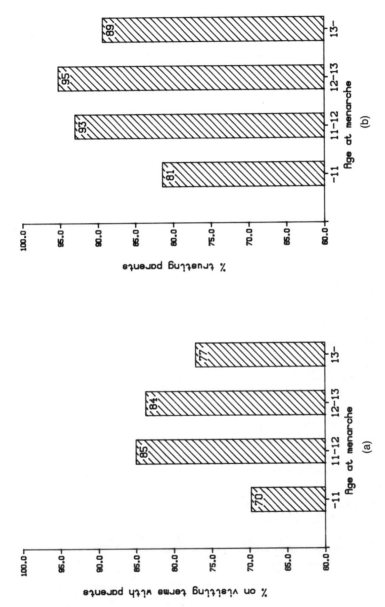

FIG. 10.3. (a) The percentage of females in four menarcheal groups who reported being on visiting terms with their parents at 25.10 years. (b) The percentage of females in four menarcheal groups who reported trusting parents for help in a difficult situation at 25.10 years.

Significant differences ($p < .05$) appeared among the menarcheal groups for the category Siblings with respect to trust. A closer analysis revealed that the latest-matured girls more often than the other three menarcheal groups of females stated that their siblings could be trusted to help them when needed. This result should be seen in light of the fact that the latest-developed girls had more siblings than did the other groups (see chapter 3).

Finally, significant differences in social support as related to menarcheal age of the subjects were found for the person category People in Clubs, Associations, etc. A monotonic positive relation, with a greater percentage of females being on visiting terms with people in clubs the later-physically matured they were, was discerned ($p < .05$).

The format of the postal questionnaire precluded the possibility of making a further inquiry into details of the subjects' parental relations. This had to be left to the more extensive study of the females in the subsample study. For this subsample of females, questions were asked addressing psychological factors in the family background. A questionnaire, the Parent–Child Relations Scale (PCR) was used, including various statements that described how the females perceived their parents' treatment and behavior toward them during their upbringing. The scale was used in reduced form (15 statements). A comparison was made among the menarcheal groups for three psychological dimensions—pedagogical attitude, emotional engagement, and authoritarian discipline—each one rated separately for father and mother. The results showed little differences among the menarcheal groups of girls with respect to how they perceived and evaluated their fathers and mothers as educators. However, a difference at the .10 level of significance was reported for the scale authoritarian discipline when the mother was rated. The earliest-matured group of females reported that their mothers had had a more authoritarian attitude toward them during upbringing than was the case for the later-developed females. No differences appeared for the emotional engagement the females perceived their mothers or fathers to have had relative to them, nor for their mothers' or fathers' pedagogical attitude toward them in their upbringing.

To summarize, the following conclusions regarding the social support at adult age as related to menarcheal timing seem pertinent. Overall, the subjects showed little signs of lacking a supportive network. Only a few reported that they had no person in their family or no relative whom they could trust would help them if necessary. In view of the fact that the earliest-matured group experienced more strained relations with their parents in adolescence, a similar tendency, albeit much weaker, was found at adult age. The females who were early-maturing reported, at adult age, seeing their parents somewhat less often than did other females, as well as showing less trust that their parents and siblings would help them in a difficult situation. Retrospective reports from the medical sample revealed a tendency that the earliest-matured females perceived their mothers to have been more authoritarian during their upbringing than did the later-maturing females.

PERSONALITY

In this section a comparison related to more psychological qualities at adult age is made between females who were early- and later-matured. It entails a comparison of adult personality, sex-role identity, and life values.

One of the few studies in the literature that examined the adult personality from the standpoint of females' biological maturation was made by Shipman (1964). This study was not designed as a follow-up or a longitudinal study, but as an investigation comparing personality dimensions of adult women who, by recall, stated their age for menarche. The subjects were 82 White females, aged 20 to 50 years, and they were tested with Cattell's 16 Personality Factor Test. To what extent the results have an adequate external generalizability can be questioned, due to certain peculiarities of the sample. Of the 82 females, 63 were "adult relatives of a diabetic child, selected on the basis of their willingness to be studied comprehensively for precursors of diabetes mellitus."

Those who had been early maturers (menarche at ages 10 and 11) rated themselves as adults as more conservative, more lax, and inexact, and more trusting of other people relative to other females. The females who reached menarche at the expected time (ages 12 and 13) were more feminine, and the late developers (menarche at 14 years or later) rated themselves as more dominant, suspicious, and more self-controlled. This characterization has to be taken at face value because no description of the results, other than correlations, were presented. The traits that were positively correlated with menarcheal age were: dominance, liberality, self-sufficiency, suspiciousness, masculinity, and self-control.

Shipman speculated about the possible developmental progressions from adolescence to adulthood that might explain these results. From a literature review, he characterized the early-developed girls as "self-conscious, confused, anxious, and ineffective in school" in adolescence and believed that in adult life these characteristics would make these females submissive, dependent and trusting—a "passive–receptive femininity." The more unclear feminine role for the later-matured girls in adolescence would make for compensation in the educational domain, consequently making this group of girls "assertive, well-informed, independent thinkers with not so much need for social approval." It is of some interest to note that the later-developing females in the present sample were more well educated as adults than the early-developed females, and also that the early-developed females were more family oriented in the sense that more of them had their own children. Whether their homemaking role should be interpreted as a "passive–receptive femininity" is an open issue.

In another investigation, Shader, Harmatz, and Tammerk (1974) administered a large test battery (TMAS, MMPI Depression Scale, Langner's Scale of Psychopathology, the Extraversion and the Neuroticism scales of EPI, the Femininity scale of CPI) to 447 women with a mean age of about 22 years. The sample was divided into two halves for cross-validation of the results. No differences related to the menarcheal age of the women were found that were valid across both sample halves.

Some interesting results about the adult life situation among early- and late-maturing females can be found in a study of Peskin (1973). An almost reversed picture of emotional, cognitive, and social adaptation for early- and late-developed females at the age of 16 was obtained for the same females when they were 30 years of age. In adolescence, the early developers were rated as more socially introverted and as showing more stressful emotional symptoms. At the age of 30, it was the later-developed females who were so characterized, whereas the early-developed women were characterized as having more intellectual interests, higher aspirations, and being more rational and responsible. In a related study, Peskin and Livson (1972) reported that the early-matured, as adults, had higher scores on overall psychological health, measured by the 100-item Q sort set, with the criterion for health being "capacity for work and for satisfying interpersonal relationships, emotional warmth, a sense of purpose, and a realistic perception of self and social reality" (Peskin, 1972, p. 156). This more favorable situation stood in sharp contrast to impulse-control data in adolescence where the early developers emerged as more "whining and explosive in temper." Although these two studies conducted on a group of females in the Guidance study are quite fascinating, they are nevertheless based on a very limited sample of 33 subjects.

Personality. In order to compare differences in adult personality among the menarcheal groups of females, a number of personality scales were given to the women in the medical sample. Seventy subjects were tested with KSP, an omnibus inventory covering various personality dimensions (Schalling, 1978):

- Somatic Anxiety
- Psychic Anxiety
- Muscular Tension
- Anxiety (SA+PA+MT)
- Social Desirability
- Impulsiveness
- Monotonic Avoidance

- Detachment
- Psychastenia
- Socialization
- Indirect Aggression
- Irritability
- Suspicion
- Guilt
- Inhibition of Aggression

The comparison of adult personality for these scales yielded few significant differences in the personality dimensions among the menarcheal groups of subjects. Significant differences were found for one scale only: the socialization scale. For this scale, made up of items from Gough's socialization scale, a difference among the menarcheal groups at the .01 level of significance was obtained. Post hoc testings showed that the earliest-matured females reported more social adjustment problems than did the latest-matured females. A closer item-by-item analysis for the socialization scale is shown in Table 10.6.

Characteristic of the socialization scale is that many of the items, in contrast to the other personality scales, refer to past events and experiences. The items in the socialization scale that differed significantly among the four menarcheal groups were also the ones picturing home and school problems at earlier ages ("I have had more than my share of things to worry about"; "Sometimes I used to feel that I would like to leave home"; "My home was less peaceful and quiet than those of most other people"; "I sometimes wanted to run away from home"; "My parents never really understood me"). For some of the items, tendencies significant at the .10 level of significance were obtained; these items also referred to the past ("The members of my family were always very close to each other" [reverse]; "In school I was sometimes sent to the principal for cutting up"; "As a youngster in school I used to give the teacher lots of trouble"). The early-developed girls, consistently for all the items just mentioned, reported more problems than did the later-matured girls.

Some comments should be made on the differentiating items. In chapter 5 it was observed that the more problematic teacher relations among the early-developed girls concerned obstinate tendencies directed from the girls toward their teachers, without these girls' necessarily perceiving that their teachers treated them incorrectly. In fact, the early-matured girls liked as many of their teachers as did the later-matured girls. This seemingly unjustified rebelliousness on the part of the early-matured girls finds some support in the socialization scale. Now, in retrospect, the females who were early-matured admitted to a higher extent than did later-maturing females that they gave their teachers lots of trouble and that they used to misbehave.

TABLE 10.6
Differences Among Females in Four Menarcheal Groups with Respect
to Socialization ($N = 71$)

Socialization items	F	p	post hoc
I have had more than my share of things to worry about	4.00	*	1>4,2
Sometimes I used to feel that I would like to leave home	3.87	*	1>3,4
My parents have often disapproved of my friends	1.17		
Life usually hands me a pretty raw deal	0.95		
My home life was always happy (rev.)	1.86		
I have often gone against my parents' wishes	1.50		
People often talk about me behind my back	1.78		
My home life was always very pleasant (rev.)	1.30		
The members of my family were always very close to each other (rev.)	2.33	a	
My home as a child was less peaceful and quiet than those of most other people	2.91	*	
In school I was sometimes sent up to the principal for cutting up	2.52	a	
I sometimes wanted to run away from home	5.77	**	1>3,4
Even when I have gotten in trouble I was usually trying to do the right thing (rev.)	0.96		
With things going as they are, it's pretty hard to keep up hope of amounting to something	1.77		
As a youngster in school I used to give the teacher lots of trouble	2.66	a	
My parents never really understood me	3.55	*	1>4
I seem to do things that I regret more often than other people do	1.63		
When I was going to school I played hookey quite often	1.36		
My parents have generally let me make my own decisions (rev.)	0.77		
I often feel as though I have done something wrong or wicked	0.53		

^{a} $p<.10$. *$p<.05$. **$p<.01$. ***$p<.001$.

At young adult age, the early-matured females also perceived in retrospect that their home had been filled with more conflicts, less harmony, and less understanding on the part of their parents than did the later-developed females. Their home life had made them want to leave home. The information the subjects gave when they were adolescents strongly supported that the early-matured girls were the menarcheal group that particularly showed runaway behavior. At the age of 14.10 years, more than one out of four of the earliest-matured females reported that they had run away from home once or on several occasions. This happened for just 6% among the rest of the girls.

Sex-Role Identity. In addition to the personality scales, sex-related characteristics of the individuals were measured by a revised version of Bem's Masculinity-Feminity scale (Bem, 1974). No differences among the four menarcheal groups of females were revealed, either with respect to masculinity [$F(3,70) = 1.61$, ns] or Femininity [$F(3,70) = 0.73$, ns]. The lack of such differences tend to designate the early-developed females' more homemaking orientation to the social domain, but less to the psychological.

Life Values. Measures of life values were obtained from an instrument labeled "The Life Wheel," developed by Shalit (1978). This is an instrument that taps aspects in everyday life that are of perceived importance. The life values have been reported to be highly age-related. The subject is instructed to describe, within a 12-graded wheel, what she thinks is important in her present life, and to describe why this is important. The information can be used both quantitatively and qualitatively. The coding was performed according to the manual provided by Shalit (1978). The following life value areas were investigated. No differences were found among the menarcheal groups with regard to the number of categories used [$F\ (3,74) = 1.74$, ns].

- Ideology
- Solidarity
- Housing conditions
- Knowledge
- Hobbies-Interests
- Cultural experiences
- Economic security
- Physical satisfaction
- Psychic satisfaction
- Recreation
- Existential
- Altruistic

No significant differences appeared among the four menarcheal groups of subjects for the different life values in the Wheel, except for the category altruistic [$F\ (3,74) = 4.61$, $p < .01$]. This category is defined from remarks such as "provide care for other, work with sick people, responsibility, etc." A considerably higher proportion of the earliest-developed subjects than of the other females reported life values that were coded in this category. A detailed presentation of the percentage of subjects in respective menarcheal groups who reported life values within the altruistic category is shown in Fig. 10.4. Having in mind the domestic situation for the earliest-matured subjects, that more of them had own children, and that the number of children were highest per person among this group of subjects, it is not surprising that they more often than other females described altruistic motives.

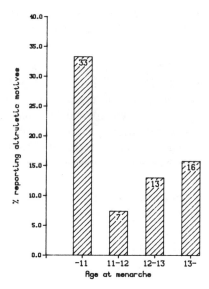

FIG. 10.4. The percentage of females in the four menarcheal groups who mentioned aspects of their life, subsequently coded in the "altruistic" life value category.

SOCIAL ADJUSTMENT

Criminal Offenses

A convincing connection between age of biological maturation and frequency of norm breaking, with the early-developed girls more involved in such activities than the later-maturing girls, has been reported for data covering the mid-adolescent years. Very few girls were registered delinquency cases in adolescence. Data for these registered offenses through age 17 showed, but not convincingly so, that more of the early-developed girls were registered for some offense.

If the differences among the menarcheal groups of girls with respect to norm violation in mid-adolescence are associated with a more general and *lasting* pattern of social maladjustment, we would expect to find these differences reflected in official data indicating violations of the law at adult age. Thus, a higher proportion of the early- than of the late-maturing females would turn up in official criminal records. To investigate if such was the case, official data on criminal offenses, covering age 18 up to age 30, were compared among the menarcheal groups. The criminal activity among females in the four menarcheal groups in this postadolescent period is shown in Table 10.7.

As Table 10.7 shows, 27 of the 466 girls (5.8%) were registered by the police for some offense between age 18 and age 30. Statistical analysis indicated that early-maturing females were overrepresented among the subjects with criminal records ($p < .05$). A word of caution should be inserted. With such a low percentage for delinquency as shown in Table 10.7, the risk of

TABLE 10.7
Registered Offenses From 18 Up to Age 30 among Females
in Four Menarcheal Groups ($N = 466$)

Age at menarche	Non-offender	One-time offender	Recidivist
−11 yrs	45 91.8%	2 4.1%	2 4.1%
11–12 yrs	96 88.9%	9 8.3%	3 2.8%
12–13 yrs	180 96.8%	6 3.2%	0 0%
13– yrs	118 95.9%	5 4.1%	0 0%
Total %	439 94.2%	22 4.7%	5 1.1%

χ^2 (6, $N = 466$) = 14.92, $p<.05$

chance fluctuations is obviously high. Although a significant relationship was observed, it might be concluded that the results fail to provide strong evidence for an assumption that registered law breaking is related to menarcheal age. The results can be compared with data presented in earlier literature, for example, Gold and Tomlin (1975; reviewed by Gold & Petronio, 1980).

Alcohol Abuse

Our original hypothesis, applied to alcohol drinking, suggested that early alcohol habits and drinking frequency in mid-adolescence were related to a girl's maturational timing, such that the earlier developed a girl was, the earlier she engaged in this type of behavior and the more advanced were her drinking habits in the mid-adolescent period. This hypothesis was strongly confirmed. At the age of 14.5 years, clear differences were established, with a greater percentage of early-matured than of the later developers having been drunk. In the short-term perspective an equalization was found in the prevalence of drunkenness among the menarcheal groups of girls. A the age of 15.10, no significant differences among the four menarcheal groups with respect to drunkenness per se were demonstrated. This finding lends support to a maturational timing interpretation of beginning to use alcohol, with the early-maturing girls starting earlier and the later-maturing girls catching up at a later time in adolescence.

On the other hand, the early-developed girls drank more frequently at all test occasions in adolescence—at 14.5 years, 14.10 years, and at 15.10 years.

No attenuation of the differences among the four menarcheal groups with regard to frequent drunkenness could be discerned. The finding that the early-developed girls were a subgroup of girls with excessive alcohol habits in adolescence might indicate that they were at particular risk for developing alcohol problems in adulthood.

In the postal inventory at adult age, when the research group of females were on the average 25.10 years old, questions covering the subjects' adult alcohol consumption were given. Two questions concerned how often the subject drank alcohol and how much was consumed on occasions when they drank the most. The first question was answered by 81% and the second by 67% of the females with menarcheal data. Table 10.8 presents the results of the comparison among the menarcheal groups for frequency of alcohol consumption, and Table 10.9 shows the similar comparison for the amount of alcohol consumed.

TABLE 10.8
Alcohol Consumption at Age 25.10 Among Females in Four Menarcheal Groups

Age at menarche	Frequency of alcohol consumption			N	%
	Never	Some-times	At least weekly		
−11 yrs	5.7	82.9	11.4	35	9.9
11–12 yrs	7.1	79.8	13.1	84	23.8
12–13 yrs	9.0	75.2	15.9	145	41.1
13– yrs	11.2	77.5	11.2	89	25.2
N	31	274	48	353	
%	8.8	77.6	13.6		

χ^2 (6, N = 353) = 2.58, ns

TABLE 10.9
Alcohol Consumed at Age 25.10 by Females in Four Menarcheal Groups

Age at menarche	Amount of alcohol consumed, grams				N	%
	0	1–30	31–60	61–		
−11 yrs	7.1	21.4	46.4	25.0	28	9.6
11–12 yrs	8.1	23.0	39.2	29.7	74	25.3
12–13 yrs	10.5	18.4	37.7	33.3	114	38.9
13– yrs	13.0	28.6	31.2	27.3	77	26.3
N	30	66	109	88	293	
%	10.2	22.5	37.2	30		

χ^2 (9, N = 293) = 5.62, ns

No systematic relationship between menarcheal age and answers to the alcohol questions were discerned. Those who drank most frequently at 25.10 years came as often from the early- as from the later-developed groups of girls. However, there was a tendency for the abstainers to come slightly more often from the later-developed group of females. Table 10.9, showing the amount of alcohol consumed, similarly showed no relationship with menarcheal age.

Complementary questions covering the occasions under which the subjects drank, whether the alcohol habits had changed during recent years, and the frequency of drinking particular types of alcoholic beverages, showed no connections with the menarcheal age of the females.

Questions on drinking habits were also asked in connection with the interview of the medical sample at the age of 25–26 years. Of the subjects in the sample, 77.2% drank alcohol occasionally or regularly. No relationship between menarcheal age and alcohol drinking was obtained [x^2 (3, $N = 55$) = 4.18, ns]. Of the subjects, five females were total abstainers. All of them belonged to the two latest-matured groups, thus agreeing with the results for the total group, which indicated some relationship between menarcheal age and being a nondrinker. Additional questions were given concerning the circumstances around alcohol drinking (on what particular occasions the subject drank, whether or not they had drunk more at any age than they intended, if drinking had affected their way of living, whether or not they had been treated for alcohol problems, how much was consumed the last week, and whether or not their drinking habits had changed during recent years). No significant relationships with menarcheal age were found, except in one case. More of the earliest-matured girls than of other girls had, when they looked back, experienced blackouts once or on several occasions (-11 years: 62.5%; 11-12 years: 30.4%; 12-13 years: 4.3%; 13- years: 29.4%).

To conclude, from the self-report data on alcohol use at adult age, no systematic relationship between alcohol use and the menarcheal age of the females was evident at 25–26 years of age, except for a slight tendency for the abstainers from alcohol to be overrepresented among the later-developed females. That the earliest-matured females had experienced more black-outs in their life than had other females is not surprising in view of their more advanced drinking habits in adolescence compared to the other girls.

In addition to self-report measures, information on alcohol abuse from official registers up to adult age was examined (Stattin & Magnusson, in press). Information was obtained from the police (drunkenness, drunk driving), the social authorities (measures taken in accordance with the Temperance Law; information on treatment at institutions for alcoholics), and open and closed psychiatric services (patient journals). This information on alcohol abuse covers the period from 14 up to 26 years of age.

No significant differences in registered alcohol abuse were found among the four menarcheal groups of girls. Only 15 of the 458 females with complete register data were registered by the age of 26. Of the subjects in the four menarcheal groups, 4.3% of the females who had attained menarche before age 11 were registered by the police, by social welfare authorities, or by psychiatrists for alcohol problems; 5.6% of the females who had had their menarche between ages 11 and 12 were likewise registered, 0.5% of the females with menarche between 12 and 13 were found in these registers, and 5% of the latest-developed group of females were registered. Observe that the lowest incidence of alcohol problems was found among the "average" maturing females (i.e., females attaining their menarche between 12 and 13 years of age). This is the same pattern as was found for self-reports of blackouts.

Drug Use

Both at the age of 14.5 and 14.10 years, and at the age of 15.10 years, more girls of the early developers than of the later-maturing had used hashish. The relationship between menarcheal age of girls and smoking hashish was significant at the .001 level, at least, at all three test occasions. No catch-up effect for the later-matured girls was discerned. Rather, the differences among the four menarcheal groups tended to increase over time.

Data on use of drugs for the total sample at adult age was not available for an analysis of the long-term stability of smoking hashish. In the interview with females in the medical sample some questions concerned use of drugs. Only one girl reported having used marijuana during the last year. She belonged to the earliest-matured group. A question was asked dealing with the subjects' earlier use of drugs. As might be expected from data covering the adolescent period, more of the earliest-matured females reported, at adult age, having earlier used drugs, cannabis, or stronger narcotics. The percentage of girls in the four groups who stated that they had used drugs at some time earlier in life is graphically depicted in Fig. 10.5 Although the differences appear strong between the earliest-matured females and the other later-developed females, they were not statistically significant $[x^2 (3, N = 78) = 3.81, ns]$. Only two girls admitted having used stronger narcotics. They belonged to each of the two earliest menarcheal groups.

In the interview the females in the medical sample were asked whether or not they had ever used tranquilizers or sleeping medicine. A total of 17.9 percent of the females admitted that they had used such medicine. No systematic relations with the age of menarche of the subjects was found. Again, the average-developed females reported the least frequent use (-11 years: 22.2%; 11-12 years: 22.2%; 12-13 years: 8.7%; 13- years: 21.1%).

Smoking

No information was collected on smoking habits in adolescence. At 25 years of age, the females in the medical sample were questioned on their present and past smoking. A slight tendency for subjects who never had started

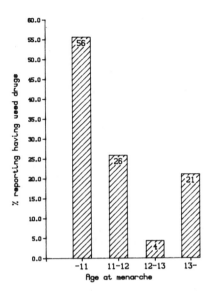

FIG. 10.5. The percentage of girls in four menarcheal groups reporting past use of narcotics ($N = 78$).

smoking to be late-matured (22.2% of the earliest-developed females had never started to smoke in comparison with 34.1% of the other females) and for girls who presently were smoking to be overrepresented among the earliest-matured group was discerned (55.6% of the earliest-developed reported that they smoked, relative to 42.1% of the other subjects). However, the differences among the subjects in the four menarcheal groups did not approach statistical significance.

A MEDIATING EFFECTS ANALYSIS

The model that has been proposed, linking the psychosocial functioning of the person to biological maturation, is one that places the key responsibility for this connection on factors operating in the individual (perceived maturity relative to agemates) and in her everyday interpersonal environment (peers outside the same-age–same-class frame). In the last chapter, the hypothesized pathway was explicitly tested for measures in mid-adolescence within the interpersonal, social transition, school adjustment, and emotional domains.

It is of central interest to investigate whether or not the same mediating model that quite successfully accounted for the observed relationships between pubertal maturation and psychosocial measures in adolescence can also be applied to explain the differences among the groups of females in adult life. If so, the implication would be that factors, within the individual and in her environment, which accounted for differences among the menarcheal groups of girls in mid-adolescence have long-term influence on the lives of the females.

In measurement terms, to the extent that perceived maturity and peer network in the adolescent years are factors that are also responsible for the long-range effects of biological maturation on adult psychosocial functioning, then the direct effect of menarcheal age on the differentiating variables at adulthood should be reduced once the self-concept and the peer network factors are controlled for.

The calculations were performed with the same rational as was done in chapter 9. First, the effects of menarcheal age on the differentiating measures in adulthood were established, with ordinary ANOVAs, for the group of subjects who had complete data for the proposed mediators (perceived maturity, association with younger, older and working peers, and heterosexual relations). The criterion for inclusion of a particular dependent variable for the subsequent analysis of covariance was that it had to differentiate at at least the .05 level among the menarcheal groups for those subjects with complete data. The following variables covering adult age were examined:

Family life

1. Childbirth

Work

2. Job position

Education

3. Level of education
4. Reported obstacles for further education at adult age: Child care
5. Reported obstacles for further education at adult age: Present employment situation

Personality and life values

6. Socialization (personality scale)
7. Altruistic needs (life value)

Social adjustment

8. Registered criminal offenses
9. Retrospective reports of blackouts in connection with drinking

The items that did not significantly differentiate among the menarcheal groups of individuals for this more circumscribed group of females, and thereby were excluded from further analyses, were the questions on perceived control over one's choice of education and occupation and the social support items.

Next, an ANCOVA was computed for each one of these criterion variables from adult life, with menarcheal age as the independent variable and the mediating variables being entered as five covariates.

The results from the ANCOVA are summarized in Table 10.10. The first columns in this table show the unadjusted effect of menarcheal age on each of the variables from the adult time period and the adjusted effect after the effect of all the five covariates had been controlled for. The last two columns show which covariate contributed to the highest effect reduction of menar-

TABLE 10.10
The Relationship of Menarcheal Age to Various Forms of Psychosocial Functioning at Adult Age With and Without Control for Self-Concept and Peer Network in Mid-adolescence

		Unadjusted effect:		Adjusted for covariates: All covariates		Strongest covariate		
	n	F	p	F	p	Peer type	F	p
Family life								
Own children? (Yes–No)	343	6.08	***	3.90	**	heterosex. rel.	4.34	**
Work								
Job position	291	3.86	**	2.58		working peers	2.49	
Education								
Level of education	344	4.23	**	2.73	*	working peers	2.64	*
Obstacles to further education:								
Child care	345	2.93	*	1.73		working peers	1.94	
Employment makes it difficult	345	3.11	*	2.21		working peers	2.28	
Personality & Life values								
Socialization (scale)	63	6.39	***	4.05	*	working peers	4.05	*
Altruistic needs	69	2.84	*	2.42		—	—	
Social adjustment								
Criminal offences: register data	381	4.29	**	4.10	**	—	—	—
Blackouts (retrospective report)	63	5.62	**	4.49	**	working peers	4.42	**

* $p<.05$. ** $p<.01$. *** $p<.001$.

cheal age, and the subsequent F value, once this covariate was introduced into the equation.

Family Life

Own Children. Comparing the unadjusted and the adjusted effect of menarcheal age on childbearing, the results in Table 10.10 show that a considerable part of the relationship between age at menarche and childbearing was conditioned by the covariates. A rather strong reduction in the strength of the menarcheal impact could be discerned once the effects of the five covariates were corrected for. At the same time, the covariates did not account for all the variance between menarcheal age and childbirth. The unadjusted impact that was significant at the .001 level, was still significant at the .01 level after the covariates had entered the equation.

Not unexpectedly, "heterosexual relations" was the covariate that most significantly related to childbirth at adult age (a significant contribution at the .05 level was also obtained for the covariate younger peers). When this peer type was entered as the single covariate, it contributed almost as much to the effect reduction for menarcheal age as all five covariates together. The women who had children at the age of 25 years had had, in mid-adolescence, more stable relations with boys (and less of them associated with younger peers) compared to women who had no children at this time in adulthood.

The finding that the outside class peer network 10 years earlier in life affected childbirth for women 25 years of age, is of interest, not only for the issue of the long-range consequences of early versus late maturation. Because the adult life situation among women in many areas is dependent on their domestic situation, the fact that the peer reference group in adolescence is already an influencing factor for adult childbirth implies that future psychological and sociological studies making detailed causal, time-linked investigations of family life might profit from more extensive surveys of the characteristics of the women's peer network much earlier in life.

Work

Job Position. The unadjusted effect of menarcheal age on job position was significant at the .01 level. The adjusted menarcheal effect, after covariate control, was reduced to nonsignificance. Thus, much of the impact that menarcheal age had on adult job position was mediated by the covariates that covered features of the adolescent life situation.

The ANCOVA identified associations with employed peers in mid-adolescence as being most significantly related to the adult job position (heterosexual relations was another covariate that had significant impact at the .05 level). The more extensive the contacts with working peers (and with boys) in mid-adolescence, the lower was the adult work position. Working peers as the single covariate was almost as efficient a mediating factor as all the covariates together.

Education

Level of Education. After all the covariates had entered the equation, the former effect of menarcheal age on adult level of education, which was significant at the .01 level at least, was markedly reduced. However, there was still a significant impact, adjusted for the covariates, at the .05 level of significance.

The adult educational level of girls was earlier found to be closely connected with their job positions. Therefore, the variables mediating the menarcheal age—education relationship should be similar to those mediating the relationship between menarcheal age and job position. As can be seen in Table 10.10, this was indeed the case. Females who in the mid-adolescent period had associated with working peers had lower adult education than did females who had had no employed peers in their circles of friends. (As for job position, heterosexual relations was another covariate that was significantly related, at the .01 level, with adult educational level.) Working peers was the essential mediating factor. Its effect singly was as high as the five covariates taken together.

It will be recalled that the covariates, working peers and heterosexual relations, were the only two significant covariates for grade-point average at the age of 15 and that association with working peers was the only significant covariate for the girls' experienced school interest at this age. Thus, girls' contacts in the mid-adolescent years with peers who were already employed were negatively connected with their school achievement and school interest in mid-adolescence, as well as with the future level of education and work position. A similar influence on the educational situation in midadolescence and the educational–vocational situation at adult age was observed for the girls' heterosexual contacts in the mid-adolescent years.

Obstacles for Further Education. The difference among the menarcheal groups of females with respect to obstacles to further education, problems with child care and reasons connected with employment, were reduced to insignificance once the effect of the covariates were corrected for. Association with working peers in mid-adolescence was the only covariate that showed a significant contribution to the two items. Females who had had working peers in their circles of friends in mid-adolescence reported more obstacles of these kinds at adult age.

Personality and Life Values

Socialization. None of the different personality scales administered at adult age, except Gough's socialization scale, differentiated among the menarcheal groups of subjects. A great part of the original differences ($p < .001$) on this scale among the four menarcheal groups of females was reduced after the covariate control. However, there still existed significant differences ($p < .05$) between early- and later-matured females after corrections for the covariates.

Above all, associating with working peers in adolescence was the covariate, that was related to the intergroup differences in socialization. The effect reduction in menarcheal age was of the same magnitude after correcting for this single covariate as it was after correcting for all five covariates. More females who engaged with employed friends in mid-adolescence perceived, at young adulthood, that their upbringing had been problematic than did females who had not associated with such peers (an effect of the covariate, perceived maturity, on socialization, at the .05 level, also could be discerned. The more mature the girls perceived themselves to be in mid-adolescence, the more upbringing problems they reported at adult age).

Altruistic Needs. None of the covariates had any significant impact on the life value scale altruistic needs. Only slight reductions in the influence of menarcheal age on altruistic needs were found after the covariate correction.

Social Adjustment

Criminal Offenses. None of the covariates were significantly related to having committed criminal offenses (albeit heterosexual relations was just above the .10 limit). Little reduction in the impact of menarcheal age could be discerned after adjustment for the five covariates.

Blackouts. Little reduction in the effect of menarcheal age after the covariate control was also obtained for retrospective reports of having had blackouts. One of the covariates, working peers, was significantly related to the retrospective measure. Association with working peers in mid-adolescence was connected with having experienced more blackouts in connection with alcohol drinking.

Comments

The main results of the analyses of the mediating impact of perceived maturity and peer network in adolescence on the menarcheal influence on different measures at adult age now can be summarized.

Did there remain any significant effect of menarcheal age on the respective psychosocial indicators once the variation in the dependent variables due to the proposed mediators was eliminated? For items within the areas of family life, work, and education, the results indicate that the menarcheal impact was to a substantial extent mediated by the covariates. The effect that pubertal development had on measures within the areas of personality and life values, and social adjustment, was rather independent of the covariates.

Which of the self-concept and the peer-type variables were the important mediators for the relationships between menarcheal age and the respective dependent measures? Of the covariates, the results indicate that it was particularly peer network conditions at mid-adolescence, rather than perceived maturity at the same period in time, that operated as intermediary factors between pubertal maturation and measures within these areas at adult age. Different peer types were mediators for different areas. Heterosexual relations was the strong mediator linking pubertal development with having own children at 25–26 years of age. The covariate working peers operated as the main mediator for measures within the areas of work and education.

THE ISSUE OF DIFFERENTIAL SUSCEPTIBILITY

The notion of *differential susceptibility* has been used to describe the phenomenon that the influence of peers on a subject's behavior is a function of the match of peer characteristics and the level of biological maturation of the individual. Greater impact on a subject's behavior of peer influences is assumed to be present when there is a match between peer characteristics and the individual's level of pubertal development than when this fit is absent. By virtue of their earlier physical maturation, greater preparedness for change is assumed to be expressed by the early-developed girls in mid-adolescence for behaviors that characteristically belong to older age strata of teenagers, employed peers, and the opposite sex. Conversely, the impact of younger peers is stronger for late-maturing girls who associate with such peers than among early-developed girls who engaged with chronologically younger peers.

The differential susceptibility issue is, essentially, an interaction hypothesis, that states that the optimal conditions for behavior change should occur when there is a fit between characteristics of the person and analogue characteristics in her interpersonal environment. This hypothesis on "individual-environment fit" (cf. Pervin, 1968) or "goodness of fit" between individual attributes and characteristics of the surrounding context (Lerner, 1987) is now applied to those measures in adulthood that differentiated the females who were early-developed from those who were late-maturing. A "critical" type of peer refers to the adolescent peer type that, in the ANCOVA just performed, had the greatest mediating impact on the dependent measure under consideration. Among girls who had not associated with the critical types of peers in mid-adolescence, we do not expect any differences between early- and later-matured women with respect to these measures in adulthood. By contrast, among women who had engaged with these types of peers in the mid-adolescent years, we expect a higher behavior rate among the females whose level of pubertal development in mid-adolescence had matched the characteristics of the peer type in question.

The hypothesis of differential susceptibility in the long-range perspective was tested for the following dependent measures and their associated "critical" adolescent peer types:

1. Number of children (heterosexual relations),
2. Job position (working peers),
3. Level of education (working peers), and
4. Gough's socialization scale (working peers).

To facilitate the comparison, the menarcheal age variable was dichotomized into one group of early-developed females (menarche: -12 years) and one group of later developed females (menarche: 12- years).

Own Children. Consider first the mean number of children among the four groups of females in Fig. 10.6. Contrary to what was expected, there emerged statistically significant differences for this family life measure between

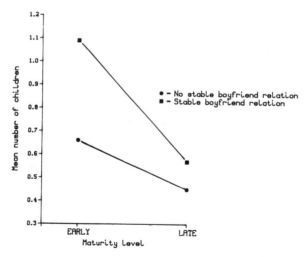

FIG. 10.6. Mean number of children at 25.10 years among early- and late-developed females, with and without stable heterosexual relations in mid-adolescence.

the early- and the later-developed females in the subsample who had not had stable relations with boys in mid-adolescence ($t = 2.11$, $p < .05$). However, and in agreement with the hypothesis, the greatest differences in mean number of children between early- and later-developed women occurred for the group of subjects who had had such stable relations with boys in the mid-adolescent years ($t = 2.89$, $p < .01$). The mean number of children was almost twice as high among the early-developed females as among the later-developed who had had stable heterosexual relations in mid-adolescence.

Job Position. Figure 10.7 shows the means of the job status variable for early- and later-developed females who had associated with versus those who had not associated with working peers in mid-adolescence. In agreement with our hypothesis, there were clear and significant differences in job status between the early- and the later-developed females who engaged with working peers in mid-adolescence ($t = 2.87$, $p < .01$). The early-developed females displayed lower job status scores than did the later-developed. In contrast, there were no significant differences among the early- and the later-developed females who did not have such peers in their circles of friends in mid-adolescence ($t = 0.04$, ns).

Level of Education. The level of education at the age of 25.10 years for early- and later-developed females who had associated with versus those who had not associated with working peers in mid-adolescence is depicted in Fig. 10.8. Again, there were significant differences between early- and later-developed females only for those who had engaged with peers who were employed in mid-adolescence ($t = 2.81$, $p < .01$, vs. $t = 1.30$, ns).

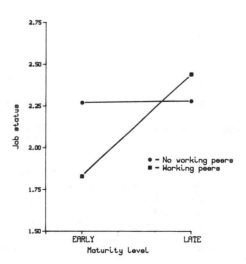

FIG. 10.7. Job status at 25.10 years among early- and late-developed females, with and without working peers in mid-adolescence.

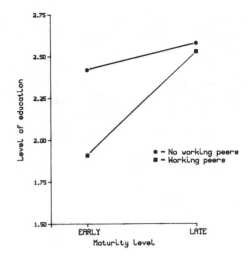

FIG. 10.8. Level of adult education among early- and late-developed females, with and without working peers in mid-adolescence.

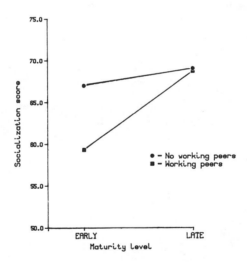

FIG. 10.9. Socialization scores for early- and late-developed females, with and without working peers in mid-adolescence.

Socialization. The mean average score for the socialization scale for the early- and the later-developed women, separately for those who had had employed friends in midadolescence and for those who had not had this friendship type is shown in Fig. 10.9. The same pattern of results were observed for this criterion variable as for the former. Among the females who had associated with working peers in mid-adolescence, the early-developed subjects had considerably lower socialization scores (less satisfying social adjustment) than did the later-developed subjects. Due to the limited number of subjects, the difference was not significant ($t = 1.73$, $p = .11$). Among the females who had not associated with working peers in mid-adolescence, small differences between the early- and the later-developed subjects were discerned ($t = 0.93$, ns).

Taken together, these results largely confirmed the hypothesis that differential susceptibility to the influence of peers in midadolescence had long-term implications for early- and later-developed females. For all four adult criterion variables, the later-developed females who had associated with the "critical" peer type in mid-adolescence had about the same mean average as the later-developed subjects who had not associated with the particular type of peers. Thus, the impact of the norms and values of the type of peers in question did not affect the later-developed girls much, even though they were in the influence sphere of these critical peers. The peer influences were mainly concentrated to the early-developed girls. The early-maturing girls who had associated with the critical peers in mid-adolescence differed considerably, with regard to own children, job position, level of education, and socialization, from the early-maturing girls who had not engaged with such peers. These results demonstrate that it is not the menarcheal age of the females as such, but the interaction of menarcheal age and peer association in mid-adolescence, which is the prime responsible factor for the effects of biological maturation on the criterion variables examined at adult age. For the subsample of females who had not associated with the critical peer types in mid-adolescence, it did not matter much whether the subject was early- or later-developed, for three of the four variables just examined, if she had not engaged in mid-adolescence with the particular type of peer mediating the association with menarcheal age.

A SUMMARY OF THE MAIN RESULTS AND SOME FURTHER ANALYSES

The overall impression of the data presented on the adult life situation can be summarized as follows. In the first place, the earliest-developed females did not pursue education to the same extent as the other women did. As many as 4 out of 10 of the early-maturing females stopped school at the earliest possible time. This proportion was twice as high as for the rest of the females. The lower education among the early-developed females was reflected in the type of employment they held in early adulthood. Early-maturing females held substantially lower job positions than did the later-matured girls.

As many of the early- as of the later-developed females were married or lived with a man at 25 years of age. However, more females in the earliest-matured group had own children at this age than was the case among later-developed subjects, and the number of children per person was higher in the earliest menarcheal group.

Looking back, the early-developing subjects at 25 years of age perceived that their educational and vocational choice had been out of their control to a higher extent than reported by the later-developing females. The social life circumstances of the earliest-matured group of females also seemed to interfere with their educational plans for the future. When asked whether they saw any obstacles for further education at adult age, more of the earliest-matured females gave as a reason that limited opportunities for day care circumscribed the possibility for such a venture. More of the earliest-matured females also stated that their present employment interfered with such educational plans. All in all, their family and occupational roles tended to set limits for the development of whatever potentialities and motivations they might have had.

A battery of personality tests were given at adult age in order to yield a broad comparison between the early and the late developers. No differences appeared among the menarcheal groups, for anxiety, aggression, hostility, impulsiveness, detachment, or psychasthenia. The lack of personality differences are quite noteworthy. The studies in the literature reporting no such long-term consequences in the personality domain is consistent with this finding.

In retrospect, the early-matured females seemed to have been more oriented toward a traditional family life than toward aspiring on higher education and a work career. In this connection it should be emphasized that there was no connection between adult sex-role identity and menarcheal age. A conclusion, therefore, has to be reached that early maturation predisposed subjects to a certain female social role, but not necessarily to a more feminine identification. The conclusion is partially supported by the data that were presented by Shipman (1964).

Finally, different indicators of social adjustment at adult age among the menarcheal groups of females were examined. Over the life span up to age 30, official register information showed that a higher proportion of the early-matured females were registered for some criminal offence compared with the later-matured subjects. The relationship, however, was weak. No systematic relationship between register information on alcohol abuse and menarcheal age was obtained. The subjects' own reports of their drinking behavior at the age of 25 years corroborated the register data of a lack of long-term consequences for alcohol. More of the early-developed females were smokers at adult age, but the relationship with menarcheal age was not significant. Taken together, no convincing relationship between menarcheal age and adult social maladjustment was established. Where differences occurred at adulthood among the menarcheal groups with respect to social behavior, these concerned the subjects' social behavior seen retrospectively. Thus, more of the early-developed females reported past incidents of blackouts in connection with alcohol drinking. Moreover, earlier use of drugs was more often reported by early-developed females as compared with their later-matured counterparts.

The Issue of Family Versus Work

In early research on identity development among females, "the career–marriage conflict" was a central issue. Much research has also been done examining features connected with a commitment toward career relative to a commitment toward homemaking. Although it is generally recognized that women tend to combine family and career (Rand & Miller, 1972), nevertheless, women are more likely than men to be faced with problems connected with the traditional female sex role of child care and homemaking versus pursuing an educational and work career (Tinsley & Faunce, 1980; Yuen, Tinsley, & Tinsley, 1980). Anyone could see this problem after plotting curves for fertility and for career opportunities over time in the same diagram. The present results, concerning the outcome in the social sectors at adult age for females who were early-developed and females who were late-maturing, show a similar tension between family life and education. Early-developed females had, as adults, own children to a higher extent than did the later-developed females, and they had comparatively lower education. They perceived problems for further education at adult age due to their home and work situation. Later-developed females, by contrast, had children less often. They were better educated and held higher job positions compared with the early-developed women. In the main, the family life orientation was more characteristic of the early developers, whereas the educational career was of more concern for later-developed females.

To the extent these results on the relationship between biological maturation and adult level of education and birth of own children can be transformed to the individual level, then we would, presumably, find an overrepresentation of two subgroups of females:

1. early-developed females who, at adult age, have their own children and low education; and
2. later-developed females without their own children and with high education.

In order to test the validity of this proposition, a configural frequency analysis was conducted for dichotomized measures. The subjects were split into one group of early-developed women (menarche before 12 years; $n = 117$) and one group of later-developed (menarche at the age of 12 or later; $n = 230$). A split was also made of the subjects into one group of women without own children ($n = 210$) and one group with one or more own children ($n = 137$).

Finally, the women were grouped into those with low education (compulsory school education or higher secondary school, practical; $n = 237$) and those with high (higher secondary school, theoretical, college, or academic education; $n = 110$). Eight different individual profiles can be produced across the three variables. They are shown in Table 10.11.

Two individual configurations occurred significantly more frequent than would be expected by chance (types). The first was the profile "early biological maturation–own children–low level of education." Fifty-one girls had such a characteristic. By chance, 32 girls would be expected to fit this pattern. The second profile was "late biological maturation–no own children–high level of education." There were a total of 65 females who were so characterized, relative to an expected number of 44. Thus, even with this relatively crude dichotomization of the variables, the results, pertaining to individual configurations of adult social life, support the hypothesis of yielding two type patterns: early-developed females with their own children at adult age and with low education, and later-developed females without their own children and with high education. To conclude, the role biological maturation played for the adult social life situation manifested in relationships between variables, also can be found when analyzing the same data in terms of distinct subgroups of persons characterized by these variables.

One individual configuration occurred less frequently than was expected by chance (antitype). This was the configuration of being later-developed with own children and with high education. Twelve females had this variable configuration. More than twice as many were expected to have this pattern.

Adult Level of Education and Job Status

Up to the present time, menarcheal age in girls has been an unknown source of influence on adult level of education in females and on adult occupational position. There is no study in the literature that has investigated whether or

TABLE 10.11
Configural Frequency Analysis of Individual Profiles Made up the Variables Biological Maturation, Own Children, and Level of Education at Adult Age

Biological maturation	Own children	Level of education	Obs	Expect	Type analysis[a]
Early	No	Low	33	48.4	
Early	No	High	24	22.5	
Early	Yes	Low	51	31.6	TYPE ($p<.01$)
Early	Yes	High	9	14.6	
Late	No	Low	88	95.1	
Late	No	High	65	44.1	TYPE ($p<.01$)
Late	Yes	Low	65	62.0	
Late	Yes	High	12	28.8	ANTITYPE ($p<.01$)

[a]Probability adjusted for multiple comparisons

not the age at which the menarche occurs in females has any long-term consequences in the educational–vocational domain. It is therefore of particular interest to compare the impact of the menarcheal age variable with the influence of common background variables typically employed in research in this domain.

One might assume, for example, that the impact of menarcheal age interacted with the intelligence of the subjects and the educational tradition in the family. These are commonly found to constitute standard predictors of educational and vocational success. Therefore, in order to examine more closely the unique contribution of menarcheal age to the variation in adult educational status and adult job position, beyond the prognosis that could be made from these standard predictors, a control analysis using a correlational methodology was applied.

Measures of parental education and the intelligence of the subjects were taken from the midadolescent years.[1] When they were in Grade 8, the subjects in the research group were tested with an intelligence battery, WIT III (Westrin, 1967), which contains four subtests: Analogies, a test of inductive ability on verbal material; Opposites, a test of verbal comprehension; Number Combinations, measuring deductive ability; and Puzzle, a test of spatial ability. At Grade 9, a questionnaire was administered to the parents of the subjects, including, among other things, a question on the father's as well as the mother's education (see Magnusson, Dunér, & Zetterblom, 1975, pp. 84–86). The educational variable was coded into seven categories:

1. unskilled laborer;
2. vocational training, vocational school, or equivalent;
3. lower secondary school, girls' school, folk high school, etc.;
4. education at intermediate level;
5. higher secondary school;
6. advanced education above matriculation but of shorter duration than university studies; and
7. university level education.

The measure of parents' education used was the educational status of the parent with the highest achieved education.

---

[1]In a previous study (Magnusson, Stattin, & Allen, 1986), data for intelligence and parents' education were analyzed for a measure of intelligence administered in Grade 3 (10 years of age) and for a measure of parental education in a parents' questionnaire administered at the same age.

TABLE 10.12
Intercorrelations among Parents' Education, Intelligence, Menarcheal Age,
Adult Level of Education, and Adult Job Position

	Intelligence	Menarcheal age	Adult level of education	Adult job position
Parent's education	.28*** (n = 413)	.08 (n = 374)	.40*** (n = 381)	.27*** (n = 320)
Intelligence		.02 (n = 456)	.44*** (n = 462)	.26*** (n = 383)
Menarcheal age			.17*** (n = 413)	.16*** (n = 347)
Adult level of education				.68*** (n = 392)

***$p<.001$.

The intercorrelations among the variables under investigation is shown in Table 10.12.[2]

As expected from other studies, parents' education and the females' intelligence, measured around 14–15 years of age, correlated significantly at the .001 level with the females' level of education and job status at the age of 25 years. The coefficients of correlation were higher for the females' educational level (IQ: $r = 0.44$; SES: $r = 0.40$) than for their job status (IQ: $r = 0.26$; SES; $r = 0.27$). The correlation coefficients between menarcheal age, on the one hand, and adult level of education and adult job status, on the other, were lower, but still significant at the .001 level (0.17 and 0.16, respectively).

In order to investigate the unique contribution of menarcheal age on the adult criterion variables, controlling for the influence of the two standard predictors of educational–vocational achievement, two partial correlation coefficients were computed. The partial correlation between menarcheal age and adult educational level simultaneously controlling for intelligence and parents' education was 0.16 ($p < .001$, $n = 372$). Observe the "raw" correlation between menarcheal age and adult level of education in Table 10.12, which is 0.17. In effect, when controlling for individual differences in intelli-

---

[2]If the mother's education is employed as the measure of parents' education, the correlations with the females' adult level of education and job status are lower: 0.33 and 0.20, respectively. Somewhat lower relationship with menarcheal age is obtained, .04, but there is no change in the relationship with the subjects' intelligence, .28. The partial correlation between menarcheal age and adult level of education controlling simultaneously for mothers' education and the subjects' intelligence is .18, $p < .001$. The analogous partial correlation between menarcheal age and adult job status is .16, $p < .01$.

gence and the educational status of the home, there was a negligible attenuation of the menarcheal age–adult education relationship. The partial correlation between menarcheal age and adult job position at 25.10 years, controlling for intelligence and parents' education, was 0.15 ($p = .004$, $n = 318$). Thus, for job status also there was little reduction in the size of the relationship after the partilization (the raw coefficient was 0.16).

In conclusion, menarcheal age in females had a significant relation to their adult level of education and adult job status. Because it is clear from Table 10.12 that the relationship between the variable, biological maturation, and the criterion variables at adult age is not linear, the impact of menarcheal age can be said to be a conservative estimate, utilizing the aforementioned linear analysis, and is stronger than is reflected in the correlation analysis. Second, the impact of the age at menarche was found to be more or less independent of the subjects' intelligence and the educational status of their home. The empirically determined observation that differences in education and occupational status at young adulthood can be traced back to biological conditions occurring 10 years before is noteworthy. There were already indications in the mid-adolescent years that pointed in this direction. However, that the influence of menarcheal age was so long-lasting for the subjects' future educational–vocational careers was surprising.

The interaction of different factors in adolescence can be said to have built up a platform for the kind of adult life to which early- and later-developed girls aspired. Of particular interest is the possibilities the knowledge of the interactional system found in adolescence offer for making prognoses on the future social life of the females. The presented research provides clear expectations regarding different types of life in adulthood on the basis of the patterns of biological maturation, self-perception, and social network in the teenage years. Seen in isolation, these factors have limited prognostic potential. Seen together, a pattern is formed that makes a certain type of social life outcome in adulthood more likely for some females, rather than for others. Already in mid-adolescence the early-matured girls encountered an interpersonal environment that shaped their future life in a direction toward a traditional family life rather than toward aspiring to higher education and a work career. Among peers already employed, they encountered a value system putting less emphasis on scholastic achievement and more on economic independence. Working peers served as models and as examples that employment after school might pay off. Data from mid-adolescence also showed that the early-developed females were not as motivated to continue studying after the compulsory school education. Early stable relations to boys further reinforced a family-oriented lifestyle.

In this connection we would like briefly to mention some results from a comparison among the four menarcheal groups of females when they were 14.5 years of age. On a question in the Norm Inventory of whether or not the girls looked forward to giving birth to and bringing up their own children, 67% among the earliest-developed wished this very much or rather much in comparison with 42% of the latest-developing girls ($x^2 = 22.48$, $df = 6$, $p <$ .05). Other evidence in the literature is in line with this more future-directed family life orientation. For example, Simmons, Blyth, and McKinney (1983) showed that early developers were allowed greater independence from their parents than were late developers and that the early-maturing girls took greater responsibility for household activities by babysitting more than did their late-developing counterparts.

Moreover, it is not an unreasonable guess that the vocational orientation in the peer group of the early-developed females had a negative effect on their parental relations in mid-adolescence, running counter to the expectations of the parents of prolonged education for their daughters. In fact, that conflicting mother–daughter relations in mid-adolescence were connected with associations with working peers as well as with contacts with the opposite sex, was confirmed in the former chapter.

In effect, in the mid-adolescent years, the early-matured girls were already more surrounded than were other girls by an interpersonal ecology that fostered positive attitudes toward starting to work and toward family formation. The empirical finding that the females' perceived maturity in midadolescence did not have comparatively as great an influence as a mediating factor as did the peer-type variables, identifies the powerful long-range mediating factors of pubertal maturation as being primarily external to the individual, as being focused in the interpersonal environment.

11
Some Final Reflections

BROAD OR NARROW IMPACT OF PUBERTAL MATURATION?

In the introductory chapter the literature on the consequences in various respects of individual variations in maturational timing were reviewed. In contrast to the rather consistent relationships that have been reported for males, with early physical development being generally beneficial, and particularly so for social relations, the role of timing of maturation among females, as reported in the empirical literature, was found to be unclear, with inconsistent findings across studies and with both positive and negative consequences attached to early maturing. It was concluded from the review that there is a strong consensus in present day research that the consequences of early and late maturation on behavior are domain specific and are narrowly connected with just a few areas, such as body image, dating, and menstrual attitudes (cf. Brooks-Gunn, 1987; Crocket & Petersen, 1987; Petersen, 1987).

The behavioral impact of menarcheal age reported in the present research, for Swedish females, has been more articulated, more coherent across domains, and more lasting than has been reported for other comparable samples of females. Several domains of adolescent development were connected with maturational timing in females.

Here we briefly summarize a sample of the findings from different areas that were found in the present research to be connected with pubertal timing in females. If not stated otherwise, data refer to the mid-adolescent period (14.5 and 14.10 years of age).

Physical Measures. From measures of skeletal age in Grade 8, it was documented that the higher the proportion of girls having reached adult status the earlier the physical timing. From height and weight measures in Grade 6, at the average age of 12.9 years, early-developed girls weighed more than the later-developed girls, they were taller, and they scored higher on the index of body mass.

Self-Defining Attributes. There was a close correspondence between pubertal timing and perceiving oneself as mature. The early-developed girls considered themselves more mature than their classmates, compared to later-matured girls' perceptions, and they also reported themselves to feel different to a higher extent. Subsequent analysis showed this to be due to the early developers' feeling more romantic than later-developed girls. The girls were also asked whether they were satisfied with themselves, and if not, what attributes they would like to change. Most girls, independent of menarcheal age, mentioned physical appearance and their shyness in contacts with others as attributes they wanted to change. On a direct question, more of the early developers than later-maturing girls reported weight problems. Furthermore, more of the early-developed girls reported that they were not satisfied with their temperament. They would like to be nicer toward their parents and have more feelings for order.

Interpersonal Relations. The early-developed girls reported more problematic parental relations, particularly with their mothers. These conflicts appeared to concern everyday life issues, while less concerning the mother as an identification object. The results indicated that the girl herself, rather than her parents, was the main instigator of these conflicts (see earlier for differences among the menarcheal girls with respect to temperament). The early-developed girls also reported more strained teacher relations. Again, the conflicts appeared to be instigated by the girls.

As reported at several instances throughout the book, the friends of the early developers were more often chronologically older, and the early developers also associated, more often than later-developed girls, with peers who had quit school and have started to work. Association with chronologically younger peers was more frequent among the later-developed girls than among the early-maturing. Among classmates of their own sex, the early-developed girls preferred other early developers, whereas the later-matured more preferred their later-developed counterparts. Finally, among the girls who engaged with a friendship group, it was more frequent that this group involved boys among the early developers, and more frequently that it contained an "only-girls" group among the late developed girls. Sociometric data on social status in the class of the ages of 9.9, 12.9, and 14.5 years of age, showed that the early-developed girls were rated lower in status first, but that their social status increased over time.

The results for the relationship between pubertal maturation and opposite-sex relations showed a strong tendency for more frequent and more advanced contacts with boys among the early developers than among the late. More of the early-maturing girls had stable boyfriend relations in mid-adolescence, and more of them had had sexual experiences, compared to later-maturing girls. Teenage abortions were also considerably more common among the early developers. For an open-ended question on the greatest problem that the girls experienced themselves to have, problems connected with boys was the most commonly given answer. The early-developed girls expressed a separation type of anxiety, being afraid that their boyfriends would leave them, whereas the later-developed girls were more worried about how to establish contact with boys.

Social Adjustment. The early developers were found to break conventional norms at home, at school, and in leisure time, more frequently than did the later-developed girls. Official data on delinquency up to age 18, to some extent, collaborated this finding. Data on conduct disturbances at school in the form of teachers' reports at the age of 9.9 and 12.9 years showed no differences among the menarcheal groups at the earlier age, but the early developers expressed more concentration difficulties and motor restlessness at the latter. From parent reports collected prior to mid-adolescence, it was found that, even at the age of 10, the parents of the early-matured girls reported more problems with their daughters than did the parents of the other girls (pocket money, TV watching, time to be in at evenings), and at the age of 13 the parents of the early-developed girls reported more problems with respect to making their daughters obey.

Data for school adjustment indicated somewhat lower achievement and considerably lower school motivation among the early-maturing girls than among the later-developed. Recorded hours of school absenteeism were particularly high for the earliest-developed group of girls. Differences between early- and later-developed girls with respect to school interest were not present at the age of 10. Some differences appeared at age 13, with the early-developed girls reporting less satisfaction. Marked differences in school motivation between early- and late-developed girls appeared at the age of 14 and 15.

The early-developed girls reported more emotional problems of psychosomatic and depressive types than did the later-developed girls. Most differentiating were the items measuring suicidal thoughts. Both the most early and the most late developed girls, particularly the former group, were overrepresented among subjects with documented contacts in adolescence with psychiatric clinics and counseling services in the town. The occurrence of extrinsic problems (school problems, antisocial behavior, and peer-related problems) among the girls who consulted these services, or were referred, was more common among the early-developed girls than among the later-developed.

In summary, timing of maturation played an important role for several areas typically investigated in adolescent research. Not only could the effects of early and late maturation be traced in several domains, there was a psychologically logical pattern of results across the various domains.

Part of the explanation of the coherent picture obtained in the present study we believe is to be found in the more detailed specification of developmental progressions, and in direct testings of these, in the present study compared with the more exploratory studies reported earlier. Part of the difference is also to be found in the present study's focus on the early part of the physical maturation continuum. From a theoretical basis it was more important to differentiate variations in early maturation than to discriminate among late developers. Also, part of the difference with earlier studies might be sought in the nature of the present sample. In contrast to many previous studies, a representative, normal sample of females was followed over time. The conditions for generalizability were quite satisfying, and little sample loss was found over time.

A Sociocultural Interpretation

Part of the difference between the present study's findings of broad maturational effects, in contrast to the more narrow effects reported in many other studies, also may be interpreted in the frame of reference of differences between subject samples in the social structure that influence upbringing conditions and constricts or enhance the impact of biological maturation on social behavior. Maturation always takes place in a cultural context, manifested in the environmental settings of a community. If the present study had been performed in another society, the overriding structure of the society would be different, and so might the consequences in everyday life of being early- or late-developed.

Up to the present time, most of the research on maturational timing effects that has been presented in the literature has been conducted on American samples of females. In view of the broad consequences of early and late maturing found for the present Swedish sample of teenagers and the more narrow reported to be connected with timing of maturation among American female samples, it is not unreasonable to assume that these differences might be due to differences with respect to how the Swedish and the American societies have organized the social institutions and formal upbringing conditions for their adolescents.

It is always difficult to ascribe differences on the level of individuals to some type of molar social processes, or vice versa. For example, although the secular trend would be assumed to be paralleled by a synchronous earlier sexual debut among youth, such a close synchronization in time does not seem to be the case (Petersen & Boxer, 1982). From the opposite perspective, although several investigations have shown that earlier pubertal maturation is connected with an earlier marriage (Ryder & Westoff, 1971; Udry & Cliquet, 1982), menarcheal age has been found to go down in age, whereas the age for marriage has been going up in many populations. Udry and Cliquet, in their cross-cultural investigation of the relationship between menarcheal age, age at marriage, and age at birth of the first child, concluded that "the cross-sectional relationship between age at menarche and age at first marriage is direct on the micro level, but inverse when the unit of analysis is populations differing in age at menarche" (p. 63). From these examples it should be clear that parallelism between processes operating on the individual and the social level cannot easily be inferred. Still, it is tempting to speculate on what features of social processes might eventually be involved behind the present broad impact of pubertal timing for Swedish females and the more narrow impact usually found among American. Before addressing the issue of the particular environmental features that might differentially constrict or enhance the influence of pubertal maturation on behavior between Swedish and American youth, a general view of the matter of broad and narrow impact should be taken.

The question can be formulated: Under what social conditions can pubertal maturation in females be expected to have a more narrow impact and under what social conditions might pubertal maturation have a broader impact on the adolescent life situation? Explicitly or implicitly we have dealt with this issue of narrow versus broad impact of menarcheal age throughout this book. In chapter 2 a general proposition underlying the present research was formulated, which suggested that the individual variations from the normal process of adolescent transition which are observed between early- and later developed girls were related to:

1. differences in perceived maturity between early- and later-developed females, and
2. differences between the early- and the later-developed girls with respect to engaging with peers outside the normal same-age–same-class frame.

Subsequent analyses, testing the validity of the propositions, provided substantial empirical support for these assumptions. The results overall suggested that the particular reason why pubertal maturation among females was manifested in social behavior was largely due to differential association with the peer network outside the same-age–same-class–same-sex frame. More frequent engagement with older friends, with working friends and with boys, with their more advanced social behavior, were observed among the early-developed girls, compared with the later-developed females, whereas the latter girls more frequently associated with younger peers, with their more childish behavior. Therefore, to the extent that one compared the results in the present research with results obtained in other studies, the availability of an interpersonal peer network outside the circle of same-class–same-age–same-sex friends is a critical comparison point.

From what has been reviewed, for Swedish females, it might be concluded that limited differences in social behavior in adolescence will appear due to variations in maturational timing among females who do not engage with peers across the age border, who do not have peers outside the school environment, or who do not have stable opposite-sex relations. A greater role of pubertal maturation is assumed to exist for the females who associate with these nonconventional type of peers. *If this line of reasoning would be transformed to the issue of sociocultural influences, one would expect a more limited and narrow impact of pubertal maturation to appear in a society in which stronger social regulations prohibit females from cutting across age and environmental borders, or which set up strong regulations for opposite-sex contacts. In contrast, the more permeable the age and setting borders in the society for the teenager, and the more permissive the attitudes are to opposite-sex contacts, the broader the influence of pubertal timing on behavior could be expected to be.* It is in this context of permeability with regard to nonconventional peer contacts that the broader impact of pubertal maturation found in the present study among Swedish youth, and the more narrow impact of pubertal development reported for several American samples, might be traced.

Similar types of arguments have been expressed by others (Dornbusch et al., 1981; Petersen, 1985; Petersen & Boxer, 1982). When discussing the role of pubertal status on behavior, Petersen (1985) attributed the limited influences found across studies to characteristics of the American society:

> We think that one factor producing these limited effects is the strong social effect of spending large amounts of time with same-age peers in school. U.S. society has no rites linked to pubertal change. Although maturational age might potentially be more relevant to the individual than chronological age, U.S. society does not recognize it in any structural way. In contrast, experiences during the school-age years are strongly structured by age and grade. Therefore, a model that includes social mediation of effects of pubertal change seems called for. (p. 212)

Dornbusch et al. (1981) advocated that if social transition issues, like dating, are strongly controlled by social customs or closely linked with age-graded development, the role of biological maturation consequently will be more limited.

The viewpoint expressed here with respect to the potentiality of the overriding social structure to constrain or enhance the impact of pubertal maturations on behavior coincides, to some extent, with the discussion of adolescent sexuality by Petersen and Boxer (1982). They differentiated two general standpoints in the literature: the "hormonal drive" perspective in which sexual activity was closely tied to the increase of sex hormone activity in puberty, and the "social control" perspective, which suggested that adolescent sexuality primarily was a social phenomenon governed by sanctions and expectations of the larger society. Integrating these perspectives, Petersen and Boxer arrived at the conclusion that biological influence on sexual behavior would be more accentuated as the social permissiveness increased (which, in fact, is a common genetic viewpoint). Specifically, with respect to the relatively higher incidence of sexual intercourse among girls than among boys in the 1970s, an increase that was much stronger than the decrease over this time period in the age in which females reached puberty, they stated the following:

> If we are correct in our hypothesis that the recent changes are due to the increased latitude in the social constraints governing the sexual behavior of girls, underlying biological influences may now be *more* apparent than previously. Thus we are not arguing that biological factors are unimportant but rather that they may be altered by social influences. Only when social restraints are minimized can purely biological influences become apparent. (p. 244)

Particularly, likely social influences were identified as the "women's movement and social sanction of and permission for sexual experiences " (p. 243).

Opposite-Sex Relations: A Comparison Between Swedish and American Youth. Among the present sample of Swedish adolescent females, variations in the timing of pubertal maturation were closely connected with opposite-sex relations and sexual behavior in mid-adolescence. Differences between the early- and the later-developed girls in mid-adolescence, with respect to parent relations, teacher relations, norm-violative activity, and unwanted pregnancy seemed to be connected with the girls' heterosexual relations at this time. As was reviewed in chapter 9, much of the difference in social relations and social adjustment among the menarcheal girls in our research group seemed to be accounted for by their contacts with the opposite sex. Among girls who did not have a steady relationship with a boyfriend, a limited impact of pubertal maturation in the aforementioned respects was obtained. In contrast, these differences among the menarcheal girls appeared strongly for the girls who were going steady. That variations in the timing of pubertal maturation were related to these features seemed in part due to the fact that the early developers were overrepresented among the girls who had steady opposite-sex relations and partly to the fact that these girls more readily adopted the more advanced behavior of their boyfriends. Against this background of the role of heterosexual relations for the effects of early and late maturation upon social behaviors and social relations in the present sample of females, it is of particular interest to compare differences between Swedish and American females with respect to the issue of the social permissiveness for opposite-sex relations in adolescence.

Sweden is generally recognized as a society with an accepting view of opposite-sex contacts and sexual behavior among youth, and the view of adolescent sexuality is probably more permissive in Sweden compared with United States, in which premarital sex is more likely to be considered as a norm violation by parents (Reiss, 1970). The legitimate character of sexual behavior among Swedish youth is reflected, among other things, in obligatory sex education early in the compulsory school and contraceptive advisory services by school nurses. In the United States, sex education enters later and comprises far from all schools (Golub, 1983). In her comparison of gender concepts of Swedish and American youth, Intons-Peterson (1988) reported that a substantial part of Swedish adolescent females described themselves as "liberated," whereas this attribute was rarely used by American females of the same age. The Swedish subjects also were found to be less gender typed compared to the American.

There are fewer socially institutionalized regulations or social rules for initiating and maintaining opposite-sex contacts in Sweden compared with U.S. society. American females face a social custom in the adolescent life phase, the dating system, that regulates heterosexual contacts and does so largely on a chronological age basis. Dating usually requires a certain expected age and certain expected behavior on part of the individuals. When adolescents reach the appropriate age, around 13–14 years, they are supposed to have dates. Although dating is less of a formal institution today than it once was, strong messages are still directed to the teenage girl from parents and from mass media about the expected manners and the behaviors that constitute a good "dating personality."

The concept of dating has little meaning for Swedish adolescents. To have a relationship with a boy means, for a Swedish girl, going steady. In U.S. society opposite-sex relations tend to be more under supervision by parents, and in that sense under certain social control. Boyfriend relationships and transient opposite-sex contacts of Swedish teenage girls are often unknown to parents.

In effect, the conditions for development of heterosexual relationships and associated sociosexual issues are, to some extent, set differently for Swedish and American females. The relatively more permissive attitudes of adolescent sexuality in Sweden, compared with the attitudes in the United States, and the fewer formal arrangements for adolescent heterosexual contacts and less supervision directed from parents for Swedish females, might make the role of sexual maturity carry relatively more weight for entering into heterosexual contacts and for associated sociosexual issues for Swedish females. For American females chronological age will be a more important determinant. Stronger restrictions and more age-normative rules for opposite-sex relations in the United States, compared with the situation in Sweden, might have the consequence of constricting the role of timing of maturation among American teenage girls and prohibit the early-developed American girls' acceleration into more advanced social and sexual behavior at an earlier time than their later-developed peers. Such crossing of the sex border, with higher exposure to advanced social behavior codes among older boys, might be a prominent factor behind the strong and consistent relationship between physical maturation and the wide range of advanced social behaviors among Swedish females found in the present investigation, compared with the more scattered findings and narrow impact generally found for American samples.

To the extent that our argument is correct, that Swedish females face more permissive attitudes toward issues of adolescent sexuality than do American females, and that such differences in societal permissiveness differentially affect the influence of menarcheal age on sexual issues, several inferences can be made.

First, we would expect differences in prevalence of sexual intercourse between Swedish adolescent females and comparable American samples. The greater permissiveness of sexual behavior in the Swedish society would be reflected in Swedish females having their first sexual intercourse at an earlier time than do Americans. That, indeed, sexual intercourse among Swedish adolescent females begins earlier in time than for comparable samples of American females has been reported in preceding chapters. As was demonstrated in chapter 7, the prevalence of sexual intercourse at the age of 15 in the present sample of Swedish females was of the same magnitude as has been reported for American females who were 3 to 4 years older in age.

These results are not isolated findings. The same low age for the first appearance of sexual intercourse among Swedish teenagers has been reported in other Swedish investigations as well. Because they have not been published in international journals, a more detailed account of the results from one of these studies is given, because in this investigation the analyses were conducted on subjects of the same age as the present females. A sample of 212 boys and girls in an urban community, 122 males and 90 females, was prospectively studied from birth to maturity within "the Clinic for the Study of Children's Development and Health" (Karlberg et al., 1968). The subjects were a probability sample, born between 1955 and 1958. Data for psychological, social, and somatic development were collected annually from the first years up to the age of 18 for the subjects. In this sample, the percentage of girls who had had sexual intercourse were: at 12 years of age 1%, at 13 years 6%, at 14 years 10%, at 15 years 33%, at 16 years 54%, at 17 years 65% and at 18 years 82% (Klackenberg-Larsson & Björkman, 1978). The median age for having had sexual intercourse was 15.8 years. This can be compared with American studies for subjects of the same age that have reported that the percentage of girls who had had sexual intercourse by age 15 was 13% to 14% (Rice, 1978). Overall, it can be concluded that, at least for subjects of the similar age as in the present study, Swedish females undoubtedly had an earlier debut with regard to sexual behavior than had American females of the same age.

Some further comments can be made with regard to the correlates of the age of first sexual intercourse. In Klackenberg-Larsson and Björkman's investigation, no differences were found with respect to social class. However, the girls who thought their mothers were strict had their first sexual intercourse at a later time than had the girls who thought their mothers as more permissive.

Moreover, in line with the argument that the less the social restraint against sexual and sociosexual behavior, the more the biological factor will impact entry into sexual behavior, it would follow that there should be a closer association between pubertal development and sexual intercourse among the more sexually liberated Swedish adolescents than among American teenage girls. The results in the present study confirmed this. A strong relationship between pubertal maturation and sexual intercourse among Swedish adolescent females was demonstrated in chapter 6. More inconsistent findings have been reported for American samples, being significant in some studies but not significant in others.

Second, and of even more importance, given the validity of the proposition that the timing of maturation has a stronger impact on the timing of sexual behavior as the social permissiveness for such behavior rises and less restrictions are put on development of opposite-sex contacts, one would expect a stronger parallel case of rate of pubertal maturation and sexual behavior among Swedish compared to American adolescents generally. This is to say that the synchronization in time between pubertal maturity and entering into sexual behavior would be stronger among Swedish adolescents, both males and females, than among American. At least we would expect that females, who enter puberty earlier than males, would have an earlier entry into sexual behavior than do males.

Among American samples of teenagers, presumed to face less social permissiveness for heterosexual behavior than that experienced by Swedish teenagers generally, adolescent males have, with few exceptions (Jessor & Jessor, 1977), been found to be more advanced sexually than their same-age female counterparts (cf. Billy, Rodgers, & Udry, 1984; Kinsey, Pomeroy, Martin, & Gebhard, 1953; Sorensen, 1973; Vener & Stewart, 1974; see Miller & Simon, 1980, for a review). Compared at any given point in time during adolescence a larger proportion of males than of females have had sexual experiences.

In contrast, in the present sample of Swedish subjects the prevalence of sexual intercourse was higher among girls than boys in mid-adolescence. At the age of 15.10 years in the ninth grade, 39% of all girls had had sexual intercourse in comparison with 30% of the boys. Similar sex differences, with girls starting earlier than boys, have been obtained for other Swedish samples. In the Swedish longitudinal study conducted by Karlberg et al. (1968) the tendency for females to be involved in sexual behavior to a greater extent than same-age males was found from early adolescent years. The figures for the prevalence of sexual intercourse for the ages 12 to 18 were : 12 years: males 0%, females 1%; 13 years: males 1%, females 6%; 14 years: males 6%, females 10%; 15 years: males 13%, females 33%; 16 years: males 35%, females 54%; 17 years: males 45%, females 65%; 18 years: males 59%, females 82%. The differences between the genders for sexual intercourse were significant at all ages from 15 to 18 years (Klackenberg-Larsson & Björkman, 1978).

In conclusion, the relatively strong relationship between biological maturation and sexual intercourse among the present sample of mid-adolescent females (an association that is stronger than has been reported generally for U.S. teenagers), and the earlier age of first sexual intercourse among Swedish females compared with Swedish males (a finding that has been found to be the reverse among American samples), both suggest that the potentiality for the factor of physical maturation for sexual issues rises as the societal permissiveness of intimate heterosexual behavior increases.

However, that pubertal maturation seems to have a more direct influence on sexuality among Swedish than among American teenagers (whether for sexual behavior or adolescent abortions) does not say anything about whether hormonal or social factors are the prime determinants for sexual and sociosexual aspects among teenagers. The "hormonal drive" versus "social control" perspective is a somewhat loose theoretical distinction. There is little evidence from the endocrinological literature suggesting that even if the external circumstances were optimal, initiation into sexual activities among females would be under the control of hormonal influences (cf. Katchadourian, 1977). Rather, what appears to be the most parsimonious explanation for the quite strong association between pubertal development and sexual experiences in adolescence among Swedish females, and the earlier sexual debut of females relative to boys among Swedish adolescents, is that in a community in which the social arrangements for heterosexual relations are more free and the attitudes toward adolescent sexuality are more liberal, the psychological significance for the girl of reaching sexual maturation becomes more salient. This is to say that it is the personal meaning of biological maturation rather than the biological process as such that carries the important behavioral implications. Perceiving themselves as more "reproductively mature" than same-age peers, issues connected with opposite-sex relations become relatively more salient for the early-developers.

Females Who Mature Early Versus Late and Sexually Experienced Versus Sexually Inexperienced Females. Having demonstrated the mediating impact of opposite-sex relations for the effects on behavior of pubertal timing for

Swedish adolescent females, results from studies comparing American female adolescents with and without heterosexual experiences become of interest. The argument can be made that for the American females who, in fact, do establish stable relations with boys, about the same differences that appear between early- and later-developed Swedish females would be obtained between American girls with and without stable heterosexual relations.

Indeed, there is an apparent *similarity* between characteristics that in the literature have been reported to differentiate girls with and girls without sexual experiences among American samples of adolescent females and the features distinguishing the early- from the later-developing girls in the present sample of Swedish females (cf. Billy, Rodgers, & Udry, 1984; Miller & Simon, 1980). In chapter 9 certain results were reviewed that were reported by Jessor and Jessor (1975, 1977) for comparisons between nonvirgin and virgin girls. The rather striking similarities in social behavior patterns of the early developers and the nonvirgin girls, on the one hand, in comparison with the late developers and the virgin girls, on the other, seems to suggest that analogous developmental progressions occur for the sexually experienced girls in American samples as for the early developers among Swedish females.

In fact, about the same kinds of differences in behavior, reported between sexually more advanced compared with sexually less advanced adolescent females, have been demonstrated in investigations in other countries as well. Schofield (1965) reported data from investigations of adolescents in England and Wales. In his study, a random sample of 15- to 19-year-old teenagers, 934 boys and 939 girls, were interviewed with regard to sexual behavior, attitudes, social behavior, and personality. Among the features of the sexually experienced girls that discriminated them from the less experienced girls, Schofield summarized the most outstanding:

> In general the experienced girls did not have less favourable backgrounds than the other girls, but there was a difference in family relations. . . . experienced girls more often reported poor relations with both the father and the mother, and there were more reports of marital difficulties among the parents. . . . In the analysis of the attitude inventory three of the girls' first-order factors were antipathy to family loyalty, dislike of home restrictions, and preferences for friends' advice. All these factors were highly correlated with sexual experience, but did not appear at all among the boys' first-order factors. It is clear that experienced girls have gone much farther than experienced boys in rejecting family influences. Relations with both parents were often strained and they were less likely to receive advice on sexual matters from their parents, and when they did get this advice, they were more likely to reject it.
>
> The religious influence is the same with girls as it is with boys. The experienced girls were less likely to go to church and less likely to hold views which were favourable to religion, but they were just as likely to come from church-going homes.

The school records of the two experienced groups were very similar. Like the boys, the girls were more likely to have been to a secondary modern or comprehensive school, and were unlikely to have taken GCE; they disliked school, had more problems, and left and an earlier age. . .the school and work record of these sexually experienced girls suggests that their physical development is more advanced than it is for the other girls in the sample. They started dating and kissing at an earlier age; they were more likely than the others to have a steady boy friend; and more often they claimed to be in love.

These [sexually experienced] girls were more likely than the other girls to go around in a mixed group, to spend more time with the group, and to meet on commercial premises. . . . The experiences girls tended to hold attitudes that indicated support for teenage freedom and had a high score on teenage ethnocentrism.

The gregarious outgoingness noted among the experienced boys is not quite so apparent among the experienced girls. They did not go to the cinema, dances or coffee bars any more than the other girls. The important difference for girls is the person they go with; experienced girls were much more likely to go to these places with a boy than with another girl.

. . . Like the boys, they went to bars more often, got drunk more often, and smoked more cigarettes. They went to more unsupervised parties, often where most of the people there were adults, and these parties lasted past midnight and sometimes all night.

Although far fewer girls than boys had ever been in trouble with the law, those who had appeared before a court were more likely to be sexually experienced. They also reported more misdeeds than the other girls. (pp. 228-230)

In agreement with the comparison between early- and later-developed girls in the present study, Schofield reported strong differences with regard to norm breaking between the sexually experienced and the sexually inexperienced girls. Although 39% of the sexually inexperienced girls reported having been drunk, drunkenness had occurred for 89% of the experienced. Only 6% of the inexperienced girls reported more frequent drunkenness in comparison with 45% of the experienced. Two out of three sexually experienced girls reported smoking compared with one out of three of the inexperienced. It occurred that experienced girls attended "all-night parties" more frequently than did sexually inexperienced girls. Although 13% of the sexually experienced girls had appeared before the court, this occurred for just 1% of the inexperienced, and 21% of the experienced girls reported frequent (often or sometimes) misconduct in comparison with 7% of the inexperienced girls.

Another area in which clear correspondences to the present results were obtained in Schofield's study concerned peer relations. It was reported that sexually experienced girls were less likely to associate with groups of exclusively girls and more likely to engage with mixed-sex groups compared with the sexually inexperienced girls. As to the places at which teenagers meet, about as many of the experienced girls as of the less experienced went to cinemas and dance halls, but it occurred considerably more often that the experienced girls were accompanied by boys to these places. The experienced girls also attended teenage parties without parents more frequently than did the inexperienced girls. Of particular interest from the point of view of boys as instigators for girls' engaging in nonconventional activities, Schofield found that when girls used alcohol in public bars, they did so to a great extent with boyfriends or in mixed-sex groups. He made the following comment:

> Most girls go to cinemas, dances, clubs, coffee bars whatever their level of sex activity; this is because they go to these places with other girls if they do not go with boy friends. It is noteworthy that the number of visits to public houses is the only one associated with sex experience, because this is an exception to the previous list of places, for girls rarely go into a bar with other girls and almost always go with boy friends. Therefore the really important factor so far as girls' sexual experience is concerned is whether she has a steady boy friend or not. (p. 181)

From the review of some of the findings reported by Schofield (1965), in addition to the results reported by Jessor and Jessor (1975, 1977), it can be concluded that pronounced individual differences with regard to a wide variety of social behaviors and interpersonal relations are connected with heterosexual relations among adolescent females. Moreover, there is a surprising similarity between the features that differentiate the sexually experienced girls from the sexually inexperienced in these studies and the features that have been found to differentiate the early- from the later-developed girls in the present research.

Therefore, it seems as if a primary reason why a broader impact of pubertal maturation has been found for the present sample of Swedish females, compared with other previous research, is due to the fact that early physical maturation in Sweden is a comparatively stronger facilitating factor for heterosexual interactions than for American females. The more permissive climate for adolescent opposite-sex contacts in Sweden compared to, for example, that found in United States, makes physical maturation a salient indicator of sexual maturation in Sweden, one that manifests itself in earlier sexual and sociosexual behavior among the girls who develop physically earlier than others. In her more frequent contacts with members of the opposite sex, the early-developed girls enters a social environment characterized by more advanced behavioral standards than is the case for the later-developed girls who will be much less within the influence sphere of male norms and attitudes. In conclusion, in view of the substantial differences, in various areas, which in the literature have been found to distinguish girls with more advanced opposite-sex relations from those with less advanced contacts with the member of the opposite sex, the broad impact of pubertal maturation found in the present research to a certain extent can be attributed to the early-developed girls' stronger opposite-sex orientation, making them face more advanced lifestyles than do the later-developed females.

THE ROLE OF THE PEER NETWORK FOR ISSUES OF TRANSITION

How transition issues in adolescence among early- and later-developed girls are affected by the peers they engage with has been a major focus in the present study. It is widely recognized that peers, among girls, are a critical source of influence behind various types of transition behaviors in adolescence. Still, we know very little about how these peer influences operate and which types of peers are actually the influential ones. With few exceptions (cf. Dunphy, 1963), little attention has been given to the implications of different types of peer relations for issues concerning the transition period from late childhood to early adulthood. The concept of "peer group" is ambiguous and seems to implicate a homogenous "in-group" where its members share about the same attitudes, values, behaviors, and general social life conditions. Obviously, this is a simplification of the situation.

In previous research, the peer network of adolescents, and its impact on social transition issues, often has been empirically investigated either in a highly general sense, as the impact of the peer culture generally, or the sampling of peers had been made mainly from one type of setting.

Previous studies have typically used either of two research paradigms. One line of research is exemplified by the research presented by Jessor and Jessor (1977), who based much of their framework for studying problem behavior among youth on the parent versus peer orientation. Subjects were asked to report their own values and behavior as well as those of their peers. The measure of peers was an indirect one, probably quite strongly influenced by the beliefs of the subjects themselves. Apart from the difficulty that the data on peer values and behavior were inferred from the perceptions of the subject rather than measured directly, it is not possible to ascertain which types of peers the subject had in mind.

In contrast, another approach, typified by research by Kandel and Lesser (1972) and Kandel (1978), involved a direct and independent measure of the behavior and values of those pupils who the subjects denoted as their best friends. Although the dyad sampling method used by Kandel is superior to the indirect method, this technique also has limitations that circumscribe the generality of the findings. Because the best friends of subjects have typically been sampled from the same school as the subjects themselves, the social behavior of the subjects' best school friend is assessed, but not necessarily the behavior of his or her friends in other situations. Pupils who have friends outside of school—for example peers who are already working full time—

cannot name them because the instructions of the sociometric measure require the names of friends in a school or classroom.

Another vast body of literature has made use of variants of the peer nomination or the sociometric assessment technique in order to highlight the issue of friendship formation and social adjustment. These studies all share more or less the same feature of assessing social status as social status in the classroom situation. Although the purpose of most of these studies has not primarily been to examine how peers influence the individual's behavior, they have yielded information about how individual's social status or membership in sociometric peer groups, such as being stars, popular, controversial, unpopular, neglected, isolated, and rejected is related to their social adjustment. The literature rather compellingly documents that subjects with problematic peer relationships in the school context, particularly the pupils who are actively rejected by their classmates, show more aggressiveness, disruptive and antisocial behavior than do other pupils (Coie, Dodge, & Coppotelli, 1982; Dodge, Coie, & Brakke, 1982; Dunnington, 1957; Hartup, Glazer, & Charlesworth, 1967; Moore, 1967; Vosk, Forehand, Parker, & Rickard, 1982).

The reliance on the sociometric methodology has certain drawbacks in that the sampling of peers is performed for more or less one type of setting. Therefore, the extent to which one might generalize findings is limited. It is trivial to comment that the best friend in the class does not necessarily refer to the best friend of the subject or that the social status in the classroom context is not the social status among peers generally. Still, much research has proceeded and still proceeds from the notion that peer influences can be validly measured as peer influences in the school context. It is somewhat ironic that the most refined instruments, such as sociometrics, and the most elegant methods for data treatment, have been used over and over again in contexts with limited ecological generalizability. Furthermore, although previous sociometric research has detailed how social standing among classmates is related to social adjustment at various ages, less progress has been made in specifying the role of peers within a more systematic developmental framework for studying issues of transition. In her invited commentary to some studies in the sociometric literature, Blyth (1983) also noted the absence of a systematic developmental perspective in this field of research, and she suggested that "we need to take a more explicitly developmental perspective, so that the variety of changes taking place within children and their social worlds can be brought together to form a more intricate and complete framework" (p. 450).

The purpose of the present study was to examine, from an age-normative developmental viewpoint, the impact of association with different types of peers for adolescent problem. Although much peer research has proceeded from the premise that the important peer influence involves peers who are associated with the subject through physical, social, and chronological age proximity, the present study proposed that contacts with peers outside of this "expected circle" of friends would have particular significance for individual differences in rate of social transition among adolescents. The analyses that were conducted confirmed this proposition to a large extent. In conclusion, the result of the present research demonstrate that the data that are necessary to characterize peer influences must be much broader than those usually found in the peer literature.

GENERAL OR DIFFERENTIAL MATURATIONAL IMPACT?

One of the standing issues that has concerned us throughout this book is how to conceptualize the impact of physical maturation in the short and long run. In the first chapter, two hypotheses for the effects of individual differences in physical growth were distinguished. One hypothesis assumed a general maturational impact. Social behavior approximates in time physical maturation. The finding that we at a particular point in time in early or mid-adolescence find behavioral differences between early- and later-matured girls is due to the fact that the early-matured girls have passed the critical phase in their pubertal sequence at which certain types of social behavior usually are acquired, whereas the late developers still have not. Because all girls will pass this point, few differences related to physical growth should appear in late adolescence when girls generally have reached about adult maturity. Nor should we, for that matter, expect that individual differences in psychosocial functioning at adult age are related to age at menarche.

The second hypothesis was the differential impact of physical growth. The unique combination of early physical maturation occurring at the most intensive transitional phase in adolescence, a self-concept that facilitates the acquisition of new social behavior, and powerful influences of the peer culture on the early-developed girls, might create life conditions for these girls that would not be duplicated later on by the later-developed girls.

Social Adjustment

The first model is actually a strict hypothesis of delayed effects for the later-matured girls. If it is a valid assumption, and applied to, for example, problem behavior in adolescence, then we would predict that (a) the higher incidence of such behavior in mid-adolescence that was documented for the early-matured females would attenuate in late adolescence due to a "catching-up" on the part of the later-developed females, and (b) when the subjects in young adulthood looked back on their earlier experiences, no differences in retrospective reports of problem behavior related to menarcheal age would emerge.

As far as the first point is concerned, there were some suggestions in chapter 7 that the prevalence of drunkenness behavior among the four menarcheal groups of girls attenuated over time. There were also some tendencies in data pointing toward more harmonious relations with parents among the early-developed girls at the age of 15.10 years of age than 17 months earlier. However, for frequent drunkenness no such catch-up effect could be discerned, and with regard to other aspects, like drug use and sexual intercourse, little equalization between the early- and the later-matured girls was revealed. Stronger differences among the menarcheal groups of girls were present at age 15.10 years than 17 months earlier. Data did not permit an analysis of the developmental trajectories for early- and later-matured girls after the age of 16, but as far as the results up to this point in time are concerned, little evidence of a systematic delay effect was found.

The argument for a general maturational impact would further imply that no differences with regard to retrospective reports of problems at adult age would be found among the menarcheal groups. However, when, at the age of 25.10 years, the females were looking back on their social experiences, there were indeed substantial differences in reports of problem behavior between early- and later-developed women. More subjects among the early-developed reported having used narcotics in the past and more reported have had blackouts in connection with alcohol drinking. They reported to a greater extent than the later-developed women that they had been rebellious at school and had experienced more conflicts at home. It was particularly the earliest-developed females who differed from the rest. If there had been a catch-up effect of the later developers in late adolescence, such differences in retrospective reports would probably not have turned up.

The lack of catch-up effects during adolescence for the later-developed females, in addition to higher incidence in retrospective reports at adult age of problem behavior in the adolescent period, seem, in the perspective of the adolescent years, to invalidate the general maturational hypothesis and designate the early-developed females as truly a problem-prone group of teenagers. The hypothesis of a *differential impact* of physical maturation seems to best represent the girls' social adjustment in adolescence.

The validity of the general and the differential models, respectively, for the impact of individual differences in growth rate is of particular interest when the focus is on development in the long-term perspective.

In this context it is interesting to note that there was no strong evidence of a long-term impact of menarcheal age. This was in strong contrast to the situation in midadolescence when the early-developed girls differentiated from other girls for a variety of social adjustment criteria. With respect to the postadolescent years, data on registered offenses showed some tendency for the delinquency cases to be overrepresented among early-developers. For self-reported drinking at young adulthood and for official data on alcohol abuse at adult age no significant relationships with menarcheal age were observed. Moreover, no significant relationships were observed between menarcheal age and drug use, and use of tranquilizers and sleeping medicine, respectively. As to cigarette use, more of the females who had never started smoking were later-developed, and more of the females who smoked presently at adult age were early developers. However, the relationship with pubertal maturation and adult smoking was not significant. The few weak relationships with menarcheal age could not invalidate the overall conclusion that the model assuming a *differential impact circumscribed only to the adolescent period* was supported when the interest is in social development.

The strong connection between early maturing and problem behavior in adolescence and the limited impact of menarcheal age in the long-term perspective must be commented on within a more general frame.

Sturm und Drang. The pendulum sometimes hovers momentarily between the extremes in developmental psychology. For a long time the view of adolescence as an unstable period with "normal" crises, emotional turbulence, and interpersonal conflicts prevailed (Erikson, 1968; A. Freud, 1958; Hall, 1904). The changes occurring in different areas—endocrine, bodily, emotional, cognitive, moral, social, and interpersonal—were assumed to require substantial coping on part of the individual, delineating the adolescent years as a particularly stressful and stormy period in the life cycle. Problems of various kinds, emotional, identity, adjustment, and interpersonal, were not unexpected.

This outlook was counterbalanced by large-scale empirical investigations showing quite the opposite (Douvan & Adelson, 1966; Kandel & Lesser, 1972; Offer, 1969; Rutter, Graham, Chadwick, & Yule, 1976). One after the other, these studies challenged the widely accepted turbulence viewpoint by showing the adolescent period as a rather peaceful and stable period for the large majority of teenagers in these years. This "new look" on the teenage years is perhaps best documented for the changes in social network conditions connected with the transition from childhood to adult life: the gaining of independence from home restrictions and the orientation toward peer and the peer culture. Empirical data showed that teenagers on the whole had a good relationship with their parents and that parents to some extent encouraged their offspring in independent behavior (Bandura, 1972; Douvan & Adelson, 1966; Kandel & Lesser, 1972; Offer, 1969). The clash between parental and peer values often implied in early research on adolescents was found less dramatic than commonly assumed. In important areas parents and peers tended to have similar values. Besides, where a problematic adolescence was found in individuals, it tended to be preceded by a problematic childhood.

The conclusion, that teenagers pass through adolescence with rather good parental relations and with few social and affective problems, is based on examinations of data for broad samples of individuals, exemplified by descriptions of the "typical" adolescent. This does not preclude the possibility that the adolescent years might be a period of emotional upheaval and of interpersonal conflicts and rebellion for subsamples of individuals (cf. Offer, 1969). Up to the present time we have had vague notions which teenagers are likely to have a problematic life situation in adolescence, and for what reasons this occurs.

In the present research we have looked into the possibility if, and to what extent, adolescent problems were inherent in the very nature of maturing itself: that is to say, the extent to which social and interpersonal problems in the teenage years might have their roots in subjects' own timing of development.

From the point of view of conventional standards, the early-matured girls could indeed be characterized as a risk group for social maladaptation in mid-adolescence. More of the subjects in this group were involved in norm-breaking activities in the mid-adolescent years, and their norm-violating activity was considerably higher compared with the other girls. As was shown in chapter 8, they were surrounded by a peer group that sanctioned norm breaking less harshly compared to the peers of later-developed females, and that, to some extent, had more positive evaluations of norm breaking. The parents of the early-developed girls expected their daughters to be ignoring their prohibitions, staying out late, and getting drunk to a higher extent that did the parents of the later-developed girls. In reiterating the criteria for being at risk for more advanced antisocial life in adulthood that Loeber (1982) reported in a comprehensive review of the associations between early socially maladaptive behavior and later antisocial behavior for boys, it can be seen that the early-developed females possessed all three: Their norm breaking started earlier, the norm-violation rate was higher than other girls', and their norm breaking occurred in more contexts—at home, at school, and in leisure time settings.

However, there was only a weak tendency for the early-matured females to have more registered criminal offenses up to age 26, and no difference in alcohol consumption was found at adulthood. What remains is the existence of a group of girls who, within a limited period of time in the adolescent years, were more socially deviant than girls in general. From the point of view of normal biological development, connected with adaptation to the social behavior among groups of peers who are, for the girls' chronological age, advanced socially, the resultant problem behavior among the early-developed girls is understandable. Chronologically speaking, these girls were expected to show the social behavior patterns that prevail among their same-age peers. Biologically, however, they were more similar to older teenagers, and in comparison with them, did not stand out socially. Consequently, what might be interpreted negatively and with a deviant connotation by the adult world, was, from the perspective of these girls, no more than a change of social behavior in the direction toward the habits and leisure-time activities more common among older groups of peers, working peers, and among boys. The acquisition of new social behavior was facilitated by a self-concept of greater maturity and a willingness to engage in behaviors that signified a more mature status. Overall, the higher norm breaking in adolescence among the early-developed girls was largely a temporary feature with limited future problem implications.

Education and Family Life

When we enter the area of education and adult family life the picture is different. In these fields there was a conspicuous relation between the educational level, the job status, and important aspects of family life at adulthood on the one hand, and individual differences in the age at menarche, on the other. Early-developed females had significantly shorter education, and they had jobs that required less theoretical education than was the case for the late-developed females. More often, females who had had their menarche early had children than those who had had their menarche later. Similar differentiating features in adolescence were found between the early- and the later-developed females. The early-developed girls were not as interested in school and they had more advanced opposite-sex relations than did their later-developed counterparts. Thus, for the areas of education, working life, and family life the results support the differential model of long-term impact of early versus late pubertal maturation on the course of development.

The overall tendency is clear: In the areas of social adjustment and personality at adult age there are only weak indications of any relationship with the age at which the menarche occurs. For the important aspects of adult life referring to the attained level of education, the kind of job having children or not, the age at menarche has an impact. These results taken together have important implications for the choice of an appropriate and effective strategy for research on individual development. They convincingly show that such research must be planned and carried through as longitudinal research over extended period of time, studying a broad spectrum of relevant factors in the individual and in the surrounding environment. The essential aspects of the course of development and its relation to biological development that have been investigated here could not have been revealed if a cross-sectional design had been applied or if the study had been limited to only single aspects of individual functioning and of the environment.

EPILOGUE

A major research strategy implication of the interactional view that forms the theoretical frame of reference for the research conducted and presented here is that longitudinal research must be conducted if the development of individuals as totalities, and not only single aspects of individual functioning, is to be understood as a process of maturation and experience. The analyses presented here clearly demonstrate the relevance of this view. There is no adequate alternative to longitudinal research for the study of the individual course of development in a life span perspective (see e.g., Baltes, Reese, & Lipsitt, 1980; Cairns, 1979; Livson & Peskin, 1980; Magnusson, 1988; McCall, 1977; Wohlwill, 1973). Because longitudinal research has become popular again during the last decade, there is a tendency to include even rather short-term observations in that category. However, the empirical analyses reported here clearly show that effective longitudinal research must cover the total critical periods of development for the function(s) under consideration, and observations must be made continuously during the critical stage of development in order to avoid neglecting important transitions.

The analyses that have been performed and reported also demonstrate the necessity of covering a broad spectrum of individual functioning in order to understand the lawful patterning of developmental change in individuals. By covering the distinct configurations of relevant factors that are characteristic of individuals and groups of individuals we can avoid the risk of drawing rapid and far-reaching conclusions about developmental process from studies of single aspects of individual functioning and of the environment. The range and restrictions of long-term consequences of early pubertal maturation among girls could be successfully investigated only by the study of a wide range of factors for a representative group of girls followed from an early age to adulthood.

It has been a challenge in the present research to approach issues connected with behavioral change. It has been discussed and analyzed here in a theoretical framework of the interplay of biological factors, social relations, and environmental factors. In our examination of the role of pubertal timing among females for issues of adolescent transitions, we benefited much from earlier research. Based on previous research evidence it was possible to leave the exploratory stage of research and apply a theory-guided approach to test certain developmental progressions.

Theoretical considerations determined the focus on early maturing. They also determined the dependent measures that were used. A developmental model was formulated specifying the conditions for the possible emergence of differences between early- and later-matured girls in various areas. This model has been investigated throughout this book. Its validity was tested using a broad range of data. Overall, the applicability of the general model for the understanding of the role of pubertal maturation for social behavior and for interpersonal relations, both in adolescence and in early adulthood, was supported in these empirical analyses. The model outlined should not be considered a final product. It is raw outline attempting to show, and to explain, the consequences of maturing early versus late. The present analyses are but the first attempts to trace the implications of pubertal timing in the female life course. The model posited has the advantage of being testable on other subject samples by other researchers, and of being investigated to provide more likely explanations. Its merit is its empirical testability of the relationship between pubertal maturation and psychosocial development. The concepts developed here concerning differential opportunity and differential susceptibility to peer influences of certain types may be confirmed or rejected with existing data or with data collected in the future. Generally, the information that the present research has yielded with respect to the interplay over time of biological growth factors with psychological and interpersonal factors and with behavior should be of value for future systematically planned, longitudinal studies uncovering with more precision the process through which physical growth intervenes in the life course of females.

References

Abernethy, E. M. (1925). Correlations in physical and mental growth. *Journal of Educational Psychology, 16,* 458, 539–546.
Adams, G. R. & Munro, G. (1979). Portrait of the North American runaway: A critical review. *Journal of Youth and Adolescence, 8,* 359–373.
Adams, J. F. (1981). Earlier menarche, greater height and weight: A stimulation-stress factor hypothesis. *Genetic Psychology Monographs, 104,* 3–22.
Adams, J. P. (1964). Adolescent personal problems as a function of age and sex. *Journal of Genetic Psychology, 104,* 207–214.
Adams, P. L. (1969). Puberty as a biosocial turning point. *Psychosomatics, 10,* 343–349.
Adams, P. L. (1972). Late sexual maturation in girls. *Medical aspects of Human Sexuality, 6,* 50–75.
Allen, C. D., & Eicher, J.B. (1973). Adolescent girls' acceptance and rejection based on appearance. *Adolescence, 8,* 125–138.
Andersson, B-E. (1969). *Studies in adolescent behavior.* Stockholm, Almqvist & Wiksell.
Andersson, O., Dunér, A., & Magnusson, D. (1980). Social adjustment among early maturing girls. *Reports from the Department of Psychology, University of Stockholm,* Report No. 35.
Andersson, T., Magnusson, D., & Dunér, A. (1983). Base Data-81: The life situation at early adulthood. *Reports from IDA, Department of Psychology, University of Stockholm,* No. 49. (In Swedish).
Apter, A., Galatzer, A., Beth-Halachmi, N., & Laron, Z. (1981). Self-image in adolescents with delayed puberty and growth retardation. *Journal of Youth and Adolescence, 10,* 501–505.
Ausubel, D. P. (1954). *Theory and problems of adolescent development.* New York: Grune & Stratton.

REFERENCES

Baltes, P.B., Reese, H. W., & Lipsitt, L. P. (1980). Life-span developmental psychology. *Annual Review of Psychology, 31,* 65–110.

Bandura, A. (1972). The stormy decade: Fact or fiction. In D. Rogers (Ed.), *Issues in adolescent psychology* (2nd ed.). New York: Appleton-Century-Crofts.

Bardwick, J. M. (1971). *Psychology of women: A study of bio-cultural conflicts.* New York: Harper & Row.

Baron, R. M., & Kenny, D. A. (1986). The moderator-mediator variable distinction in social psychological research: Conceptual, strategic, and statistical considerations. *Journal of Personality and Social Psychology, 51,* 11–73–1182.

Bayley, N., & Tuddenham, R. (1944). Adolescent changes in body build. *43rd yearbook. National Society for the Study of Education.* Chicago: University of Chicago Press.

Bean, J. A., Leeper, J.D., Wallace, R. B., Sherman, B. M., & Jagger, H. (1979). Variations in the reporting of menstrual histories. *American Journal of Epidemiology, 109,* 181–185.

Bem, S. L. (1974). The measurement of psychological androgyny. *Journal of Consulting and Clinical Psychology, 42,* 155–162.

Benedek, E. P., Poznanski, E., & Mason, S. (1979). A note on the female adolescent's psychological reactions to breast development. *Journal of the American Academy of Child Psychiatry, 18,* 537–545.

Bergman, L. R. (1973). Parent's education and mean change in intelligence. *Scandinavian Journal of Psychology, 14,* 273–281.

Bergman, L. R., & El-Khouri, B. (1986). Some exact tests of single cell frequencies in two-way contingency tables. *Reports from the Department of Psychology, University of Stockholm,* No. 645.

Bergman, L. R., & Magnusson, D. (1984a). Patterns of adjustment problems at age 10: An empirical and methodological study. *Reports from the Department of Psychology, University of Stockholm,* No. 615.

Bergman, L. R., & Magnusson, D. (1984b). Patterns of adjustment problems at age 13: An empirical and methodological study. *Reports from the Department of Psychology, University of Stockholm,* No. 620.

Bergman, L. R., & Magnusson, D. (1986). Type A behavior: A longitudinal study from childhood to adulthood. *Psychosomatic Medicine,* 134–142.

Bergman, L. R., & Magnusson, D. (1987). A person approach to the study of the development of adjustment problems: An empirical example and some research strategical considerations. In D. Magnusson & A. Öhman (Eds.), *Psychopathology: An interactional perspective.* New York: Academic Press.

Bergsten-Brucefors, A. (1974). Mensdebutåldern hos svenska skolflickor [Age at menarche in Swedish girls]. *Rapport från Pedagogiska Institutionen vid Lärarhögskolan i Stockholm,* No. 105.

Bergsten-Brucefors, A. (1976). A note on the accuracy of recalled age at menarche. *Annals of Human Biology, 3,* 71–73.

Berzonski, M. D. (1981). *Adolescent development.* New York: Macmillan.

Billy, J. O. G., Rodgers, J. L., & Udry, J. R. (1984). Adolescent sexual behavior and friendship choise. *Social Forces, 62,* 653–678.

Bloch, H. A., & Niederhoffer, A. (1958). *The gang: A study in adolescent behavior.* New York: Philosophical Library.

REFERENCES

Blos, P. (1962). *On adolescence.* New York: Macmillan.
Blyth, D. A. (1983). Surviving and thriving in the social world: A commentary on six bew studies of popular, rejected, and neglected children. *Merril-Palmer Quarterly, 29,* 449–458.
Blyth, D. A., Hill, J. P., & Thiel, K. S. (1982). Early adolescents' significant others: Grade and gender differences in perceived relationships with familial and nonfamilial adults and young people. *Journal of Youth and Adolescence, 11,* 425–450.
Blyth, D. A., Simmons, R. G., & Zakin, D. F. (1985). Satisfaction with body image for early adolescent females: The impact of pubertal timing within different school environments. *Journal of Youth and Adolescence, 14,* 207–227.
Bowerman, C. E., & Kinch, J. W. (1959). Changes in family and peer orientation of children between the fourth and tenth grades. *Social Forces, 57,* 206–211.
Bronfenbrenner, U. (1984). *The ecology of the family as a context for human development: Research perspectives.* Unpublished manuscript.
Brooks-Gunn, J. (1984). The psychological significance of different pubertal events to young girls. *Journal of Early Adolescence, 4,* 315–327.
Brooks-Gunn, J. (1987). Pubertal processes and girls' psychological adaptation. In R. M. Lerner & T. T. Foch (Eds.), *Biological-psychosocial interactions in early adolescence.* Hillsdale, NJ: Lawrence Erlbaum Associates.
Brooks-Gunn, J., & Petersen, A. C. (1983). *Girls at puberty. Biological and psychosocial perspectives.* New York: Plenum.
Brooks-Gunn, J., & Petersen, A. C. (1984). Problems in studying and defining pubertal events. *Journal of Youth and Adolescence, 13,* 181–196.
Brooks-Gunn, J., Petersen, A. C., & Eichorn, D. H. (1985). The study of maturational timing effects in adolescence. *Journal of Youth and Adolescence, 14,* 149–161.
Brooks-Gunn, J., & Ruble, D. N. (1982a). The development of menstrual-related beliefs and behaviors during early adolescence. *Child Development, 53,* 1567–1577.
Brooks-Gunn, J., & Ruble, D. N. (1982b). Developmental processes in the experience of menarche. In A. Baum & J. E. Singer (Eds.), *Handbook of psychology and health: Issues in child health and adolescent health* (Vol. 2). Hillsdale, NJ: Lawrence Erlbaum Associates.
Brooks-Gunn, J., & Ruble, D. N. (1983). The experience of menarche from a developmental perspective. In J. Brooks-Gunn & A. C. Petersen (Eds.), *Girls at puberty. Biological and psychosocial perspectives.* New York: Plenum.
Brooks-Gunn, J., & Warren, M. P. (1985a). Measuring physical status and timing in early adolescence: A developmental perspective. *Journal of Youth and Adolescence, 14,* 163–189.
Brooks-Gunn, J., & Warren, M. P. (1985b). Effects of delayed menarche in different contexts: Dance and nondance students. *Journal of Youth and Adolescence, 14,* 285–300.
Brooks-Gunn, J., & Warren, M. P. (1987, April). *Biological contributions to depressive and aggressive affect in young adolescent girls.* Paper presented at the Society for Research in Child Development meetings in a symposium on Hormone Status at Puberty. Baltimore.
Brooks-Gunn, J., Warren, M. P., Rosso, J., & Gargiulo, J. (1987). Validity of self-report measures of girls' pubertal status. *Child Development, 58,* 829–841.

Brooks-Gunn, J., Warren, M. P., Samelson, M., & Fox, R. (1986). Physical similarity of and disclosure of menarcheal status to friends: Effects of grade and pubertal status. *Journal of Early Adolescence, 6,* 3–14.
Brown, B. B. (1982). The extent and effects of peer pressure among high school students: A retrospective analysis. *Journal of Youth and Adolescence, 11,* 121–133.
Bruch, H. (1941). Obesity in relation to puberty. *Journal of Pediatrics, 19,* 365–375.
Brundtlang, G. H., & Walloe, L. (1973). Menarcheal age in Norway: Halt in the trend towards earlier maturation. *Nature, 241,* 478–479.
Bruner, J. S. (1966). *Studies on cognitive growth.* New York: Wiley.
Buck, C., & Stavraky, K. (1967). The relationship between age at menarche and age at marriage among childbearing women. *Human Biology, 39,* 93–102.
Bullough, V. L. (1983). Menarche and teenage pregnancy: A misuse of historical data. In S. Golub (Ed.), *Menarche. The transition from girl to woman.* Lexington, MA: Lexington Books, D. C. Heath.
Burns, R. B. (1979). *The self-concept: in theory, measurement, development and behaviour.* London: Longman.
Bush, D. E., Simmons, R. G., Hutchinson, B., & Blyth, D. A. (1977–1978). Adolescent perceptions of sex-roles in 1968 and 1975. *Public Opinion Quarterly, 41,* 459–474.
Cahalan, B., Cicin, I. H., & Crossley, H. M. (1969). *American drinking practices* (Rutgers Center of Alcohol Studies). New Brunswick: College and University Press.
Cairns, R. B. (1979). *Social development: The origins and plasticity of interchanges.* San Francisco: Freeman.
Campbell, A. C. (1980). Friendship as a factor in male and female delinquency. In H. C. Foot, A. J. Chapman, & J. R. Smith (Eds.), *Friendship and social relations in children.* New York: Wiley.
Cavior, N., & Dokecki, P. R. (1973). Physical attractiveness, perceived attitude similarity and academic achievement as contributors to interpersonal attraction among adolescents. *Developmental Psychology, 9,* 44–54.
Clarke, A. E., & Ruble, D. N. (1978). Young adolescents' beliefs concerning menstruation. *Child Development, 49,* 231–234.
Clausen, J. A. (1975). The social meaning of differential physical and sexual maturation. In S. E. Dragastin & G. H. Elder (Eds.), *Adolescence in the life cycle: Psychological change and social context.* Washington, DC: Hempisphere.
Clifford, E. (1971). Body satisfaction in adolescence. *Perceptual and Motor Skills, 33,* 119–125.
Coie, J. D., Dodge, K. A., & Coppotelli, H. (1982). Dimensions and types of social status: A cross-age perspective. *Developmental Psychology, 18,* 557–570.
Coleman, J. C. (1980). Friendship and the peer group in adolescence. In J. Adelson (Ed.), *Handbook of adolescent psychology.* New York: Wiley.
Coleman, J. S. (1961). *The adolescent society.* New York: The Free Press.
Collins, J. K., & Propert, D. S. (1983). A developmental study of body recognition in adolescent girls. *Adolescence, 18,* 767–774.
Comite, F., Pescovitz, O. H., Sonis, W. A., Hench, K., McNemar, A., Klein, R. P., Loriaux, D. L., & Cutler, G. B. (1987). Premature adolescence: Neuroendocrine and psychosocial studies. In R. M. Lerner & T. T. Foch (Eds.), *Biological-*

psychosocial interactions in early adolescence. Hillsdale, NJ: Lawrence Erlbaum Associates.

Conklin, E. S. (1933). *Principles of adolescent psychology.* New York: Holt.

Constanzo, P. R., & Shaw, M. E. (1966). Conformity as a function of age level. *Child Development, 37,* 967–975.

Covington, J. (1982). Adolescent deviation and age. *Journal of Youth and Adolescence, 11,* 329–344.

Crafoord, K. (1986). Flickproblem och problemflickor i tidiga tonåren. [Problem behavior and problem-girls in early youth]. *Individuell Utveckling och Miljö, Stockholms Universitet,* Rapport Nr 65.

Crocket, L., Losoff, M., & Petersen, A. C. (1984). Perceptions of the peer group and friendship in early adolescence. *Journal of Early Adolescence, 4,* 155–181.

Crocket, L., & Petersen, A. C. (1987). Pubertal status and psychosocial development: Findings from the Early Adolescence Study. In R. M. Lerner & T. T. Foch (Eds.), *Biological-psychosocial interactions in early adolescence: A life-span perspective.* Hillsdale, NJ: Lawrence Erlbaum Associates.

Csikszentmihalyi, M., & Larson, R. (1984). *Being adolescent: Conflict and growth in the teenage years.* New York: Basic Books.

Damon, A., & Bajema, C. J. (1974). Age at menarche: Accuracy of recall after thirty-nine years. *Human Biology, 46,* 381–384.

Damon, A., Damon, S. T., Reed, R. B., & Valadian, I. (1969). Age at menarche of mothers and daughters, with a note on accuracy of recall. *Human Biology, 41,* 161–175.

Danza, R. (1983). Menarche: Its effects on mother-daughter and father-daughter interactions. In S. Golub (Ed.), *Menarche. The transition from girl to woman.* Lexington, MA: Lexington Books, D. C. Heath.

Davidson, H. H., & Gottlieb, L. S. (1955). The emotional maturity of pre- and postmenarcheal girls. *Journal of Genetic Psychology, 86,* 261–266.

Davies, B. L. (1977). Attitudes towards school among early and late maturing adolescent girls. *Journal of Genetic Psychology, 131,* 261–265.

de Anda, D. (1983). Pregnancy in early and late adolescence. *Journal of Youth and Adolescence, 12,* 33–42.

Deutsch, H. (1944). *The psychology of women.* (Vol. 1). New York: Grune & Stratton.

Diers, C. J. (1974). Historical trends in the age at menarche and menopause. *Psychological Reports, 34,* 931–937.

Dion, K. K., Berscheid, E., & Walster, E. (1972). What is beautiful is good. *Journal of Personality and Social Psychology, 24,* 285–290.

Dodge, K. A., Coie, J. D., & Brakke, N. P. (1982). Behavioral patterns of socially rejected and neglected preadolescents: The role of social approach and aggression. *Journal of Abnormal Child Psychology, 10,* 389–409.

Donovan, J. E., & Jessor, R. (1978). Adolescent problem drinking: Psychosocial correlates in a national sample study. *Journal of Studies on Alcohol, 9,* 1506–1524.

Dornbusch, S. M., Carlsmith, J. M., Duncan, P. D., Gross, R. T., Martin J. A., Ritter, P. L., & Siegel-Gorelick, B. (1984). Sexual maturation, social class, and the desire to be thin among adolescent females. *Journal of Developmental and Behavioral Pediatrics, 5,* 308–314.

Dornbusch, S. M., Carlsmith, J. M., Gross, R. T., Martin, J. A., Jennings, D., Rosenberg, A., & Duke, P. M. (1981). Sexual development, age, and dating: A comparison of biological and social influences upon one set of behaviors. *Child Development, 52*, 179–185.

Dornbusch, S. M., Gross, R. T., Duncan, P. D., & Ritter, P. L. (1987). Stanford studies of adolescence using the National Health Examination Survey. In R. M. Lerner & T. T. Foch (Eds.), *Biological-psychosocial interactions in early adolescence*. Hillsdale, NJ: Lawrence Erlbaum Associates.

Douglas, J. W. B. (1964). *The home and the school*. London: Macgibbon & Kee.

Douglas, J. W. B., & Ross, J. M. (1964). Age of puberty related to educational ability, attainment and school leaving age. *Journal of Child Psychology and Psychiatry, 5*, 185–195.

Douvan, E. (1970). New sources of conflict in females at adolescence and early adulthood. In J. M. Bardwick (Ed.), *Feminine personality and conflict*. Monterey, CA: Brooks/Cole.

Douvan, E., & Adelson, J. (1966). *The adolescent experience*. New York: Wiley.

Downs, W. R. (1985). Using panel data to examine sex differences in causal relationships among adolescent alcohol use, norms, and peer alcohol use, *Journal of Youth and Adolescence, 14*, 469–486.

Dreyer, A. S., Hulac, V., & Rigler, D. (1971). Differential adjustment to pubescence and cognitive style patterns. *Developmental Psychology, 4*, 456–462.

Duck, S. W. (1975). Personality similarity and friendship choises by adolescents. *European Journal of Social Psychology, 5*, 351–365.

Duke, D. L. (1978). Why don't girls misbehave more than boys in school? *Journal of Youth and Adolescence, 7*, 141–157.

Duke, D. L., & Duke, P. M. (1978). The prediction of delinquency in girls. *Journal of Research and Development in Education, 11*, 18–33.

Duke, P. M., Carlsmith, J. M., Jennings, D., Martin, J. A., Dornbusch, S. M., Gross, R. T., & Siegel-Gorelick, B. (1982). Educational correlates of early and late sexual maturation in adolescence. *Journal of Pediatrics, 100*, 633–637.

Duke, P. M., Litt, I. F., & Gross, R. T. (1980). Adolescents' self-assessment of sexual maturation, *Pediatrics, 66*, 918–920.

Duncan, P. D., Ritter, P. L., Dornbusch, S. M., Gross, R. T., & Carlsmith, J. M. (1985). The effects of pubertal timing on body image, school behavior, and deviance. *Journal of Youth and Adolescence, 14*, 227–235.

Dunér, A. (1972). *Vad ska det bliva? Undersökningar om studie- och yrkesvalsprocessen* [Studies on the educational and vocational career process]. Stockholm: Allmänna Förlaget. (English summary)

Dunnington, M. J. (1957). Behavioral differences of sociometric status groups in a nursery school. *Child Development, 28*, 93–102.

Dunphy, D. C. (1963). The social structure of urban adolescent peer groups. *Sociometry, 26*, 230–246.

Dusek, J. B., & Flaherty, J. F. (1981). The development of the self-concept during the adolescent years. *Monographs of the Society for Research in Child Development, 46*, 1–70.

Dwyer, J., & Mayer, J. (1968–1969). Psychological effects of variations in physical appearance during adolescence. *Adolescence, 3*, 353–380.

Ehrhardt, A. A., Meyer-Bahlburg, H. F. L., Bell, J. R., Cohen, S. F., Healey, J. M., Stiel, R., Feldman, J. F., Morishima, A., & New, M. I. (1984). Idiopathic precocious puberty in girls: Psychiatric follow-up in adolescence, *Journal of the American Academy of Child Psychiatry, 23*, 23–33.

Eichorn, D. H. (1963). Biological correlates of behavior. In H. W. Stevenson, J. Kagan, & H. G. Richey (Eds.), *The sixty-second Yearbook of the National society for the study of education.* Chicago: University of Chicago Press.

Eichorn, D. H. (1975). Asynchronizations in adolescent development. In S. Dragastin & N. S. Endler (Eds.), *Adolescence in the life cycle.* New York: Wiley.

Ellis-Schwabe, M. E., & Thornburg, H. D. (1986). Conflict areas between parents and their adolescents. *Journal of Psychology, 120,* 59–68.

Ellison, P. T. (1982). Skeletal growth, fatness, and menarcheal age: A comparison of two hypotheses. *Human Biology, 54,* 269–281.

English, H. B. (1957). Chronological divisions of the life span. *Journal of Educational Psychology, 48,* 436–439.

Epstein, S. (1980). The self-concept: A review and the proposal of an integrated theory of personality. In E. Staub (Ed), *Personality: Basic issues and current research.* Englewoods Cliffs, NJ: Prentice-Hall.

Erikson, E. H. (1968). *Identity, youth and crisis.* New York: Norton.

Farrington, D. P. (1986). Age and crime. In M. Tonry & N. Morris (Eds.), *Crime and justice* (Vol. 7). Chicago: University of Chicago Press.

Faust, M. S. (1960). Developmental maturity as a determinant in prestige of adolescent girls. *Child Development, 31,* 173–184.

Faust, M. S. (1983). Alternative constructions of adolescence growth. In J. Brooks-Gunn & A. C. Petersen (Eds.), *Girls at puberty. Biological and psychosocial perspectives.* New York: Plenum.

Feldhusen, J. F., Thurston, J. R., & Benning, J. J. (1973). A longitudinal study of delinquency and other aspects of children's behavior. *International Journal of Criminology and Penology, 1,* 341–351.

Fischbein, S. (1977). Intra-pair similarity in physical growth of monozygotic and of dizygotic twins during puberty. *Annals of Human Biology, 4,* 417–430.

Fishbein, M. (1967). Attitude and the prediction of behavior. In M. Fishbein (Ed.), *Readings in attitude theory and measurement.* New York: Wiley.

Fishbein, M., & Ajzen, I. (1975). *Belief, attitude, intention, and behavior.* Reading, MA: Addison-Wesley.

Foner, A. (1975). Age in society. Structure and change. *American Behavioral Scientist, 19,* 144–165.

Freud, A. (1958). Adolescence. *Psychoanalytic study of the child, 13,* 255–278.

Friedman, S. B., & Sarles, R. M. (1980). "Out of control" behavior in adolescents. *Pediatric Clinics of North America, 27,* 97–107.

Frisch, R. E., & Revelle, R. (1970). Height and weight at menarche and a hypothesis of critical body weights and adolescent events. *Science, 169,* 397–399.

Frisk, M. (1968). *Tonårsproblem. En studie av läroverksungdom.* [Teenage problems]. Helsingfors: Samfundet Folkhälsan. (Dissertation, not in English).

Frisk, M. (1975). Puberty: Emotional maturation and behaviour. In S. B. Berenberg (Ed.), *Puberty, Biological and psychosocial components.* Leiden: H. E. Stenfert Kroese B. V.

Frisk, M., Tenhunen, T., Widholm, O., & Hortling, M. (1966). Psychological problems in adolescents showing advanced or delayed physical maturation. *Adolescence, 1*, 126–135.

Furstenberg, F. (1976). *Unplanned parenthood: The social consequences of teenage child bearing.* New York: The Free Press.

Furu, M. (1976). Menarcheal age in Stockholm girls, 1967. *Annals of Human Biology, 3*, 587–590.

Gaddis, A., & Brooks-Gunn, J. (1985). The male experience of pubertal change. *Journal of Youth and Adolescence, 14*, 61–69.

Gagnon, J. H. (1983). Age at menarche and sexual conduct in adolescence and young adulthood. In S. Golub (Ed.), *Menarche. The transition from girl to woman.* Lexington, MA: Lexington Books, D. C. Heath.

Garbarino, J., Burston, N., Raber, S., Russel, R., & Crouter, A. (1978). The social maps of children approaching adolescence: Studying the ecology of youth development. *Journal of Youth and Adolescence, 7*, 417–428.

Garn, S. M. (1980). Continuities and change in maturational timing. In O. G. Brim, Jr. & J. Kagan, (Eds.), *Constancy and change in human development.* Cambridge, MA: Harvard University Press.

Garwood, S. G., & Allen, L. (1979). Self-concept and identified problem differences between pre- and postmenarcheal adolescents. *Journal of Clinical Psychology, 35*, 528–537.

Gates, A. I. (1924). The nature and educational significance of physical status and of mental, physiological, social and emotional maturity. *Journal of Educational Psychology, 15*, 329–358.

Gibbs, J. T. (1980). Depression and suicidal behavior among delinquent females. *Journal of Youth and Adolescence, 10*, 159–167.

Gilligan, C. (1982). *In a different voice. Psychological theory and women's development.* Cambridge, MA: Harvard University Press.

Glass, G. V., Peckham, P. D., & Sanders, J. R. (1972). Consequences of failure to meet assumptions underlying the fixed effects of analyses of variance and covariance. *Review of Educational Research, 43*, 237–288.

Gold, M., & Petronio, R. J. (1980). Delinquent behavior in adolescence. In J. Adelson (Ed.), *Handbook of adolescent psychology.* New York: Wiley.

Goldberg, S., Blumberg, S. L., & Kriger, A. (1982). Menarche and interests in infants: Biological and social influences. *Child Development, 53*, 1544–1550.

Goldstein, H. (1979). *The design and analysis of longitudinal studies.* New York: Academic Press.

Gollin, E. S. (1958). Organizational characteristics of social judgment: A developmental investigation. *Journal of Personality, 26*, 139–154.

Golub, S. (1983). Implications for women's health and well-being. In S. Golub (Ed.), *Menarche. The transition from girl to woman.* Lexington, MA: Lexington Books, D. C. Heath.

Golub, S. (1984). Menarche: The beginning of menstrual life. *Women & Health, 9*, 17–36.

Gottschalk, L. A., Titchener, J. L., Piker, H. N., & Stewart, S. S. (1964). Psychosocial factors associated with pregnancy in adolescent girls: A preliminary report. *Journal of Nervous and Mental Disease, 138*, 524–534.

Greif, E. B., & Ulman, K. J. (1982). The psychological impact of menarche on early adolescent females: A review of the literature. *Child Development, 53,* 1413–1430.

Greulich, W. W. (1944). Physical changes in adolescence. In Adolescence. *43rd Yearbook of the National Society for the Study of Education.* Chicago: University of Chicago Press.

Gross, R. T., & Duke, P. M. (1980). The effect of early versus late physical maturation on adolescent behavior. *Pediatric Clinics of North America, 27,* 71–77.

Hall, G. S. (1904). *Adolescence: Its pathology and its relations to physiology, anthropology, sociology, sex, crime, religion and education* (Vols. I and II). New York: Appleton.

Harper, J., & Collins, J. K. (1972). The secular trend in the age of menarche in Australian schoolgirls. *Australian Paediatric Journal, 8,* 44–48.

Hart, M., & Sarnoff, C. A. (1971). The impact of the menarche. A study of two stages of organization. *Journal of the American Journal of Child Psychiatry, 10,* 257–271.

Hartup, W. W. (1970). Peer interaction and social organization. In P. H. Mussen (Ed.), *Carmichael's manual of child psychology.* New York: Wiley.

Hartup, W. W. (1976). Cross-age versus same-age peer interaction: Ethological and cross-cultural perspectives. In V. L. Allen (Ed.), *Children as teachers: Theory and research on tutoring.* New York: Academic Press.

Hartup, W. W., Glazer, J. A., & Charlesworth, R. (1967). Peer reinforcement and sociometric status. *Child Development, 38,* 1017–1024.

Hauge, R. (1971). *Kriminalitet som ungdomsfenomen* [Crime as a youth characterizing phenomenon]. Stockholm: Aldus.

Havighurst, R. J., Bowman, P. H., Liddle, G. P., Matthews, C. V., & Pierce, J. V. (1962). *Growing up in River City.* New York: Wiley.

Hebbelinck, M. (1977). Biological aspects of development at adolescence. In J. P. Hill & F. J. Mönks (Eds.), *Adolescence and youth in prospect.* Guilford, England: IPC Science & Technology.

Helgasson, T., & Asmundsson, G. (1975). Behavior and social characteristics of young asocial alcohol abusers. *Neuropsychobiology, 1,* 109–120.

Henricson, M. (1973). *Tonåringar och normer* [Teenagers and norms]. Stockholm: Skolöverstyrelsen.

Henton, C. L. (1961). The effect of socio-economic and emotional factors on the onset of menarche among negro and white girls. *Journal of Genetic Psychology, 98,* 255–264.

Hess, B. (1981). Friendship. In M. W. Riley, M. Johnson, & A. Foner (Eds.), *Aging and society, Vol. 3. A sociology of age stratification.* New York: Russell Sage Foundation.

Hiernaux, J. (1972). Ethnic differences in growth and development. *Eugenics Quarterly, 15,* 13–21.

Higham, E. (1980). Variations in adolescent psychohormonal development. In J. Adelson (Ed.), *Handbook of adolescent psychology.* New York: Wiley.

Hill, J. P., & Holmbeck, G. N. (1987). Familial adaptation to biological change during adolescence. In R. M. Lerner & T. T. Foch (Eds.), *Biological-psychosocial interactions in early adolescence.* Hillsdale, NJ: Lawrence Erlbaum Associates.

Hill, J. P., Holmbeck, G. N., Marlow, L., Green, T. M., & Lynch, M. E. (1985). Menarcheal status and parent-child relations in families of seventh-grade girls. *Journal of Youth and Adolescence, 14,* 301–316.

Hill, J. P., & Lynch, M. E. (1983). The intensification of gender-related role expectations during early adolescence. In J. Brooks-Gunn & A. C. Petersen (Eds.), *Girls at puberty. Biological and psychosocial perspectives.* New York: Plenum.

Hirschi, T., & Gottfredson, M. (1983). Age and the explanation of crime. *American Journal of Sociology, 89,* 552–584.

Hunter, F. T., & Youniss, J. (1982). Changes in functions of three relations during adolescence. *Developmental Psychology, 18,* 806–811.

Intons-Peterson, M. J. (1988). *Gender concepts of Swedish and American youth.* Hillsdale, NJ: Lawrence Erlbaum Associates.

James W. H. (1973). Age of menarche, family size and birth order. *American Journal of Obstetrics and Gynecology, 116,* 292–293.

Jaquish, G. A., & Savin-Williams, R. C. (1981). Biological and ecological factors in the expression of adolescent self-esteem. *Journal of Youth and Adolescence, 10,* 473–485.

Jenicek, M. & Demirjian, A. (1974). Age at menarche in French Canadian urban girls. *Annals of Human Biology, 1,* 339–346.

Jenkins, G. G. (1931). Factors involved in children's friendships. *Journal of Educational Psychology, 22,* 440–448.

Jenkins, R. R. (1983). Future directions in research. In J. Brooks-Gunn & A. C. Petersen (Eds.), *Girls at puberty. Biological and psychosocial perspectives.* New York: Plenum Press.

Jensen, A. R. (1969). Reducing the heredity-environment uncertainty. In Environment, heredity and Intelligence. *Reprint Series No. 2. compiled from Harvard Educational Review,* 1969, 209–243.

Jersild, A. T. (1952). *In search of self.* New York: Teachers College, Columbia University, Bureau of Publications.

Jessor, S. L., & Jessor, R. (1975). Transition from virginity to nonvirginity among youth: A social-psychological study over time. *Developmental Psychology, 11,* 473–484.

Jessor, S. L., & Jessor, R. (1977). *Problem behavior and psychological development: A longitudinal study of youth.* New York: Academic Press.

Jones, H. E. (1949). *Adolescence in our society. The family in a democratic society. Anniversary papers of the Community Service Society of New York.* New York: Columbia University Press.

Jones, M. C. (1948). Adolescent friendships. *American Psychologist, 3,* 352.

Jones, M. C. (1957). The later careers of boys who were early- or late-maturing. *Child Development, 28,* 113–128.

Jones, M. C. (1958). A study of socialization patterns at the high-school level. *Journal of Genetic Psychology, 93,* 87–111.

Jones, M. C. (1965). Psychological correlates of somatic development. *Child Development, 36,* 899–911.

Jones, M. C., & Bayley, N. (1950). Physical maturing among boys as related to behavior. *Journal of Educational Psychology, 41,* 129–148.

Jones, M. C., & Mussen, P. H. (1958). Self-conceptions, motivations, and interpersonal attitudes of early- and late-maturing girls. *Child Development, 29,* 492–500.

Kandel, D. B. (1978). Similarity in real-life friendship pairs. *Journal of Personality and Social Psychology, 36,* 306–312.

Kandel, D. B., & Lesser, G. S. (1972). *Youth in two worlds. United States and Denmark.* San Francisco: Jossey-Bass.

Kandel, D. B., Kessler, R. C., & Margulies, R. Z. (1978). Antecedents of adolescent initiation into stages of drug use: A developmental analysis. *Journal of Youth and Adolescence, 7,* 13–40.

Kantero, R-L., & Widholm, O. (1971). The age at menarche in Finnish girls in 1969. *Acta Obstetricia & Gynecologia Scandinavia. Supplement 14,* 7–18.

Karlberg, P., Klackenberg, G., Engström, I., Klackenberg-Larsson, I., Lichtenstein, H., Stensson, J., & Svennberg, I. (1968). The development of children in a Swedish Urban community. A prospective longitudinal study. *Acta Paediatrica Scandinavica,* Suppl. 187.

Katchadourian, H. (1977). *The biology of adolescence.* San Francisco: Freeman.

Katz, E., & Stodtland, A. (1959). The functional approach to attitude measurement. In S. Koch (Ed.), *Psychology: A study of a science* (Vol. III). New York: McGraw-Hill.

Kelly, H., & Menking, S. (1979). Recalled breast development experiences and young adult breast satisfaction and breast display behavior. *Psychology: A Quarterly Journal of Human Behavior, 16,* 17–24.

Kestenberg, J. S. (1961). Menarche. In S. Lorand & H. I. Schneer (Eds.), *Adolescents: Psychoanalytic approach to problems and therapy.* New York: Hoeber.

Kestenberg, J. S. (1967a). Phases of adolescence with suggestions for a correlation of psychic and hormonal organizations. Part 1: Antecedents of adolescent organizations in childhood. *Journal of the American Academy of Child Psychiatry, 6,* 426–463.

Kestenberg, J. S. (1967b). Phases of adolescence with suggestions for a correlation of psychic and hormonal organizations. Part 2: Prepuberty diffusion and reintegration. *Journal of the American Academy of Child Psychiatry, 6,* 577–614.

Kestenberg, J. S. (1968). Phases of adolescence with suggestions for a correlation of psychic and hormonal organizations. Part 3: Puberty growth, differentiation, and consolidation. *Journal of the American Academy of Child Psychiatry, 7,* 108–151.

Kiernan, K. E. (1977). Age at puberty in relation to age at marriage and parenthood: A national longitudinal study. *Annals of Human Biology, 4,* 301–308.

Kinsey, A. C., Pomeroy, W. B., & Martin C. E. (1948). *Sexual behavior in the human male.* Philadelphia: W. B. Saunders.

Kinsey, A. C., Pomeroy, W. B., Martin, C. E., & Gebhard, P. H. (1953). *Sexual behavior in the human female.* Philadelphia: Saunders.

Kirkegaard-Sorensen, L., & Mednick, S. A. (1977). A prospective study of predictors of criminality. In S. A. Mednick & K. O. Christiansen (Eds.), *Biosocial bases of criminal behavior.* New York: Gardner Press.

Klackenberg-Larsson, I., & Björkman, K. (1978) Tonåringars sexdebut [The début of sexual intercourse among teenagers]. *Sociol-medicinsk tidskrift, 2,* 333–341.

Koestler, A. (1978). *Janus: a summing up.* London: Hutchinson.

Koff, E., Rierdan, J., & Jacobson, B. A. (1981). The personal and interpersonal significance of menarche. *Journal of the American Academy of Child Psychiatry, 20,* 148–158.

Koff, E., Rierdan, J., & Sheingold, K. (1982). Memories of menarche: Age, preparation, and prior knowledge as determinants of initial menstrual experience. *Journal of Youth and Adolescence, 11,* 1–9.

Koff, E., Rierdan, J., & Silverstone, E. (1978). Changes in representation of body image as a function of menarcheal status. *Developmental Psychology, 14,* 635–642.

Kratcoski, P. C., & Kratcoski, J. E. (1975). Changing patterns in the delinquent activities of boys and girls: A self-reported delinquency analysis. *Adolescence, 10,* 83–91.
Kugelmass, S., & Breznitz, S. (1968). Intentionality in moral judgment: Adolescent development. *Child Development, 39,* 249–256.
Lenerz, K., Kucher, J. S., East, P. L., Lerner, J. V., & Lerner, R. M. (1987). Early adolescents' physical organismic characteristics and psychosocial functioning: Findings from the Pennsylvania Early Adolescent Transitions Study (PEATS). In R. M. Lerner & T. T. Foch (Eds.), *Biological-psychosocial interactions in early adolescence.* Hillsdale, NJ: Lawrence Erlbaum Associates.
Lerner, R. M. (1987). A life-span perspective for early adolescence. In R. M. Lerner & T. T. Foch (Eds.), *Biological-psychosocial interactions in early adolescence.* Hillsdale, NJ: Lawrence Erlbaum Associates.
Lerner, R. M., & Busch-Rossnagel, N. A. (1981). Individuals as producers of their development: Conceptual and empirical bases. In R. M. Lerner & N. A. Busch-Rossnagel (Eds.), *Individuals as producers of their development: A life-span perspective.* New York: Academic Press.
Lerner, R. M., & Foch, T. T. (1987). *Biological-psychosocial interactions in early adolescence.* Hillsdale, NJ: Lawrence Erlbaum Associates.
Lerner, R. M., & Karabenick, S. A. (1974). Physical attractiveness, body attitudes, and self-concept in late adolescents. *Journal of Youth and Adolescence, 3,* 307–316.
Lewis, R. A. (1963). Parents and peers: Socialization agents in the coital behavior of young adults. *Journal of Sex Research, 9,* 156–170.
Lindahl, M., & Bergh, T. (1988, June). Kvinnor och idrott [Women and sports]. Paper presented at the conference on Women Athletes, Bosön.
Lindgren, G. (1975). Pubertet och psykosocial status I. [Puberty and psychosocial status]. *Rapport från Pedagogiska Institutionen vid Lärarhögskolan i Stockholm.* (Not available in English)
Lindgren, G. (1976). Height, weight and menarche in Swedish urban school children in relation to socio-economic and regional factors. *Annals of Human Biology, 3,* 501–528.
Livesley, W. J., & Bromley, D. B. (1973). *Person perception in childhood and adolescence.* London: Wiley.
Livson, N., & Bronson, W. C. (1961). An exploration of patterns of impulse control in early adolescence. *Child Development, 32,* 75–88.
Livson, N., & McNeill, D. (1962). The accuracy of recalled age of menarche. *Human Biology, 34,* 218–221.
Livson, N., & Peskin, H. (1980). Perspectives on adolescence from longitudinal research. In J. Adelson (Ed.), *Handbook of adolescent psychology.* New York: Wiley.
Ljung, B-O, Bergsten-Brucefors, A., & Lindgren, G. (1974). The secular trend in physical growth in Sweden. *Annals of Human Biology, 1,* 245–256.
Loeber, R. (1982). The stability of antisocial and delinquent child behavior: A review. *Child Development, 53,* 1431–1446.
Loeber, R., & Dishion, T. J. (1983). Early predictors of male delinquency: A review. *Psychological Bulletin, 94,* 68–99.
Loeber, R., & Schmaling, K. B. (1985). Empirical evidence for overt and covert patterns of antisocial conduct problems: A metanalysis. *Journal of Abnormal Child Psychology, 13,* 337–353.

Logan, D. D. (1980). The menarche experience in twenty-three foreign countries. *Adolescence*, 15, 247–256.

Logan, D. D., Calder, J. A., & Cohen, B. L. (1980). Toward a contemporary tradition for menarche. *Journal of Youth and Adolescence*, 9, 263–269.

MacFarlane, J. W., Allen, L., & Honzik, M. P. (1954). *A developmental study of the behavior problems of normal children between twenty-one months and fourteen years.* Berkeley: University of California Press.

Magnusson, D. (1985). Early conduct and biological factors in the developmental background of adult delinquency. *The British Psychological Society, Newsletter*, 13, 4–17.

Magnusson, D. (1987). Adult delinquency in the light of conduct and physiology at an early age. In D. Magnusson & A. Öhman (Eds.), *Psychopathology: An interactional perspective.* New York: Academic Press.

Magnusson, D. (1988). Individual development from an interactional perspective: A longitudinal study. Vol. 1 in D. Magnusson (Ed.), *Paths through life.* Hillsdale, NJ: Lawrence Erlbaum Associates.

Magnusson, D., & Allen, V. L. (1983a). *Human development. An interactional perspective.* New York: Academic Press.

Magnusson, D., & Allen, V. L. (1983b). Implications and applications of an interactional perspective for human development. In D. Magnusson & V. L. Allen (Eds.), *Human development. An interactional perspective.* New York: Academic Press.

Magnusson, D., & Dunér, A. (1981). Individual development and environment: A longitudinal study in Sweden. In S. A. Mednick & A. E. Baert (Eds.), *Prospective longitudinal research: An empirical basis for the primary prevention of psychosocial disorders.* Oxford: Oxford University Press.

Magnusson, D., Dunér, A., & Zetterblom, G. (1975). *Adjustment: A longitudinal study.* New York: Wiley.

Magnusson, D., & Endler, N. S. (1977). Interactional psychology: Present status and future prospects. In D. Magnusson & N. S. Endler (Eds.), *Personality at the crossroads.* Hillsdale, NJ: Lawrence Erlbaum Associates.

Magnusson, D., Stattin, H., & Allen, V. L. (1985). Biological maturation and social development: A longitudinal study of some adjustment processes from midadolescence to adulthood. *Journal of Youth and Adolescence*, 14, 267–283.

Magnusson, D., Stattin, H., & Allen, V. L. (1986). Differential maturation among girls and its relation to social adjustment: A longitudinal perspective. In P. B. Baltes, D. L. Featherman, & R. M. Lerner (Eds.), *Life-span development and behavior* (Vol. 7, pp. 735–172). Hillsdale, NJ: Lawrence Erlbaum Associates.

Magnusson, D., Stattin, H. & Dunér, A. (1983). Aggression and criminality in a longitudinal perspective. In K. T. Van Dusen & S. A. Mednick (Eds.), *Prospective studies of crime and delinquency.* Boston: Kluwer-Nijhoff Publishing.

Maidman, L. L. (1984). From deference to confidence: Changing attitudes toward menarche. *Journal of Sex and Marital Therapy*, 10, 137–139.

Marcia, J. E. (1980). Identity in adolescence. In J. Adelson (Ed.), *Handbook of adolescent psychology.* New York: Wiley.

Marklund, U. (1985). Flickor på högstadiet—En diskussion om tonårsflickors alkoholvanor [The drinking habits of teenage girls]. In P. Sennerfeldt (Ed.), *Hälsa är helhet.* Stockholm: Skolöverstyrelsen, Rapport 85:15.

Mathes, E. W., & Kahn, A. (1975). Physical attractiveness, happiness, neuroticism, and self-esteem. *Journal of Psychology, 90*, 27–30.
McCall, R. B. (1977). Challenges to a science of developmental psychology. *Child Development, 55*, 17–29.
McCord, J. A. (1983). A longitudinal study of aggression and antisocial behavior. In K. T. Van Dusen & S. A. Mednick (Eds.), *Prospective studies of crime and delinquency*. Boston: Kluwer-Nijhoff Publishing.
McKeever, P. (1984). The perpetuation of menstrual shame: Implications and directions. *Women and Health, 9*, 33–47.
McNeill, D., & Livson, N. (1963). Maturation rate and body build in women. *Child Development, 34*, 25–32.
Mead, M. (1952). Adolescence in primitive and in modern society. In G. E. Swanson, T. M. Newcombe, & E. L. Hartley (Eds.), *Readings in social psychology*. New York: Holt.
Meyer-Balhburg, H. F. L., Ehrhardt, A. A., Bell, J. R., Cohen, S. F., Healey, J. M., Feldman, J. F., Morishima, A., Baker, S. W., & New, M. I. (1985). Idiopathic precocious puberty in girls: Psychosexual development. *Journal of Youth and Adolescence, 14*, 339–353.
Miller, P. Y. (1979). Female delinquency: Fact and fiction. In M. Sugar (Ed.), *Female adolescent development*. New York: Brunner/Mazel.
Miller, P. Y., & Simon, W. (1980). The development of sexuality in adolescence. In J. Adelson (Ed.), *Handbook of adolescent psychology*. New York: Wiley.
Milow, V. J. (1983). Menstrual education: Past, present, and future. In S. Golub (Ed.), *Menarche. The transition from girl to woman*. Lexington, MA: Lexington Books, D. C. Heath.
Miserandino, M. (1986). Summary of analyses of the Solna data set: Factors affecting puberty. *Report from the Solna Project. University of Stockholm* (7/21).
Mitchell, S., & Rosa, P. (1981). Boyhood behavior problems as precursors of criminality: A fifteen-year follow-up study. *Journal of Child Psychology, 22*, 19–33.
Money, J., & Clopper, R. R. (1974). Psychosocial and psychosexual aspects of errors of pubertal onset and development. *Human Biology, 46*, 173–181.
Money, J., & Walker, P. A. (1971). Psychosexual development, maternalism, nonpromiscuity, and body-image in 15 females with precocious puberty, *Archives of Sexual Behavior, 1*, 45–60.
Montemayor, R. (1982). The relationship between parent-adolescent conflict and the amount of time adolescents spend alone and with parents and peers. *Child Development, 53*, 1512–1519.
Montemayor, R., & Van Komen, R. (1980). Age segregation of adolescents in and out of school. *Journal of Youth and Adolescence, 9*, 371–381.
Moore, S. G. (1967). Correlates of peer acceptance in nursery school children. In W. W. Hartup & N. L. Smothergill (Eds.), *The young child*. Washington, DC: National Association for the Education of Young Children.
More, D. M. (1953). Developmental concordance and disconcordance during puberty and early adolescence. *Monographs of the Society for Research in Child Development, 18*, (1).
Mulligan, G., Douglas, J. W. B., Hammond, W. A., & Tizard, J. (1963). Delinquency and symptoms of maladjustment: The findings from a longitudinal study. *Proceedings from the Royal Society of Medicine, 56*, 1083–1086.

Murphey, E. H., Silber, E., Coelho, G. V., Hamburg, D. A., & Greenberg, I. (1963). Development of autonomy and parent-child interaction in late adolescence. *American Journal of Orthopsychiatry, 33*, 643–652.

Mussen, P. H., Conger, J. J., & Kagan, J. (1963). *Child development and personality.* New York: Harper & Row.

Muuss, R. E. (1970). Adolescent development and the secular trend. *Adolescence, 5*, 267–284.

Nesselroade, J., & Baltes, P. B. (1974). Adolescent personality development and historical change: 1970–1972. *Monographs of the Society for Research in Child Development, 39*(1, Serial No. 154).

Newcombe, N., & Dubas, J. S. (1987). Individual differences in cognitive ability: Are they related to timing of puberty? In R. M. Lerner & T. T. Foch (Eds.), *Biological-psychosocial interactions in early adolescence.* Hillsdale, NJ: Lawrence Erlbaum Associates.

Neugarten, B. L., & Datan, N. (1973). Sociological perspectives on the life cycle. In P. Baltes & K. W. Schaie (Eds.), *Life-span developmental psychology: Personality and socialization.* New York: Academic Press.

Nicolson, A. B., & Hanley, C. (1953). Indices of physiological maturity: Derivation and interrelationships. *Child Development, 24*, 3–38.

Niles, F. S. (1979). The adolescent girls' perception of parents and peers. *Adolescence, 14*, 591–597.

Nisbet, J. (1953). Family environment and intelligence. *Eugenics Review, 45*, 31–40.

Notman, M. T. (1983). Menarche: A psychoanalytic perspective. In S. Golub (Ed.), *Menarche. The transition from girl to woman.* Lexington, MA: Lexington Books, D. C. Heath.

Offer, D. (1969). *The psychological world of the teenager.* New York: Basic Books.

Olofsson, B. (1971). *Vad var det vi sa. Om kriminellt och konformt beteende bland skolpojkar* [On delinquent and conforming behavior among boys]. Stockholm: Utbildningsförlaget. (English summary).

Osgood, C. E., Suci, G., & Tannenbaum, P. (1957). *The measurement of meaning.* Chicago: University of Illinois Press.

Paige, J. E., & Paige, J. M. (1981). *The politics of reproductive ritual.* Berkeley: University of California Press.

Peevers, B. H., & Secord, P. F. (1973). Developmental changes in attribution of descriptive concepts to persons. *Journal of Personality and Social Psychology, 27*, 120–128.

Pervin, L. A. (1968). Performance and satisfaction as a function of individual-environment fit. *Psychological Bulletin, 69*, 56–68.

Peskin, H. (1967). Pubertal onset and ego functioning. *Journal of Abnormal Psychology, 72*, 1–15.

Peskin, H. (1972). Multiple prediction of adult psychological health from preadolescent and adolescent behavior. *Journal of Consulting and Clinical Psychology, 38*, 155–160.

Peskin, H. (1973). Influence of the developmental schedule of puberty on learning and ego functioning, *Journal of Youth and Adolescence, 2*, 273–290.

Peskin, H., & Livson, N. (1972). Pre- and postpubertal personality and adult psychological functioning. *Seminars in Psychiatry, 4*, 343–353.

Petersen, A. C. (1979). Female pubertal development. In M. Sugar (Ed.), *Female adolescent development.* New York: Brunner/Mazel.
Petersen, A. C. (1980). Biopsychosocial processes in the development of sex-related differences. In J. E. Parsons (Ed.), *The psychobiology of sex differences and sex roles.* Washington: Hemisphere.
Petersen, A. C. (1983). Menarche: Meaning of measures and measuring meaning. In S. Golub (Ed.), *Menarche. The transition from girl to woman.* Lexington, MA: Lexington Books, D. C. Heath.
Petersen, A. C. (1984). The Early Adolescent Study: An overview. *Journal of Early Adolescence, 4,* 1–4.
Petersen, A. C. (1985). Pubertal development as a cause of disturbance: Myths, realities, and unanswered questions. *Genetic Psychology Monographs, 111,* 205–232.
Petersen, A. C. (1987). The nature of biological-psychosocial interactions: The sample case of early adolescence. In R. M. Lerner & T. T. Foch (Eds.), *Biological-psychosocial interactions in early adolescence.* Hillsdale, NJ: Lawrence Erlbaum Associates.
Petersen, A. C., & Boxer, A. M. (1982). Adolescent sexuality. In T. J. Coates, A. C. Petersen, & C. Perry (Eds.), *Promoting adolescent health: A dialog on research and practice.* New York: Academic Press.
Petersen, A.C., & Crocket, L. (1985). Pubertal timing and grade effects on adjustment. *Journal of Youth and Adolescence, 14,* 191–206.
Petersen, A.C., Schulenberg, J. E., Abramowitz, R. H., Offer, D., & Jarcho, H. D. (1984). A self-report questionnaire for young adolescents (SIQYA): Reliability and validity studies. *Journal of Youth and Adolescence, 13,* 93–111.
Petersen, A.C., & Spiga, R. (1982). Adolescence and stress. In L. Goldberg and S. Breznitz (Eds.), *Handbook of stress: Theoretical and clinical aspects.* New York: Macmillan.
Petersen, A.C., & Taylor, B. (1980). The biological approach to adolescence. In J. Adelson (Ed.), *Handbook of adolescent psychology.* New York: Wiley.
Petersen, A.C., & Wittig, M. A. (1979). Differential cognitive development in adolescent girls. In M. Sugar (Ed.), *Female adolescent development.* New York: Brunner/Mazel.
Piaget, J. (1950). *Psychology of intelligence.* New York: Harcourt, Brace & World.
Piaget, J. (1969). *The mechanisms of perception.* London: Routledge & Kegan Paul.
Place, D. M. (1975). The dating experience for adolescent girls. *Adolescence, 10,* 157–174.
Pomerantz, S. C. (1979). Sex differences in the relative importance of self-esteem, physical self-satisfaction, and identity in predicting adolescent satisfaction. *Journal of Youth and Adolescence, 8,* 51–61.
Poppleton, P. K. (1968). Puberty, family size and the educational progress of girls. *British Journal of Educational Psychology, 38,* 286–292.
Poppleton, P. K., & Brown, P. E. (1966). The secular trend in puberty: Has stability been achieved? *British Journal of Educational Psychology, 36,* 95–100.
Presser, H. B. (1978). Age at menarche, socio-sexual behavior, and fertility. *Social Biology, 25,* 94–101.
Pressey, S. L., & Pressey, L. C. (1933). Development of the interest-attitude tests. *Journal of Applied Psychology, 17,* 1–16.

Pulkkinen, L. (1983a). Youthful smoking and drinking in a longitudinal perspective. *Journal of Youth and Adolescence, 12,* 253–283.

Pulkkinen, L. (1983b). The search for alternatives to aggression. In A. P. Goldstein & M. H. Segall (Eds.), *Aggression in global perspectives.* New York: Pergamon Press.

Rand, L. M., & Miller, A. L. (1972). A developmental cross-sectioning of women's career and marriage attitudes and life spans. *Journal of Vocational Behavior, 2,* 317–331.

Reiss, I. L. (1970). Premarital sex as deviant behavior: An application of current approaches to deviance. *American Sociological Review, 35,* 78–87.

Rierdan, J., & Koff, E. (1980a). The psychological impact of menarche: Integrative versus disruptive changes. *Journal of Youth and Adolescence, 9,* 49–58.

Rierdan, J., & Koff, E. (1980b). Representation of the female body by early and late adolescent girls. *Journal of Youth and Adolescence, 9,* 339–346.

Rierdan, J., & Koff. E. (1985). Timing of menarche and initial menstrual experience. *Journal of Youth and Adolescence, 14,* 237–244.

Rice, F. P. (1978). *The adolescent. Development, relationships, and culture.* Boston: Allyn & Bacon.

Riley, M. W., Johnson, M., & Foner, A. (1972). *Aging and society. A sociology of age stratification* (Vol. 3). New York: Russel Sage.

Ritvo, S. (1977). Adolescent to woman. In H. P. Blum (Ed.), *Female psychology. Contemporary psychoanalytic views.* New York: International Universities Press.

Roberts, D. F., & Dann, T. C. (1967). Influence on menarcheal age in girls in a Welsh college. *British Journal of Preventive Social Medicine, 21,* 170–176.

Roberts, D. F., Rozner, L. M., & Swan, A. V. (1971). Age at menarche, physique and environment in industrial north east England. *Acta Paediatricia Scandinavia, 60,* 158–164.

Robins, L. N. (1966). *Deviant children grown up: A sociological and psychiatric study of sociopathic personality.* Baltimore, MD: Williams & Wilkins.

Robins, L. N. (1986). The consequences of conduct disorder in girls. In D. Olweus, J. Block, & M. Radke-Yarrow (Ed.), *Development of antisocial and prosocial behavior.* New York: Academic Press.

Rommetveit, R. (1954). *Social norms and roles.* Oslo: Universitetsforlaget.

Rona, R., & Pereira, G. (1974). Factors that influence age at menarche in girls in Santiago, Chile. *Human Biology, 46,* 33–42.

Rosenbaum, M-B. (1979). The changing body image of the adolescent girl. In M. Sugar (Ed.), *Female adolescent development.* New York: Brunner/Mazel.

Rosenberg, F. R., & Simmons, R. G. (1975). Sex differences in the self-concept in adolescence. *Sex Roles, 1,* 147–159.

Ruble, D. N., & Brooks-Gunn, J. (1982). The experience of menarche. *Child Development, 53,* 1557–1566.

Rutter, M., Graham, P., Chadwick, O. F. D., & Yule, W. (1976). Adolescent turmoil: Fact or fiction? *Journal of Child Psychology & Psychiatry, 17,* 35–56.

Ryder, N. B., & Westoff, C. F. (1971). *Reproduction in the United States, 1965.* Princeton: Princeton University Press.

Sandler, D. P., Wilcox, A. J., & Horney, L. F. (1984). Age at menarche and subsequent reproductive events. *American Journal of Epidemiology, 119,* 765–774.

Savin-Williams, R. C. (1979). Dominance hierarchies in groups of early adolescents. *Child Development, 1979, 50,* 923–935.

Savin-Williams, R. C. (1980). Social interactions of adolescent females in natural groups. In H. C. Foot, A. J. Chapman, & J. R. Smith (Eds.), *Friendship and social relations in children.* New York: Wiley.

Savin-Williams, R. C., & Small, S. A. (1986). The timing of puberty and its relationship to adolescent and parent perceptions of family interactions. *Developmental Psychology, 22,* 342–347.

Schalling D. (1978). Psychopathy-related personality variables and the psychophysiology of socialization. In R. D. Hare & D. Schalling (Eds.), *Psychopathic behavior. Approaches to Research.* Chicester: Wiley.

Schofield, J. W. (1981). Complementary and conflicting identities: Images and interaction in an interracial school. In S. R. Asher & J. M. Gottman (Eds.), *The development of children's friendships.* New York: Cambridge University Press.

Schofield, M. (1965). *The sexual behaviour of young people.* London: Longmans.

Schonfeld, W. A. (1963). Body-image in adolescents: A psychiatric concept for the pediatrician. *Pediatrics, 31,* 845–855.

Schonfeld, W. A. (1971). Adolescent development: Biological, psychological, and sociological determinants. In S. C. Feinstein, P. L. Giovacchini, & A. A. Miller (Eds.), *Adolescent psychiatry. Development and clinical studies* (Vol. 1). New York: Basic Books.

Shader, R. I., Harmatz, J. S., & Tammerk, H-A. (1974). Menarcheal age and personality: The choice of a statistical test of relationship. *Psychosomatic Medicine, 36,* 321–326.

Shainess, N. (1961). A re-evaluation of some aspects of femininity through a study of menstruation: A preliminary report. *Comprehensive Psychiatry, 2,* 20–25.

Shainess, N. (1962). Psychiatric evaluation of premenstrual tension. *New York State Journal of Medicine, 62,* 3573–3579.

Shalit, B. (1978). Shalit Perceptual Organization and Reduction Questionnaire (SPORQ). Stockholm: FOA, Report, No. 1–3.

Shipman, W. (1964). Age of menarche and adult personality. *Archives for General Psychology, 10,* 155–159.

Shuttleworth, F. K. (1937). Sexual maturation and the physical growth of girls age six to nineteen. *Monographs of the Society for Research in Child Development, 2.*

Silbereisen, R. K., & Noack, P. (1986). On the constructive role of problem behavior in adolescence. *Technical University of Berlin, 72/86.*

Simmons, K., & Greulich, W. W. (1943). Menarcheal age and the height, the weight, and skeletal age of girls seven to seventeen years. *Journal of Pediatrics, 22,* 518–548.

Simmons, R. G., Blyth, D. A., & McKinney, K. L. (1983). The social and psychological effects of puberty on white females. In J. Brooks-Gunn & A. C. Petersen (Eds.), *Girls at puberty. Biological and Psychosocial perspectives.* New York: Plenum Press.

Simmons, R. G., Blyth, D. A., Van Cleave, E. F., & Bush, D. E. (1979). Entry into early adolescence: The impact of school structure, puberty, and early dating on self-esteem. *American Sociological Review, 44,* 948–967.

Simmons, R. G., Carlton-Ford, S. L., & Blyth, D. A. (1987). Predicting how a child will cope with the transition to junior high school. In R. M. Lerner & T. T. Foch (Eds.), *Biological-psychological interaction in early adolescence.* Hillsdale, NJ: Lawerence Erlbaum Associates.

Simmons, R. G., & Rosenberg, F. R. (1975). Sex, sex roles, and self-image. *Journal of Youth and Adolescence, 4,* 229–258.
Smith, W., & Powell, E. K. (1956). Responses to projective material by pre- and post-menarcheal subjects. *Perceptual and Motor Skills, 6,* 155–158.
Snedecor, G. W., & Cochran, W. G. (1967). *Statistical methods* (6th ed.). Ames, IA: The Iowa State University Press.
Snyder, E. E. (1972). High school student perceptions of prestige criteria. *Adolescence, 6,* 129–136.
Sorensen, R. C. (1973). *Adolescent sexuality in contemporary America: Personal values and sexual behavior: Ages 13–19.* New York: World Publishing.
Stattin, H. (1979). *Juvenile delinquency and changes in relative achievement: A longitudinal analysis.* Report from the Department of Psychology, University of Stockholm, no. 30 (Summary in English)
Stattin, H., & Magnusson, D. (1984). The role of early aggressive behavior for the frequency, the seriousness and the types of later criminal offences. *Reports from the Department of Psychology. University of Stockholm,* No. 618.
Stattin, H., & Magnusson, D. (in press). Social transition in adolescence: A biosocial perspective. In A. de Ribaupierre (Ed.), *Transition mechanisms in child development: The longitudinal perspective.* Cambridge: Cambridge University Press.
Stattin, H., Magnusson, D., & Reichel, H. (in press). Criminality from childhood to adulthood: A longitudinal study of the development of criminal behavior. Part 1: Criminal activity at different ages. *British Journal of Criminology.*
Steinberg, L. (1985, March). *The ABCs of transformations in the family at adolescence. Changes in affect, behavior, and cognition.* Paper presented at the third biennal Conference on Adolescence Research, Tucson, AZ.
Steinberg, L. (1987). Impact of puberty on family relations: Effects of pubertal status and pubertal timing. *Developmental Psychology, 23,* 451–460.
Stolz, H. R., & Stolz, L. M. (1944). Adolescent problems related to somatic variations. *Yearbook of the National Society for the Study of Education, 43,* 80–99.
Stone, C. P., & Barker, R. G. (1934). On the relationships between menarcheal age and certain aspects of personality, intelligence and physique in college women. *Journal of Genetic Psychology, 19,* 121–134.
Stone, C. P., & Barker, R. G. (1937). Aspects of personality and intelligence in post menarcheal and premenarcheal girls of the same chronological age. *Journal of Comparative Psychology, 23,* 439–445.
Stone, C. P., & Barker, R. G. (1939). The attitudes and interests of premenarcheal and post menarcheal girls. *Journal of Genetic Psychology, 54,* 27–71.
Stonequist, E. V. (1937). *The marginal man.* New York: Scribner's & Sons.
Stubbs, M. L. (1982). Period piece. *Adolescence, 17,* 45–55.
Susman, E. J., Nottelman, E. D., Inoff-Germain, G. E., Dorn, L. D., Cutler, Jr., G. B., Loriaux, D. L., & Chrousos, G. P. (1985). The relation of relative hormonal levels and physical development and social-emotional behavior in young adolescents. *Journal of Youth and Adolescence, 14,* 245–264.
Sutherland, E. H., & Cressey, D. R. (1970). *Criminology* (8th ed.). Philadelphia: Lippincott.
Tanner, J. M. (1962). *Growth at adolescence* (2nd ed.). Oxford: Blackwell.
Tanner, J. M. (1965). The trend towards earlier physical maturation. In J. E. Meade

& A. S. Parker (Eds.), *Biological aspects of social problems*. Edinburgh: Oliver & Boyd.

Tanner, J. M. (1966). The secular trend towards earlier physical maturation. *Tijdschrift voor Sociale Geneeskunde, 44*, 524–538.

Tanner, J. M. (1970). Physical growth. In P. H. Mussen (Ed.), *Carmichael's manual of child psychology* (3rd ed.). New York: Wiley.

Tanner, J. M. (1978). *Fetus into man*. Cambridge, MA: Harvard University Press.

Tanner, J. M., Whitehouse, R. H., Marshall, W. A., Healy, M. J. R., & Goldstein, H. (1975). *Assessment of skeletal maturity and prediction of adult height: TW2 method*. London: Academic Press.

Tinsley, D. J., & Faunce, P. S. (1980). Enabling, facilitating, and precipitating factors associated with women's career orientation. *Journal of Vocational Behavior, 17*, 183–194.

Tobin-Richards, M. H., Boxer, A. M., & Petersen, A. C. (1983). The psychological significance of pubertal change. Sex differences in perceptions of self during early adolescence. In J. Brooks-Gunn & A. C. Petersen (Eds.), *Girls at puberty, Biological and psychosocial persepectives*. New York: Plenum.

Triandis, H. C. (1967). Towards an analysis of the components of interpersonal attitudes. In M. Sherif & C. W. Sherif (Eds.), *Attitude, ego-involvement and change*. New York: Wiley.

Trost, A-C. (1982). *Abortion and emotional disturbances*. International Library, Västerås (Swedish text with a summary in English).

Udry, J. R. (1979). Age at menarche, at first intercourse, and at first pregnancy. *Journal of Biosocial Science, 11*, 433–441.

Udry, J. R., & Cliquet, R. L. (1982). A cross-cultural examination of the relationship between ages a menarche, marriage, and first birth. *Demography, 19*, 53–63.

Van Dyne, E. V. (1940). Personality traits and friendship formation in adolescent girls. *Journal of Social Psychology, 12*, 291–303.

Vener, A. M., & Stewart, C. S. (1974). Adolescent sexual behavior in middle America revisited: 1970–1973. *Journal of Marriage and the Family, 36*, 728–735.

Viteles, M. S. (1929). The influence of age of pubescence upon the physical and mental status of normal school students. *Journal of Educational Psychology, 20*, 360–368.

Vosk, B., Forehand, R., Parker, J. B., & Rickard, K. (1982). A multimethod comparison of popular and unpopular children. *Developmental Psychology, 18*, 571–575.

Wadsworth, M. (1979). *Roots of delinquency. Infancy, adolescence and crime*. Oxford: Martin Robertson.

Walker, L. S., & Green, J. W. (1986). The social context of adolescent self-esteem. *Journal of Youth and Adolescence, 15*, 315–322.

Weatherley, D. (1964). Self-perceived rate of physical maturation and personality in late adolescence. *Child Development, 35*, 1197–1210.

Wechsler, H., & Thum, D. (1973). Teenage drinking, drug use, and social correlates. *Quarterly Journal of Studies on Alcohol, 34*, 1220–1227.

Weiss, P. A. (1969). The living system: determinism stratified. In A. Koestler & J. R. Smythies (Eds.), *Beyond reductionism*. New York: Macmillan.

Werner, H. (1948). *Comparative psychology of mental development*. New York: International Universities Press.

West, D. J., & Farrington, D. P. (1973). *Who becomes delinquent? Second report of the Cambridge study in delinquent development.* London: Heinemann.

Westney, O. E., Jenkins, R. R., & Benjamin, C. A. (1983). Sociosexual development of preadolescents. In J. Brooks-Gunn & A. C. Petersen (Eds.), *Girls at puberty. Biological and psychosocial perspectives.* New York: Plenum Press.

Westrin, P. A. (1967). *Wit III manual.* Stockholm: Skandinaviska testförlaget.

Whisnant, L., Brett, E., & Zegans, L. (1975). Implicit messages concerning menstruation in commercial educational materials prepared for young adolescent girls. *American Journal of Psychiatry, 132,* 815–820.

Whisnant, L., Brett, E., & Zegans, L. (1979). Adolescent girls and menstruation. *Adolescent Psychiatry, 7,* 157–171.

Whisnant, L. & Zegans, L. (1975). A study of attitudes toward menarche in white middle-class american adolescent girls. *American Journal of Psychiatry, 132,* 809–814.

Williams, L. R. (1983). Beliefs and attitudes of young girls regarding menstruation. In S. Golub (Ed.), *Menarche. The transition from girl to woman.* Lexington, MA: Lexington Books, D. C. Heath.

Wohlwill, J. F. (1973). *The study of behavioral development.* London: Academic Press.

Woods, N. F., Dery, G. K., & Most, A. (1982). Recollections of menarche, current menstrual attitudes, and perimenstrual symptoms, *Psychosomatic Medicine, 44,* 285–293.

Wylie, R. (1979). *The self-concept* (2nd ed.). Lincoln, NE: University of Nebraska Press.

Yuen, R. K. W., Tinsley, D. J., & Tinsley, H. E. A. (1980). The vocational needs and background characteristics of homemaker-oriented women and career-oriented women. *Vocational Guidance Quarterly, 28,* 250–256.

Young, H. B. (1963). Ageing and adolescence. *Developmental Medicine and Child Neurology, 5,* 451–460.

Young, J. W., & Ferguson, L. R. (1979). Developmental changes through adolescence in the spontaneous nomination of reference groups as a function of decision content. *Journal of Youth and Adolescence, 8,* 239–252.

Zacharias, L., Rand, W. M., & Wurtman, R. J. (1976). A prospective study of sexual development and growth in American girls: The statistics of menarche. *Obstetrical and Gynecological Survey, 31,* 325–337.

Zakin, D. F., Blyth, D. A., & Simmons, R. G. (1984). Physical attractiveness as a mediator of the impact of early pubertal changes of girls. *Journal of Youth and Adolescence, 13,* 439–450.

Zelnik, M., & Kantner, J. F. (1972). The probability of premarital intercourse. *Social Science Research, 1,* 335–341.

Zelnik, M., Kantner, J. F., & Ford, K. (1981). *Sex and pregnancy in adolescence.* Beverly Hills: Sage.

Zucker, R. A., & Devoe, C. I. (1975). Life history characteristics associated with problem drinking and antisocial behavior in adolescent girls: A comparison with male findings. In R. D. Wirth, G. Winokur, & M. Roff (Eds.), *Life history research in psychopathology.* Minneapolis: University of Minnesota Press.

Author Index

A

Abernethy, E.M., 12, 53, 202
Abramowitz, R.H., 100, 101, 103
Adams, G.R., 230
Adams, J.F., 5
Adams, J.P., 124
Adams, P.L., 53, 101
Adelson, J., 59, 130, 131, 185, 365
Ajzen, I., 252
Allen, C.D., 100
Allen, L., 18, 26, 31, 32, 33, 35, 56, 112, 114, 138, 140, 182, 210, 293, 303
Allen, V.L., 2, 37, 70, 78, 343
Ames, R., 13
Andersson, B-E., 186
Andersson, O., 33, 59, 201, 303
Andersson, T., 305
Apter, A., 19, 53
Asmundsson, G., 195
Ausubel, D.P., 1, 11, 34, 60, 68

B

Bajema, C.J., 82
Baker, S.W., 115, 137, 139
Baltes, P.B., 54, 180, 368
Bandura, A., 131, 251, 365
Bardwick, J.M., 102, 108
Barker, R.G., 29, 33, 61, 62, 112, 118, 138, 208, 209, 210, 212, 248, 249, 250, 251, 293, 295, 301, 303
Baron, R.M., 263
Bayley, N., 12, 24, 37, 50
Bean, J.A., 81
Bell, J.R., 115, 137, 139, 189
Bem, S.L., 322
Benedek, E.P., 19, 20
Benjamin, C.A., 19, 293
Benning, J.J., 222
Bergh, T., 6
Bergman, L.R., 8, 79, 161
Bergsten-Brucefors, A., 10, 81, 82
Berscheid, E., 101
Berzonsky, M.D., 33, 71, 131, 211, 302
Beth-Halachmi, N., 19, 53
Billy, J.O.G., 293, 356, 358
Björkman, K., 355, 356, 357
Bloch, H.A., 183
Blos, P., 31, 56, 58
Blumberg, S.L., 30, 56, 112, 303
Blyth, D.A., 15, 18, 19, 24, 25, 26, 31, 32, 33, 45, 51, 57, 58, 62, 69, 71, 89, 101, 103, 104, 112, 113, 114, 116, 129, 133, 138, 140, 186, 189, 201, 293, 294, 295, 296, 297, 302, 303, 346, 362
Bowerman, C.E., 130
Bowman, P.H., 222
Boxer, A.M., 13, 43, 51, 113, 119, 135, 350, 352, 353
Brakke, N.P., 362

AUTHOR INDEX

Brett, E., 36, 107
Breznitz, S., 101
Bromley, D.B., 101
Bronfenbrenner, U., 302
Bronson, W.C., 19, 210
Brooks-Gunn, J., 1, 5, 13, 14, 18, 19, 20, 21, 24, 29, 30, 31, 35, 36, 45, 51, 52, 70, 106, 107, 108, 109, 110, 112, 113, 114, 134, 138, 141, 210, 347
Brown, B.B., 101, 102, 290
Brown, P.E., 10
Bruch, H., 5
Brundtlang, G.H., 10
Bruner, J.S., 101
Buck, C., 304
Bullough, V.L., 10
Burns, R.B., 16, 99
Burston, N., 140
Bush, D.E., 19, 26, 32, 138, 186, 201, 293, 296, 303

C

Cahalan, B., 195
Cairns, R.B., 368
Calder, J.A., 108
Campbell, A.C., 68, 292
Carlsmith, J.M., 12, 17, 30, 55, 104, 113, 115, 116, 135, 136, 188, 201, 293, 352
Carlton-Ford, S.L., 19, 32, 33, 62, 103, 114, 116, 189, 293, 294
Cavior, N., 101
Chadwick, O.F.D., 92, 164, 221, 365
Charlesworth, R., 362
Chrousos, G.P., 210, 212
Cicin, I.H., 195
Clarke, A.E., 106, 109
Clausen, J.A., 21, 22, 32, 103, 116
Clifford, E., 113, 114
Cliquet, R.L., 305, 351
Clopper, R.R., 51, 53, 135
Cochran, W.G., 273
Coelho, G.V., 295
Cohen, B.L., 108
Cohen, S.F., 115, 137, 139, 189
Coie, J.D., 362
Coleman, J.C., 131
Coleman, J.S., 69, 124, 128, 130, 186
Collins, J.K., 100, 133
Comite, F., 189

Conger, J.J., 104
Conklin, E.S., 108, 110
Constanzo, P.R., 36
Coppotelli, H., 362
Covington, J., 182
Crafoord, K., 214, 215
Cressey, D.R., 72
Crocket, L., 13, 18, 33, 44, 58, 100, 113, 124, 130, 133, 138, 140, 141, 185, 186, 201, 209, 210, 293, 302, 347
Crossley, H.M., 195
Crouter, A., 140
Csikszentmihalyi, M., 130, 131
Cutler, G.B., 189, 210, 212

D

Damon, A., 6, 81, 82
Damon, S.T., 6, 81
Dann, T.C., 5, 6
Danza, R., 250
Datan, N., 17, 54, 56
Davidson, H.H., 56, 114, 210
Davies, B.L., 30, 32, 33, 201, 303
de Anda, D., 58, 218
Demirjian, A., 90
Dery, G.K., 35, 109, 110
Deutsch, H., 17, 31, 35, 105, 108, 110, 185
Devoe, C.I., 195
Diers, C.J., 10, 90, 100
Dion, K.K., 101
Dishion, T.J., 179, 222
Dodge, K.A., 362
Dokecki, P.R., 101
Donovan, J.E., 195
Dorn, L.D., 210, 212
Dornbusch, S.M., 12, 17, 30, 55, 104, 113, 115, 116, 124, 135, 137, 188, 201, 293, 295, 352
Douglas, J.W.B., 90, 202, 222
Douvan, E., 59, 102, 104, 130, 131, 185, 365
Downs, W.R., 187, 292, 297
Dreyer, A.S., 19
Dubas, J.S., 89
Duck, S.W., 69, 101
Duke, D.L., 222
Duke, P.M., 12, 17, 30, 51, 55, 113, 135, 137, 201, 222, 293, 352
Duncan, P.D., 104, 113, 115, 116, 124, 188, 201, 296

Dunér, A., 8, 33, 59, 78, 79, 148, 152, 190, 197, 199, 201, 253, 303, 305, 343
Dunnington, M.J., 362
Dunphy, D.C., 69, 71, 130, 361
Dusek, J.B., 103, 104, 185
Dwyer, J., 101, 102, 104, 113

E

East, P.L., 100, 101, 188
Ehrhardt, A.A., 115, 137, 139, 189
Eicher, J.B., 100
Eichorn, D.H., 13, 14, 24, 32, 34, 67, 133
El-Khouri, B., 161
Ellison, P.T., 5
Ellis-Schwabe, M.E., 59, 294
Endler, N.S., 9
English, H.B., 1
Engström, I., 355, 357
Epstein, S., 117, 118
Erikson, E.H., 184, 365
Everett, E.G., 133

F

Farrington, D.P., 59, 222
Faunce, P.S., 341
Faust, M.S., 18, 20, 24, 30, 36, 37, 52, 70, 125, 132, 133, 134, 139, 153, 174, 269, 303
Feldhusen, J.F., 222
Feldman, J.F., 115, 137, 139, 189
Ferguson, L.R., 131
Fischbein, S., 6
Fishbein, M., 252
Flaherty, J.F., 103, 104, 185
Foch, T.T., 7, 9
Foner, A., 54
Ford, K., 137, 218, 228
Forehand, R., 362
Fox, R., 19, 51, 134
Freud, A., 365
Friedman, S.B., 221
Frisch, R.E., 5
Frisk, M., 13, 32, 33, 51, 59, 140, 141, 189, 201, 210, 269, 294, 303
Furstenberg, F., 302
Furu, M., 36, 90

G

Gaddis, A., 13
Gagnon, J.H., 135, 136, 137, 138
Galatzer, A., 19, 53
Garbarino, J., 140
Gargiulo, J., 51
Garn, S.M., 5, 6, 10
Garwood, S.G., 18, 31, 32, 33, 56, 112, 114, 138, 140, 210, 293, 303
Gates, A.I., 12
Gebhard, P.H., 134, 356
Gibbs, J.T., 222
Gilligan, C., 185
Glass, G.V., 273
Glazer, J.A., 362
Gold, M., 183, 189, 325
Goldberg, S., 30, 56, 112, 303
Goldstein, H., 2, 44, 84
Gollin, E.S., 101
Golub, S., 15, 32, 105, 354
Gottfredson, M., 59
Gottlieb, L.S., 56, 114, 210
Gottschalk, L.A., 33, 218
Graham, P., 92, 164, 221, 365
Green, J.W., 102
Green, T.M., 32, 33, 35, 141, 142
Greenberg, I., 295
Greif, E.B., 14, 23, 108
Greulich, W.W., 5, 81, 84
Gross, R. T., 12, 17, 30, 51, 55, 104, 113, 114, 115, 116, 124, 135, 136, 188, 201, 293, 295, 352

H

Hall, G.S., 365
Hamburg, D.A., 295
Hammond, W.A., 222
Hanley, C., 83, 84
Harmatz, J.S., 27, 320
Harper, J., 133
Hart, M., 31, 38, 56
Hartup, W.W., 59, 69, 130, 362
Hauge, R., 256
Havighurst, R.J., 222
Healey, J.M., 115, 137, 139, 189
Healy, M.J.R., 44, 84
Hebbelinck, M., 11
Helgasson, T., 195

Hench, K., 189
Henricson, M., 58, 182
Henton, C.L., 110
Hess, B., 69
Hiernaux, J., 6
Higham, E., 36, 53
Hill, J.P., 32, 33, 35, 58, 68, 69, 71, 116, 141, 142, 315
Hirschi, T., 59
Holmbeck, G.N., 32, 33, 35, 141, 142, 315
Honzik, M.P., 26, 35, 182, 210
Horney, L.F., 304, 305
Hortling, M., 32, 33, 51, 140, 141, 189, 201, 210, 269, 294, 303
Hulac, V., 19
Hunter, F.T., 131
Hutchinson, B., 186

I

Inoff-Germain, G.E., 210, 212
Intons-Peterson, M.J., 354

J

Jacobson, B.A., 106, 108, 141
Jagger, H., 81
James, W.H., 5
Jaquish, G.A., 115
Jarcho, H.D., 101, 103
Jenicek, M., 90
Jenkins, G.G., 69
Jenkins, R.R., 19, 293
Jennings, D., 12, 17, 30, 55, 135, 136, 201, 293, 352
Jensen, A.R., 6
Jersild, A.T., 43
Jessor, R., 54, 59, 179, 180, 182, 183, 184, 187, 195, 199, 228, 257, 290, 291, 292, 293, 297, 356, 357, 360, 361
Jessor, S.L., 54, 59, 179, 180, 182, 183, 184, 187, 199, 228, 257, 290, 291, 292, 293, 297, 356, 357, 360, 361
Johnson, M., 54
Jones, H.E., 24, 32, 33, 37, 50, 52, 53, 125, 132
Jones, M.C., 12, 13, 30, 33, 37, 53, 70, 115, 133, 138, 140, 141, 176, 201, 293, 301, 303

K

Kagan, J., 104
Kahn, A., 101, 102
Kandel, D.B., 69, 131, 160, 177, 269, 297, 361, 365
Kantero, R-L., 5, 6
Kantner, J.F., 137, 218, 228
Karabenick, S.A., 102
Karlberg, P., 355, 357
Katchadourian, H., 47, 357
Katz, E., 252
Kelly, H., 53
Kenny, D.A., 263
Kessler, R.C., 177, 297
Kestenberg, J.S., 17, 31, 38, 56, 62, 105, 106
Kiernan, K.E., 30, 33, 304
Kinch, J.W., 130
Kinsey, A.C., 13, 134, 136, 356
Kirkegaard-Sorensen, L., 222
Klackenberg, G., 355, 357
Klackenberg-Larsson, I., 355, 356, 357
Klein, R.P., 189
Koestler, A., 8
Koff, E., 24, 31, 35, 44, 45, 51, 56, 57, 103, 105, 106, 107, 108, 110, 117, 119, 141, 210, 212
Kratcoski, J.E., 230
Kratcoski, P.C., 230
Kriger, A., 30, 56, 112, 303
Kucher, J.S., 100, 101, 188
Kugelmass, S., 101

L

Laron, Z., 19, 53
Larson, R., 130, 131
Leeper, J.D., 81
Lenerz, K., 100, 101, 188
Lerner, J.V., 100, 101, 188
Lerner, R.M., 7, 9, 100, 101, 102, 188, 335
Lesser, G.S., 69, 131, 160, 177, 269, 361, 365
Lewis, R.A., 269
Lichtenstein, H., 355, 357
Liddle, G.P., 222
Lindahl, M., 6
Lindgren, G., 10, 49, 50, 81, 82, 90
Lipsitt, L.P., 54, 368
Litt, I.F., 51
Livesley, W.J., 101

AUTHOR INDEX

Livson, N., 5, 14, 19, 34, 81, 210, 320, 368
Ljung, B-O., 10
Loeber, R., 179, 181, 222, 366
Logan, D.D., 108, 109
Loriaux, D.L., 189, 210, 212
Losoff, M., 58, 100, 124, 130, 185, 186
Lynch, M.E., 32, 33, 35, 68, 116, 141, 142

M

MacFarlane, J.W., 24, 26, 35, 182, 210
Magnusson, D., 2, 7, 8, 9, 27, 33, 37, 59, 70, 78, 79, 97, 148, 152, 181, 190, 196, 197, 201, 222, 253, 303, 305, 327, 343, 368
Maidman, L.L., 107
Marcia, J.E., 128, 184, 185
Margulies, R.Z., 177, 297
Marklund, U., 221, 268
Marlow, L., 32, 33, 35, 141, 142
Marshall, W.A., 44, 84
Martin, C.E., 13, 134, 356
Martin, J.A., 12, 17, 30, 55, 104, 115, 116, 135, 136, 201, 293, 352
Mason, S., 19, 20
Mathes, E.W., 101, 102
Matthews, C.V., 222
Mayer, J., 101, 102, 104, 113
McCall, R.B., 368
McCord, J.A., 222
McKeever, P., 107, 108
McKinney, K.L., 15, 18, 24, 25, 31, 33, 45, 89, 101, 104, 112, 114, 133, 138, 140, 189, 201, 293, 295, 296, 297, 302, 303, 346
McNeill, D., 5, 81
McNemar, A., 189
Mead, M., 6, 107
Mednick, S.A., 222
Meyer-Balhburg, H.F.L., 115, 137, 139, 189
Menking, S., 53
Miller, A.L., 341
Miller, P.Y., 128, 136, 179, 186, 187, 293, 296, 356, 357
Milow, V.J., 107
Miserandino, M., 5
Mitchell, S., 222
Money, J., 51, 53, 70, 110, 134, 135, 212
Montemayor, R., 69, 130, 131, 141, 186
Moore, S.G., 362
More, D.M., 16, 30

Morishima, A., 115, 137, 139, 189
Most, A., 35, 109, 110
Mulligan, G., 222
Munro, G., 230
Murphey, E.H., 295
Mussen, P.H., 13, 30, 33, 37, 53, 104, 115, 133, 138, 140, 141, 176, 201, 293, 301, 303
Muuss, R.E., 11

N

Nesselroade, J., 180
New, M.I., 115, 137, 139, 189
Newcombe, N., 89
Neugarten, B.L., 17, 54, 56
Nicolson, A.B., 83, 84
Niederhoffer, A., 183
Niles, F.S., 131
Nisbet, J., 90
Noack, P., 183
Notman, M.T., 31
Nottelman, E.D., 210, 212

O

Offer, D., 59, 101, 103, 131, 365, 366
Olofsson, B., 8
Osgood, C.E., 125

P

Paige, J.E., 107
Paige, J.M., 107
Parker, J.B., 362
Peckham, P.D., 273
Peevers, B.H., 101
Pereira, G., 90
Pervin, L.A., 335
Pescovitz, O.H., 189
Peskin, H., 2, 14, 34, 35, 104, 210, 302, 320, 368
Petersen, A.C., 1, 5, 13, 14, 18, 19, 20, 21, 24, 27, 28, 31, 33, 36, 41, 43, 44, 51, 52, 56, 58, 62, 63, 64, 100, 103, 109, 110, 113, 119, 124, 130, 133, 135, 138, 140, 141, 185, 186, 199, 201, 209, 210, 293, 303, 347, 350, 352, 353

Petronio, R.J., 183, 189, 325
Piaget, J., 101
Pierce, J.V., 222
Piker, H.N., 33, 218
Place, D.M., 101
Pomerantz, S.C., 101
Pomeroy, W.B., 13, 134, 356
Poppleton, P.K., 10, 90, 202
Powell, E.K., 33, 57, 140
Poznanski, E., 19, 20
Presser, H.B., 33, 218, 304
Pressey, L.C., 34
Pressey, S.L., 34
Propert, D.S., 100
Pulkkinen, L., 182, 187, 292, 297

R

Raber, S., 140
Rand, W.M., 6, 89, 90
Rand, L.M., 341
Reed, R.B., 6, 81
Reese, H.W., 54, 368
Reichel, H., 8, 79, 97, 181, 196
Reiss, I.L., 354
Revelle, R., 5
Rice, F.P., 11, 15, 53, 228, 356
Rierdan, J., 24, 31, 35, 44, 45, 51, 56, 103, 105, 106, 107, 108, 110, 117, 119, 141, 210, 212
Rickard, K., 362
Rigler, D., 19
Riley, M.W., 54
Ritter, P.L., 104, 113, 115, 116, 124, 188, 201, 296
Ritvo, S., 31, 36
Roberts, D.F., 56
Robins, L.N., 181, 222
Rodgers, J.L., 293, 356, 358
Rommetveit, R., 255
Rona, R., 90
Rosa, P., 222
Rosenbaum, M-B., 100
Rosenberg, A., 17, 30, 55, 135, 136, 293, 352
Rosenberg, F.R., 68, 100, 101, 102, 103, 104, 185
Ross, J.M., 202
Rosso, J., 51
Rozner, L.M., 5

Ruble, D.N., 24, 35, 36, 45, 70, 106, 107, 108, 109, 110, 114, 141
Russel, R., 140
Rutter, M., 92, 164, 221, 365
Ryder, N.B., 351

S

Samelson, M., 19, 51, 134
Sanders, J.R., 273
Sandler, D.P., 304, 305
Sarles, R.M., 221
Sarnoff, C.A., 31, 38, 56
Savin-Williams, R.C., 33, 68, 115, 134, 142
Schalling, D., 320
Schmaling, K.B., 181
Schofield, J.W., 102
Schofield, M., 71, 156, 293, 294, 296, 358, 359, 360
Schonfeld, W.A., 53, 103
Schulenberg, J.E., 100, 103
Secord, P.F., 101
Shader, R.I., 27, 320
Shainess, N., 35, 108
Shalit, B., 323
Shaw, M.E., 36
Sheingold, K., 35, 107
Sherman, B.M., 81
Shipman, W., 302, 319, 340
Shuttleworth, F.K., 83
Siegel-Gorelick, B., 12, 104, 115, 116, 201
Silber, E., 295
Silbereisen, R.K., 189
Silverstone, E., 24, 31, 44, 45, 56, 105, 110, 117
Simmons, K., 84
Simmons, R.G., 15, 18, 19, 24, 25, 26, 31, 32, 33, 45, 51, 57, 62, 68, 89, 100, 101, 102, 103, 104, 112, 113, 114, 116, 129, 133, 138, 140, 185, 186, 189, 201, 293, 294, 295, 296, 297, 302, 303, 346
Simon, W., 136, 293, 356, 357
Small, S.A., 33, 142
Smith, W., 33, 57, 140
Snedecor, G.W., 273
Snyder, E.E., 101
Sonis, W.A., 189
Sorensen, R.C., 228, 356
Spiga, R., 36

AUTHOR INDEX

Stattin, H., 8, 37, 70, 79, 97, 181, 196, 222, 327, 343
Stavraky, K., 304
Steinberg, L., 32, 33, 142, 243, 251
Stensson, J., 355, 357
Stewart, C.S., 228, 356
Stewart, S.S., 33, 218
Stiel, R., 115, 189
Stodtland, A., 252
Stolz, L.M., 18, 36, 104
Stolz, H.R., 18, 24, 36, 104
Stone, C.P., 29, 33, 61, 62, 112, 118, 138, 208, 209, 210, 212, 248, 249, 250, 251, 293, 295, 301, 303
Stonequist, E.V., 209
Stubbs, M.L., 106, 107
Suci, G., 125
Susman, E.J., 210, 212
Sutherland, E.H., 72
Svennberg, I., 355, 357
Swan, A.V., 5

T

Tammerk, H-A., 27, 320
Tannenbaum, P., 125
Tanner, J.M., 5, 6, 10, 44, 48, 60, 82, 84, 90
Taylor, B., 21, 27, 28, 41, 62, 63, 64
Tenhunen, T., 32, 33, 51, 140, 141, 189, 201, 210, 269, 294, 303
Thiel, K.S., 58, 69, 71
Thornburg, H.D., 59, 294
Thum, D., 195
Thurston, J.R., 222
Tinsley, D.J., 341
Tinsley, H.E.A., 341
Titchener, J.L., 33, 218
Tizard, J., 222
Tobin-Richards, M.H., 13, 19, 43, 51, 113, 119
Triandis, H.C., 252
Trost, A-C., 218
Tuddenham, R., 37, 50

U

Udry, J.R., 137, 293, 304, 305, 351, 356, 358
Ulman, K.J., 14, 23, 108

V

Valadian, I., 6, 81
Van Cleave, E.F., 19, 26, 32, 138, 201, 293, 296, 303
Van Dyne, E.V., 69
Van Komen, R., 69
Vener, A.M., 228, 356
Viteles, M.S., 201, 202
Vosk, B., 362

W

Wadsworth, M., 189
Walker, L.S., 102
Walker, P.A., 70, 110, 134, 135, 212
Wallace, R.B., 81
Walloe, L., 10
Walster, E., 101
Warren, M.P., 19, 21, 51, 113, 134, 210
Weatherley, D., 13, 15, 16, 27, 36, 51, 81, 119, 301
Wechsler, H., 195
Weiss, P.A., 8
Werner, H., 101
West, D.J., 222
Westney, O.E., 19, 293
Westoff, C.F., 351
Westrin, P.A., 89, 343
Whisnant, L., 36, 106, 107, 108, 141
Whitehouse, R.H., 44, 84
Widholm, O., 5, 6, 32, 33, 51, 140, 141, 189, 201, 210, 269, 294, 303
Wilcox, A.J., 304, 305
Williams, L.R., 31, 107
Wittig, M.A., 199
Wohlwill, J.F., 368
Woods, N.F., 35, 108, 109, 110
Wurtman, R.J., 6, 89, 90
Wylie, R., 103

Y

Yuen, R.K.W., 341
Young, H.B., 2
Young, J.W., 131
Youniss, J., 131
Yule, W., 92, 164, 221, 365

Z

Zacharias, L., 6, 89, 90
Zakin, D.F., 19, 51, 57, 103, 113, 114, 129, 133

Zegans, L., 36, 106, 107, 108, 141
Zelnick, M., 137, 218, 228
Zetterblom, G., 8, 59, 78, 79, 148, 152, 190, 197, 253, 343
Zucker, R.A., 195

Subject Index

A

Adjustment, see also Emotions and Norm-breaking, 363–367
  extrinsic, 216–217
  intrinsic, 216–217
Adolescence, 1, 130–131
Affective reactions, see Emotions
Age-graded perspective, 2, 54–61
  and norm-breaking, 182–184
Aggressiveness, 222, 321
Alcohol
  abuse, 327
  drinking habits, 194–196, 234, 238–239, 325–327, 334
  peers as transmitters of, 283–285
Anxiety, 212, 320
Autonomy, 131, 182, 295–296

B

Body image, 15–16, 18, 36, 112–114, 122–125
Body mass, 87
Bone age, see Skeletal age
Breast development, 20, 48

C

Causal model, 5, 19, 39–41, 61–66, 262–265
Childbirth, 304–307, 332, 336, 367
Chronological age effects, 2, 17, 46–47, 54–61
Classroom behavior, see Teacher ratings
Conduct disturbances, 197–199
Configural frequency analysis (CFA), 161, 342
Criminal activity, 92, 97, 181, 196, 324–325, 334

D

Data collection, 8–9, 78–81, 93–96
Dating, 135–137, 354
Delayed puberty, see Late maturation
Delinquency, see Criminal activity
Depression, see also Emotions, 211–212, 268, 270, 272, 278
Development, 54–56, 250–251, 365–367
  unitary, 29–31, 60–61
  asynchronic, 32–36
Developmental tasks, 1
Deviancy hypothesis, 36–37
Differential opportunity, 71–73
Differential susceptibility, 71–73, 280–288, 299, 335–339
Drop out, 24–26, 92, 97
Drug use, see also Norm-breaking, 231, 328

E

Education, 310–313, 333, 337–338, 342–345, 367
Emotions, 32, 73–74, 209–213, 221–222, 267–268, 270, 272, 277–278

F

Family life, 57–59, 303–307, 312–313, 315–318, 331–332, 337, 342–345, 367
Family size, 90
Family vs. work, 341–343
Father relations, 164, 237
Femininity, 125, 184–187, 322

G

Grade point average, 202, 268, 270, 272, 285–287

H

Height, 49–50, 84–86, 112–114
Heterosexual relations, see Opposite-sex relations
Holistic approach, 7–9
Hormones, 47–48, 353
Hyperactivity, 211, 320

I

Intelligence, 89, 344
  and parents' education, 344
Intensive studies, 83
Interactional approach, 4–9, 60–66, 335
Intraindividual changes, 34

J

Job position, see Work conditions

L

Level of education, see Education
Life values, 323, 334
Longitudinal research, 23–24, 368
  long-term effects, 301–346, 363–367
  short-term effects, 224–247

M

Marriage, 304–307, 351
Maturation
  early, 32–37, 50–53, 63–66, 73–74
  late, 52–53
Mediating analysis, 262–265, 272–280, 329–331
Medical examination, 83
Menarche, 44, 105–111, 118
  and body image, 122–125
  and psychological adaptation, 105–111
  and self-evaluation, 114–117, 126–127
  and self-image, 31, 111–114, 115–117, 120–122
  and sex-role identification, 125
  correlates of, 5, 89–91
  cultural beliefs about, 107
  genetic influences, 6
  individual differences, 48–49, 81
  measurement of, 80–81
  preparations for, 34–36, 108–109
  reactions to, 108–110
  reliability, 81–83
  secular trend, 9–10
Menstrual problems, 6, 108
Methodology, 21–27, 44–47, 98, 273
  data treatment, 26–27
Mother relations, 140–142, 163–164, 237, 277, 281–282

N

Nonconventional peers
  definition of, 71
  influence on behavior of, 74–77, 335–339, 361–363
Norm-breaking, 10–11, 73, 179–199, 251–254, 257–258, 266–268, 276, 283, 290–293, 363–367
  and parent support, 253–258

and pubertal development, 188–189, 192–196, 276
and self-concept, 184–187
evaluations of, 254–256
expectations of, 257
overt vs. covert, 181–182
peers as transmitters of, 258–259, 283, 361, 366
sanctioning of, 256
sex differences, 180–182

O

Occupation, 307–310
  domain, 308
Official records, 92, 97, 196, 213
Opposite-sex relations, 15, 19, 32–33, 60, 70, 102, 112, 134–139, 153–155, 157–158, 174–176, 266, 269, 290–298, 353–361

P

Parent-child relations, 32–33, 57–59, 67, 130–131, 139–142, 165–166, 177, 236–240, 249–251, 266, 269, 276, 293–296, 315–319
Parent vs. peer orientation, 130–131, 168–171, 242–244
Peer conformity, 101–102, 241
Peer relations, 67–77, 100–101, 132–134, 150–152, 172, 240–242, 276, 315–318, 361–363, 366
  and norm-breaking, 187–188, 269–271, 361–367
  cross pressures, 73, 209
Peer characterization, 68–71, 75, 143–150, 158–162, 172–174, 187
  classmates, 69, 75, 146
  friendship-group, 148–149
  number of, 144
  older, 70–71, 75–76, 144–146, 298–299
  same-age, 69, 144–146, 147–148
  working, 70–71, 75–76, 146, 298–299
  younger, 70–71, 75–76, 144–145
Perceived control, 314
Perceived maturity, see Self-perception
Personality, 319–322
Physical appearance, 100–102, 124
Popularity

among peers, 52, 132–133, 152–153, 174, 266, 269, 362
among boys, see Opposite-sex relations
Pregnancy, 217–219, 268, 271, 276
Problem behavior, see also Norm-breaking, 33, 59, 219–220
Psychiatric disorders, 92, 97, 213–218
Psychoanalytic theory, 28, 30–31, 34
Psychosomatic reactions, 211, 235–236
Pubertal change, 47–51
Pubertal impact on behavior, 10–11, 22, 36–41, 61–66, 352–357, 360
  broad vs. narrow, 17–21, 347–352
  females, 14–21
  general vs. differential, 37–40, 363–367
  males, 12–14
  positive vs. negative, 18–21, 102–105
  short-term vs. long-term, 22, 39–41
  sociocultural interpretation of, 350–353
Pubertal status vs. timing, 44–47

R

Readiness, 64, 72, 77
Reliability, 81–89
Reciprocal influences, 7, 77, 248–251
Runaway behavior, 230

S

Sample, 24–26, 78–80
  representativeness of, 92, 97
School, 199–209, 220–221, 238, 240, 268, 270, 272, 278, 285, 296–298
  absent from, 204
  achievement, 202–204, 268, 270, 272, 278, 285
  motivation, 199–202, 204–206, 268, 270, 272, 278, 285
  parent reports of, 205–208
Secular trend, 9–11
  effects on behavior, 10–11
Self-concept, 99–129, 265–268
  and norm breaking, 184–187, 266–268
  and peer characteristics, 265–266
  self-evaluation (esteem), 99, 114–117, 126
    and physical appearance, 102
    and body image, 103, 129

self-perception (image), 18, 29, 51–53, 58, 66, 99, 111–117, 120–122, 225–227
  stability, 117–118
Sex role identification, 103, 111–113, 322
Sexual intercourse, 134–139, 154–155, 228–230, 355–357
Sexuality, attitude toward, 155–157
Siblings, 90–91, 318
Skeletal age, 83–84
Smoking, 328–329
Social maladjustment, see Norm breaking
Social comparison, 51–53
Social prestige, see Popularity
Social status, see Popularity
Social support, 315–318
Socialization, 321–322, 334, 338
Socioeconomic conditions, 89–90

Sociometric measures, 175–177, 361–362
Stress, 5

T

Theory, 28–37, 60–66, 74–77, 262–265
Teacher ratings of pupils, 197–199, 208–209
Teacher relations, 166–168, 266, 269, 276
Teenage pregnancy, 217–219, 268, 271, 276

W

Weight, 5, 50, 86–87, 112–116, 123–124
Work conditions, 307–312, 331–333, 337, 342–345